Urban Environment 1

Control and Management of Urban Water Environment

Edited by

HIROSHI TSUNO AND XIA HUANG

Kyoto University Press

Kyoto University Press
Yoshida-South Campus, Kyoto University
69 Yoshida-Konoe-Cho, Sakyo-ku
Kyoto 606-8315, Japan
www.kyoto-up.or.jp

ISBN 978-4-87698-371-1 (Paper)
First edition published 2013
Revised edition published 2015

Distributor

USA and Canada
International Specialized Book Services (ISBS)
920 NE 58th Avenue, Suite 300
Portland, Oregon 97213-3786
USA
Telephone: (800) 944-6190
Fax: (503) 280-8832
Email: orders@isbs.com
Web: http://www.isbs.com

Typeset by: Cactus Communications K.K.
Printed by: Kwix Co. Ltd.

Contents

Foreword

We are now in the era when people require more convenient and comfortable urban areas for their living places, which has been called 'Era of the Environment'. We need to ensure that the living environment where we live is safe and peaceful, by solving pollution issues and environmental deterioration arising from our daily lives and factory activities under a society rich in materials. For this purpose, it is more important to not only solve our pollution issues, but to also establish a society of resource recycling that aspires to a symbiosis with nature, saves energy and resources at the personal level. This is the very purpose of this program, in which we conducted research exchanges and cooperative research activities on the following four themes between universities in Japan and China over a ten-year period from April 2001 to March 2011:

1. Control of Water Resources and Consumption, and Water Pollution
2. Air Pollution Control and Innovation of Prevention Technology
3. Management of Solid Wastes and Establishment of Recycle System
4. Development of Comprehensive Management and Design Manuals on Urban Infrastructure.

Group 1 papers details research on water regime management in mega cities suffering from water resource shortages and a deterioration of quality. Group 2 researches focus on the management of air pollution and global warming arising from a fossil-based society. Group 3 researches describe how to manage the solid waste issue and how to recycle solid waste products. Finally, Group 4 studies detail the control and management of infrastructure from an energy saving perspective.

This book is based on the achievements of the above-mentioned research groups, and aims to be a reference for graduate students and strategists.

In Japan, we have suffered from industrial, urban, and global pollution. This has expanded from the regional level to the national level. We strive to achieve a sustainable urban society formed under the cooperation between developed and developing countries, using our experience, knowledge, and technologies.

In the People's Republic of China, which is a growing economic powerhouse, the approach to environmental conservation and technological development has drastically changed, compared with the situation in 2001 when this program started. Particularly after 2006, the National People's Congress have adopted the Eleventh Five-Year Plan in which they have placed importance on environmental conservation more than ever before. Now, after the Beijing Olympic Games in 2008 and Shanghai Expo in 2010, China's environmental issues have been receiving international attention.

Based on the fruits of this book, many developing countries should understand the triggers and methods to promote efficient and secure infrastructure development avoiding the detour that Japan experienced, as China is attempting to do. We must at least establish laws that control certain behaviors to discuss a sound societal development and a reformation of the present consumption patterns, which are based on personal desire. In addition, we aspire to establish road maps to produce an acceptable social environment whose efficiency and comfort will be applicable to many countries, both developed and developing.

We would like to express our sincere acknowledgement to Isao Somiya and Nobuo Takeda who are Professor Emeritus of Kyoto University and past coordinator in Japan.

We profoundly thank the contributors to this book and also thankful to the JSPS-MOE Core University Program on Urban Environment and Kyoto University Global COE program "Global Center for Education and Research on Human Security Engineering for Asian Mega Cities" for financial assistance towards the completion of this book.

August, 2014

Hiroshi TSUNO
(Coordinator in Japan, Professor Emeritus,
Kyoto University)

Jiming HAO
(Coordinator in China, Professor,
School of Environment,
Tsinghua University)

Preface

With the increasing society urbanization and the abnormal weather caused by global warming, water pollution and water shortage have become serious problems both in Japan and China. Aquatic ecological environment has been also under pressure. The increasing global population and economic booming make the problems more serious. These problems are common all over the world, both in developed and developing countries. And more, there are many areas where people cannot access safe water which is free from sanitary risks. It is also said that the 21st century is called "Water Century."

In order to solve these issues, it is quite important to establish "control and management of urban water environment" which enable supply and preservation of safe and comfortable water based on reasonable and appropriate distribution and use of water, development of new urban water resources, water quality preservation, protection of environment and so on. It should be conducted under save of energy and recycle of resources and contribute to establish the sustainable development society.

This volume is written by researchers of universities in Japan and China, where weather and geological conditions, and society and economic environments are different each other, with incorporating new research results and discoveries through our collaboration during these 10 years.

In Chapter 1, issues about water resources are discussed. Water, which is inevitable for survival and growth of all living things including human beings, can take role only when satisfies 3 conditions, those are enough quantity, appropriate quality and existence at required place. Topics and issues about present and future state of water resources in Japan and China as well as over the world are explained. Readers can recognize the common and different conditions between two countries. In Chapter 2, water quality issues are discussed. Characteristics of water and its contributions to our life, water pollution issues, history of water pollution and countermeasures to it in Japan and China, and emerging water pollution issues are included. International viewpoints are also discussed.

In Chapter 3, issues associated with urban water, mainly drinking water, are discussed including framework and risks on security and management of urban water, and purification technologies and systems under sustainable development. In Chapter 4, issues about control and management of municipal wastewater are discussed with including its framework and technologies. Especially technologies and water quality levels required for reuse of the wastewater which is expected as a new urban water source are also discussed.

In Chapter 5, management of lakes which are very important as resources for urban water is discussed with introducing a new concept "integrated management." Lakes management is commonly one of the most important issues all over the world. In Chapter 6, monitoring of micro pollutants with mussels is discussed from the viewpoints of safety of water for intended and unintended reuse of water with incorporating the new research results.

In this volume, we discuss the Japanese experiences in which Japanese suffered from and overcame the serious and sad water pollution at the age without scientific knowledge, and Chinese experiences in which Chinese try to solve the issues associated with water resources and water pollution under the rapid economic development, and grope about for control and management of urban water environment to establish more safe and comfortable urban water environment as well as sustainable development society. We are very happy if this volume contributes for graduate students in universities and business persons as a reference book. We appreciate all Japanese and Chinese contributors (names listed separately), and proof contributions by Prof. Peng LIANG (Chapter 2), Prof. Xiang LIU (Chapter 5), and Prof. Jin WU (Chapter 6) of Tsinghua University. Last, but not the least, we would like to express our sincere appreciation to Dr. Tadao MIZUNO for his editorial assistance.

August, 2013

Hiroshi TSUNO
(Japanese side leader of group 1,
Professor Emeritus in Kyoto University)

Xia HUANG
(Chinese side leader of group 1,
Professor in Tsinghua University)

Contributors

Hiroshi TSUNO, *Professor Emeritus, Graduate School of Engineering, Kyoto University; Professor, Department of Human Life and Environment, Osaka Sangyo University*

Editor

Chapter 1 (Technical editor), 1.2.1, 1.2.2, 1.2.3, 1.2.4, 1.2.5, 1.4.1, 1.4.2

Chapter 2 (Technical editor), 2.1.1, 2.1.2, 2.2.1, 2.5.1, 2.5.2

Chapter 4, 4.3.2, 4.3.3

Chapter 6 (Technical editor), 6.1, 6.2, 6.3.1, 6.3.2, 6.3.3, 6.4.1, 6.4.2

Xia HUANG, *Professor, School of Environment, Tsinghua University, China*

Editor

Chapter 1, 1.4

Chapter 4 (Technical editor), 4.2.3.4

Chapter 6 (Technical editor), 6.1, 6.2, 6.3.1, 6.4

Shinya ECHIGO, *Associate Professor, Department of Environmental Engineering, Graduate School of Engineering, Kyoto University*

Chapter 3 (Technical editor), 3.2.1, 3.2.2, 3.2.3, 3.2.4

Taira HIDAKA, *Senior Researcher, Materials and Resources Research Group, Public Works Research Institute*

Chapter 4, 4.2.1, 4.2.2

Sadahiko ITOH, *Professor, Department of Environmental Engineering, Graduate School of Engineering, Kyoto University*

Chapter 3, 3.1.1, 3.1.2, 3.1.3, 3.1.4, 3.3.1

Ilho KIM, *Senior Researcher, Water Resources & Environment Research Department, Korea Institute of Construction Technology*

Chapter 4, 4.2.3.1, 4.2.3.2, 4.2.3.3

Yuzuru MATSUOKA, *Professor, Department of Environmental Engineering, Graduate School of Engineering, Kyoto University*	**Chapter 1**, 1.1.1, 1.1.2, 1.1.3
Norihide NAKADA, *Assistant Professor, Department of Environmental Engineering, Graduate School of Engineering, Kyoto University*	**Chapter 2**, 2.4.1, 2.4.3 **Chapter 4**, 4.1.1, 4.1.2, 4.1.3, 4.1.4, 4.1.5
Fumitake NISHIMURA, *Associate Professor, Department of Environmental Engineering, Graduate School of Engineering, Kyoto University*	**Chapter 2**, 2.3.1, 2.3.3, 2.4.2, 2.4.5
Yuichi Sato, *Researcher, Lake Biwa Environmental Research Institute*	**Chapter 5**, 5.2.1, 5.2.2, 5.2.3, 5.2.4
Yoshihisa SHIMIZU, *Professor, Department of Environmental Engineering, Graduate School of Engineering, Kyoto University*	**Chapter 5 (Technical editor)**, 5.1, 5.1.1, 5.1.2, 5.1.3, 5.3, 5.3.1, 5.3.2, 5.4
Yugo TAKABE, *Assistant Professor, Department of Environmental Engineering, Graduate School of Engineering, Kyoto University*	**Chapter 2**, 2.4.2 **Chapter 6**, 6.3.3, 6.4.1
Hiroaki TANAKA, *Professor, Department of Environmental Engineering, Graduate School of Engineering, Kyoto University*	**Chapter 4 (Technical editor)**, 4.2.3.5, 4.3.4
Hui WANG, *Professor, School of Environment, Tsinghua University, China*	**Chapter 2 (Technical editor)**, 2.3.2, 2.3.4, 2.3.5, 2.3.6
Xianghua WEN, *Professor, School of Environment, Tsinghua University, China*	**Chapter 1 (Technical editor)**, 1.3.1, 1.3.2, 1.3.3, 1.3.4
Naoyuki YAMASHITA, *Lecturer, Department of Environmental Engineering, Graduate School of Engineering, Kyoto University*	**Chapter 2**, 2.4.4, 2.4.6 **Chapter 4**, 4.3.1
Jiane Zuo, *Professor, School of Environment, Tsinghua University, China*	**Chapter 2 (Technical editor)**, 2.2.2

1

Present and Future State of Water Resources

1.1 PRESENT AND FUTURE STATE OF GLOBAL WATER RESOURCE

1.1.1 *People's water demand and global water resource*

Water is an absolute necessity for human life and activity. The Preamble to the Declaration adopted at the United Nations Water Conference held in 1977 at Mar del Plata in Argentina (UN 1977) states the following about the right of human beings to have access to water: "All peoples, whatever their stage of development and their social and economic conditions, have the right to have access to drinking water in quantities and of a quality equal to their basic needs." The United Nations Convention on the Rights of the Child adopted in 1989 (UN 1989) also clearly states the right to receive clean drinking water.

These documents limit human being's water rights to water for drinking and other daily living needs. What about water for food production, economic activities and preserving the environment? The United Nations Convention on the Law of the Non-navigational Uses of International Watercourses adopted in 1997 (UN 1997) defined the fundamental necessary amount of water as including not only drinking water but also the water needed to produce food to the extent necessary to prevent starvation.

In this way, the right to water is one of the most important basic human rights. Due to the increase in population and progressive urbanization and industrialization in recent years, however, the demand for water has increased and water quality is continuing to decline. Within the next several decades, there is expected to be a dramatic increase in the number of regions suffering from an imbalance in water supply and demand. This section will focus primarily on this topic and will discuss the structure of water demand, past changes in and future prospects for global water supply and demand, and ways to resolve water problems on a global scale.

1.1.2 *Outline of global water resources*

1.1.2.1 "Blue water" and "green water"

Almost all of the water used by people for daily life needs and environmental preservation comes from freshwater sources. Although seawater is also desalinated or used directly as cooling water, the quantities involved are not very great. Freshwater sources can be divided into two categories: green water and blue water. Green water is water that comes from rainwater that seeps into the soil and then evaporates and is ingested by living organisms and the like. Blue water is water that is stored in rivers and lakes or as groundwater. As this type of water is easy to use, it is the focus of water resource management efforts. The total quantity of blue water in existence is estimated at approximately 10 million km^3, mainly in the form of groundwater. The total quantity of green water in existence is estimated at approximately 20,000 km^3, primarily water in the soil. From a use perspective, however, it is not the stocks of water but the replenishment quantity that is important. The replenishment quantities of blue water and green water are estimated to be approximately 40,000 km^3/year and 60,000 km^3/year, respectively. As a general concept, the replenishment quantity of blue water is roughly equivalent to the reserves of water resources.

Figure 1-1 may help the readers understand the "blue" and "green" water concept.

Most of the blue water flows via rivers into the ocean. However, approximately 8% is stored in inland basins and the like and evaporates without flowing into the ocean. From a stock perspective, groundwater accounts for most of the blue water. From a flow perspective, however, groundwater is estimated at approximately 2,600 km^3/year, only about 5% of the flow of rivers.

The quantity of rainfall in a certain area that runs off as surface water and replenishes underground aquifers, amended by the pure exchange amount between the aquifers and surface water (overlap), is known as the internal renewable water resources (IRWR). The rainwater that flows into the region via rivers and groundwater flow even though it actually fell in adjacent areas is known as external renewable water resources (ERWR). ERWR also include the portion of water that is provided based on regional arrangements between regions for the water in rivers and lakes that are located on regional boundaries. If there is an arrangement or the like promising a certain quantity of outflow (the flow in the opposite direction from the inflow), it must be subtracted from the total quantity of water resources in order to guarantee that amount. IRWR and ERWR together represent the total renewable water resources (TRWR).

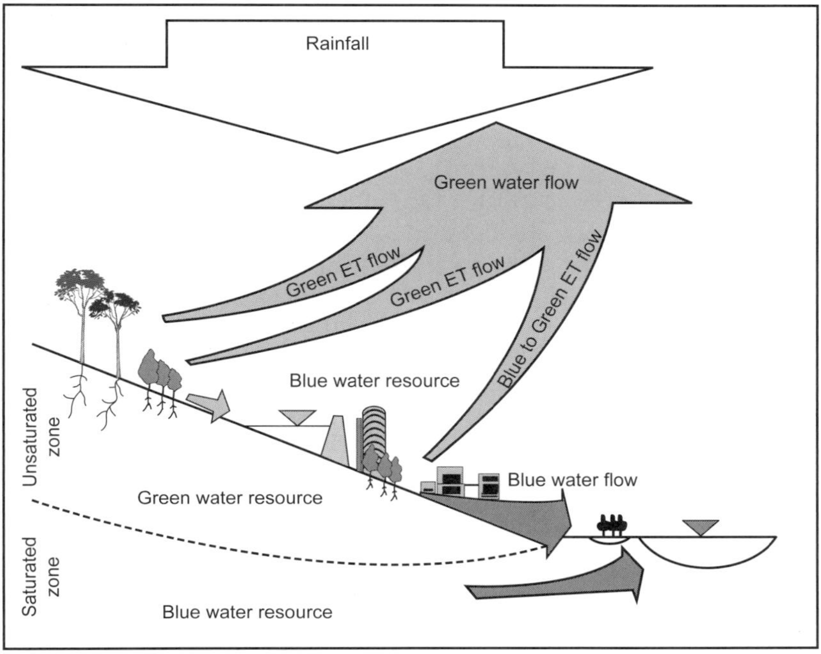

FIGURE 1-1. Conceptualization of a widened green-blue approach to water-resource planning and management. Rainfall, the undifferentiated freshwater resource, is partitioned in a green-water resource as moisture in the unsaturated zone and in a blue-water resource in aquifers, lakes, wetlands, and impoundments e.g., dams. These resources generate flows, as green-water flow from terrestrial biomass producing systems crops, forests, grasslands, and savannas and blue-water flow in rivers, through wetlands, and through base flow from groundwater.

Calculating these amounts will result in either of two values for water resource quantity, depending on whether the calculations took into account arrangements between regions and restrictions imposed due to customary practices. Values that consider only the natural conditions are described as "natural" values. Values that also consider arrangements between regions and so on are described as "actual" values.

From a water use perspective, the important thing with regard to renewable water resources is the water intake quantity that is achievable with existing technologies at the time when the water is needed. This is referred to as the utilizable water supply (UWS). The net amount of water that includes rivers, reservoirs, groundwater and so on that can be obtained by existing facilities is called the primary water supply (PWS). The International Water Management Institute (IWMI), Sri Lanka, 2000 estimated the global average of the ratio between UWS and RAWR to be 36% and estimates the values for Japan, China and India at 33%, 30% and 38%, respectively. The value for the United States, Russia and other countries that have low river regime coefficients (the ratio between maximum and minimum flow) was estimated to be 60%. L'vovich *et al.*, (1995) estimated the ratio between UWS and RAWR to be 27% when reservoir facilities are not taken into consideration, and estimated the value for UWS as 13,000 km^3/year in which 2,180 km^3/year is from man-made reservoirs. PWS and intake quantity are different concepts. In terms of the actual water intake, if the wastewater from water use facilities upstream becomes intake once again, in many cases this is counted when calculating the intake quantity,

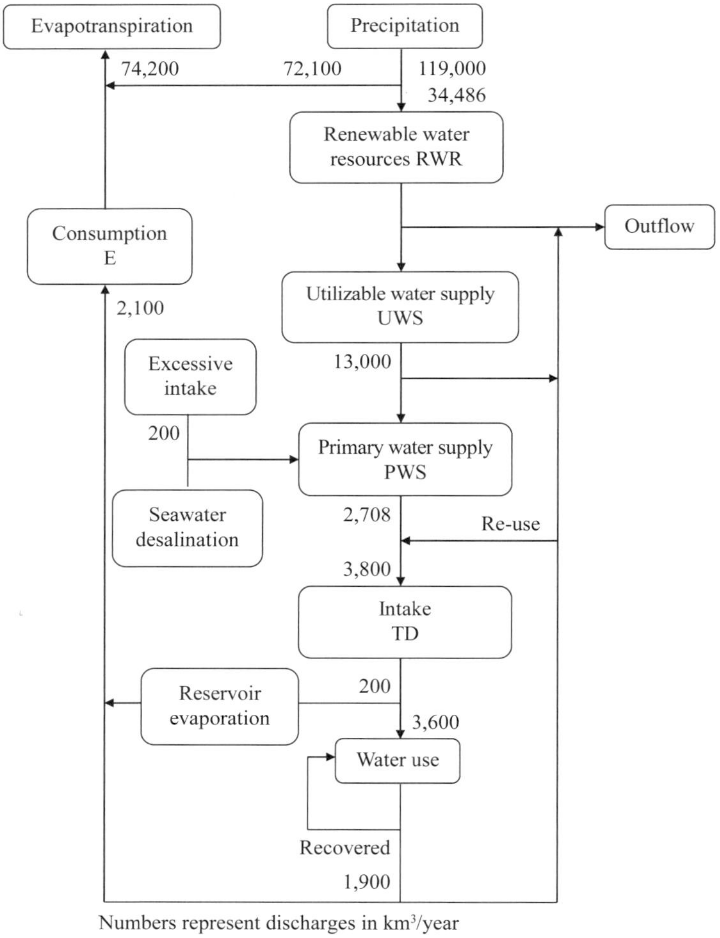

FIGURE 1-2. Global water resources and estimated flow volumes in 1995.

so the total intake that represents the sum of these values tends to be greater than the PWS. IWMI (2000) estimated the global average for the ratio between PWS and UWS in 1995 at 18% and put the RAWR at 34,486 km^3/year, so the value for PWS is calculated as 34,486 km^3/year × 0.36 × 0.18 = 2,235 km^3/year. This corresponds to 60% of the 1995 total intake of 3,800 km^3/year. Moreover, the ratio between PWS and UWS is not necessarily always a value less than 1. In 1995, the PWS values for Saudi Arabia, Pakistan and Jordan were 6.8, 1.4 and 1.1 times the values for UWS, respectively. The PWS values exceed the UWS values due to excessive pumping of groundwater. **Figure 1-2** shows the relationship of RWR, UWS and PWS and estimated global water use in 1995.

Green water is used to maintain ecosystems and for agricultural production (in the form of rainwater) and so on. Falkenmark (2000) estimated the amount of green water that is supplied to agricultural land at 6,700 km^3/year, and the amounts supplied to grasslands, forests and wetlands as 15,100 km^3/year, 40,000 km^3/year and 1,400 km^3/year, respectively.

1.1.2.2 Consumptive use and non-consumptive use

Water is a stable compound that remains virtually unchanged by circulation within the environment and use by human beings. Except for cases in which the water evaporates and is dissipated in the atmosphere or cannot be reused due to pollution, water is not extinguished as a result of use. To put it another way, there are two modes of water use: non-consumptive use, in which the water is discharged after use and can be reused by another user, and consumptive use, in which it is not possible. The sum of both of these values is the required water intake quantity. The quantity of intake that is not consumed is returned and reused at downstream.

The portion of the intake quantity that is consumed varies greatly depending on the mode of use, but at present it is estimated to be approximately 50%. **Table 1-1** shows freshwater intake and consumption quantities for all regions of the world for the year 1995 according to use. The total intake quantity was 3,788 km^3/year and the total consumption quantity was 2,074 km^3/year. The proportion of intake used for agricultural, industrial and domestic purposes was 70%, 21% and 10%, respectively. As a proportion of consumption, the values were 93%, 4.4% and 2.6%, respectively. The rate of consumption for agricultural water (the proportion of water intake that is consumed) was approximately 70%, while this figure was a little larger than 10% in the case of industrial and domestic purposes. The industrial water shown in **Table 1-1** does not include water for hydroelectric power plants, but it does include power plant cooling water from freshwater sources.

Table 1-1 also shows intake sources. Here groundwater includes both shallow groundwater with a high degree of exchangeability with river water, and deep groundwater with low regeneracy. Globally, 15% of total intake comes from groundwater, and the total groundwater intake is estimated at 600–700 km^3/year

TABLE 1-1. Freshwater intake and consumption in the world.

	Freshwater intake			Freshwater consumption		
	Intake (km³/year)	Percentage in total	Proportion of groundwater	Consumption (km³/year)	Percentage in total	Consumption/ intake (%)
Domestic	344	10%	50%	49.8	3%	14.5%
Industrial	752	20%	20%	82.6	4%	11.0%
Agricultural	2,504	70%	10%	1,753	93%	70.0%
Evaporation from reservoirs	188			188		–
Total	3,788		15%	2,074		54.8%
Per capita	661 m³/ (year•Person)			361 m³/ (year•Person)		

Estimated with Shiklomanov (1999).

TABLE 1-2. Aquifers in which excessive groundwater intake has become a problem.

Region/Country	Aquifer	Recharge quantity (km³/year)	Uptake quantity (km³/year)
Algeria and Tunisia	Sahara basin	0.58	0.74
Saudi Arabia	Saq	0.3	1.43
China	Hebei plain	35	19
Canary Islands	Tenerife	0.22	0.22
Gaza	Coastal	0.31	0.5
United States	Ogallala	6~8	22.2
United States	Arizona	0.37	3.78

From Margat (1996).

(Margat, J., 1994). Of this amount, nearly half is accounted for by two countries: India (180 km³/year) and the United States (110 km³/year). The arid regions, for example Libya and Saudi Arabia, are greatly dependent on groundwater. In these regions, groundwater accounts for more than 90% of the total intake , and this figure is 59% in the Haihe river basin and 24% in the Yellow river basin of China. In Japan, the rate of dependence on groundwater is 12%.

In recent years, groundwater intake has increased dramatically. In the United States, groundwater intake has increased 3% per year for the past 30 years, and in Saudi Arabia it has increased 15% per year. As a result, the amount of groundwater is decreasing at the rate of 200 km³ each year (Postel, 1992). As noted in the examples shown in **Table 1-2**, in some regions the amount greatly exceeds the groundwater recharge, resulting in major problems such as

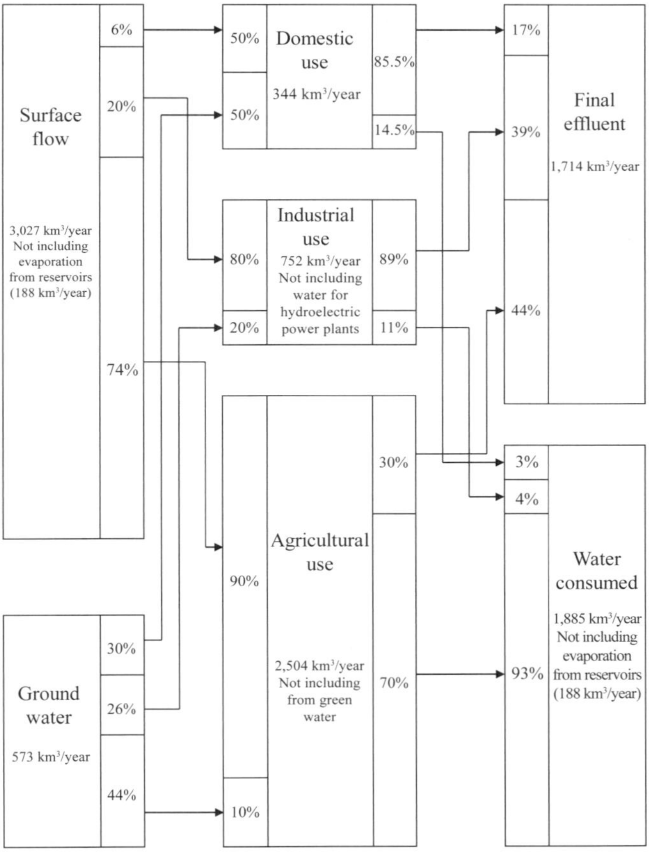

FIGURE 1-3. Global intake and consumption of blue water.

land subsidence, seawater penetration and so on. **Figure 1-3** shows the state of water use in 1995 with a focus on blue water.

1.1.2.3 The quantity of global water and its changes

The flow of blue water and green water is approximately 100,000 km³/year, however, their stocks are different. **Table 1-3** shows them. Groundwater accounts for 99.7% of freshwater stocks, with the remainder comprising water in soil (0.16%), lake water (0.11%) and river water (0.02%). When the retention times are calculated from these values and the inflow from various water regions, the results are retention times of several thousand years for groundwater, 17 years for lakes and 17 days for rivers. Reservoirs built by human beings now have a capacity of more than 8,400 km³ (Vorosmarty, C.

TABLE 1-3. Water located near the surface of the earth.

Type of water	Water volume (10^6 km^3)	Percentage in total water volume (%)	Percentage in total freshwater volume (except glaciers, %)
	Salt water		
Seawater	1,338.00	96.5	
Groundwater	12.87	0.929	
Lake water	0.0854	0.00616	
	Freshwater		
Glaciers, etc.	24.06	1.74	227.8
Subterranean ice in permafrost regions	0.300	0.022	2.84
Water in soil	0.0165	0.001	0.156
Water in living organisms	0.00112	0.00008	0.0106
Groundwater	10.53	0.760	99.7
Lake water	0.0115	0.001	0.109
River water	0.00212	0.000	0.0201
Water in atmosphere	0.0129	0.001	
Total	1,385.89	100.0	100.0

From Shiklomanov (1997).

et al., 1997), and this is estimated to make the retention time of river water tripled.

Figure 1-4 depicts the movement of water near the surface of the earth. Some 600,000 km^3/year of water exchange takes place between the atmosphere and the earth's surface, and more than 10% of this water flows via rivers into the ocean, and 10% of it is taken for use by human beings. In this Figure, the inflow to the atmosphere and to the oceans excluding groundwater is equal to the outflow. The quantity of water in the atmosphere and oceans has not changed, but in actuality a one-way movement has been produced due to human activities.

Houghton *et al.*, (2001) summarized the decrease of freshwater volume as **Table 1-4**, based on previous studies. Due to excessive pumping of groundwater, an extra 40–360 km^3/year is discharged into the oceans. The drying up of lakes is 11 km^3/year in the case of the Caspian Sea and 65 km^3/year in the case of the Aral Sea, but a portion of these amounts is added to seawater through evapotranspiration via the atmosphere, and the effect of this globally is approximately 0–70 km^3/year. Reservoir construction was actively

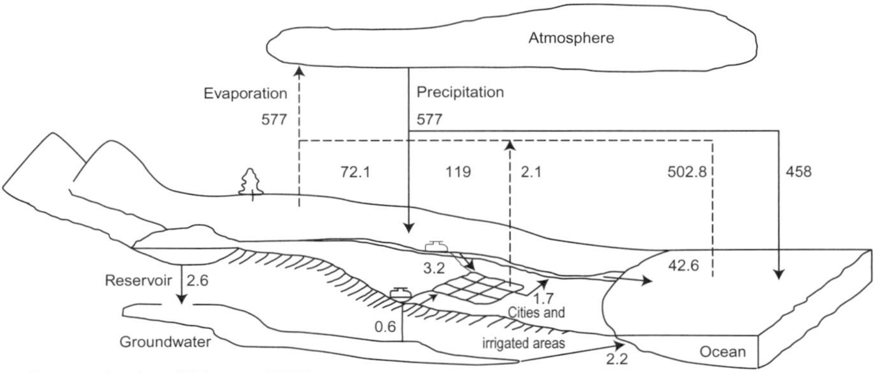

FIGURE 1-4. Movement of water near the earth's surface.

TABLE 1-4. Reduction in freshwater volume due to human activities.

Human activities	**year 1990 (km^3/year)**	**Average 1910–1990 (km^3/year)**
Excessive pumping of groundwater	40 ~ 360	0 ~ 180
Drying up of lakes	0 ~ 70	0 ~ 40
Changes in reservoirs	−540 ~ −250	−330 ~ −145
Changes in the rate of runoff due to urbanization	0 ~ 140	0 ~ 40
Deforestation	40~50	40
Changes in the volume of evapotranspiration due to increased area of irrigation	−40 ~ 0	−40 ~ 0
Increase in volume of groundwater due to leakage of irrigation water	−180 ~ 0	−70 ~ 0
Total	−680 ~ 370	−400 ~ 150

From Houghton *et al.*, (2001).

conducted during the past 50 years, and this reduced the inflow into the oceans by 250–540 km^3/year. The increased rate of runoff due to urbanization and the reduction in the moisture content of plants and soil due to deforestation are also thought to change the volume of freshwater in land areas. **Table 1-4** does not include the effect of changes in continental snow and ice due to climate change.

1.1.3 *Water demand*

1.1.3.1 Lifestyle and water demand

Human beings use water in the course of their daily lives: water for drinking, for cooking and for maintaining good hygiene. For each 1 kcal of dietary intake, approximately 1–1.5 ml of drinking water is needed. In the case of a person with a daily food intake of 2,000–3,000 kcal, between 2 and 5 liters of water is needed. While water is indispensable for people to maintain their health (to ensure good hygiene and especially for flushing human waste), the quantity of water that is needed varies greatly depending on the level of sanitation. In the case of a pit latrine, 1 to 2 liters of water is needed each time. In the case of a constantly running flush toilet, 75 liters or more per day per person is needed (Kalbermatten *et al.*, 1982; Yolles, 1993). Moreover, the combustion type and circulating type toilets developed in recent years require almost no water at all for flushing human waste. In the case of a conventional flush toilet system, however, it has been said that human health will be affected if the quantity of water goes below 20 liters per person per day (Esrey, 1986). Moreover, water for hand-washing use and bathing is also needed to maintain sanitation. In Japan, 20 liters (per person per day) and 65 liters (per person per day) of water is used for hand-washing use and bathing, respectively (MLIT, 2001). In other developed countries, the combined amount is approximately 45–100 liters (per person per day) (Gleick, 1996). A World Bank study (Kalbermatten *et al.*, 1982) found that water consumption was 5–15 liters or more (per person per day) in the case of bathing and 15–25 liters or more (per person per day) in the case of showering. The amount of water used for cooking varies greatly depending on the region. In Japan, the amount has been determined to be approximately 55 liters (per person per day) (MLIT, 2001). However, several studies that included both developed and developing nations have determined the minimum quantity to be 10–20 liters (per person per day). Combining both of these amounts, Gleick (1996) considered 50 liters (per person per day) to be the Basic Requirement of Water (BWR). And in actual, this level of water distribution is the level at which it is known that health is greatly improved (World Bank, 1992). The BWR for drinking water and toilet use alone is 25 liters (per person per day), and this is at the lower limit of the 20–40 liters (per person per day) that is the target of Agenda 21, the International Drinking Water Supply and Sanitation Decade (IDWSSD) and other campaigns. The World Health Organization (WHO) has set the near term target at 20 liters (per person per day), 10 for drinking and cooking and another 10 for bodily cleansing (primary washing hands). The WHO goal is the distribution of an amount of water that exceeds this level to all persons in the world. If this 20 liters of water (per person per day) cannot be obtained indoors or from the vicinity, people must travel a long distance to obtain water. However, the farther away it is, the more likely they compromise with less amount, so approximately 1 km is thought to be the limit (**Table 1-5**).

The World Health Organization (WHO) and UNICEF joint study (WHO/UNICEF/WSSCC (2000)) estimated that, as of 2000, 1.1 billion people had no indoor plumbing and no accessible running water or safe (in terms of sanitation) well water within 1 kilometer, and therefore had difficulty in obtaining 20 liters of water (per person per day). Furthermore, Lvovsky (2001) compiled various estimates and estimated the adverse health impact in terms of proportion of total sickness caused by inadequate water supply and sanitation as 4–10% (developing nations) and 1% (advanced industrialized nations). **Table 1-6** shows the proportion of total sickness caused by this and other environmental factors. As the table shows, the lack of safe water and sanitation is the most serious factor in terms of sickness.

Figure 1-5 shows the above mentioned relationship between lack of water supply and sanitation and impact on health with the focus on developing nations. The Figure shows three influence paths:

1. Insufficient quantity of water
2. Supply of polluted water
3. Faulty disposal of human wastes

According to a study by Esrey *et al.*, (1986), the proportion of impact of these three routes for (1), (2) and (3) was 25%, 16% and 22%, respectively. The remaining 37% is accounted for by the combined impact of (1) and (2).

20–50 liters (per person per day) is the minimum level of water needed for people to live a healthy life. However, domestic water consumption increases with the increase in per capita gross domestic product (GDP) that results from water system diffusion and economic development. In Canada, the United States and Japan, the value is 100 m^3 per person per year (equivalent to 250 liters per person per day) or greater. In Germany, the Netherlands and other countries, however, the level is stable at approximately 40 m^3 per person per year (equivalent to 100–120

TABLE 1-5. Distance traveled to obtain domestic water and quantity of water obtained.

Water source	Water quantity (liters per person per day)
Public water faucet located more than 1 km away	<10
Public water faucet located 1 km or less away	20
Indoor water faucet and flush toilet	>60

From Gleick (1996).

TABLE 1-6. Percentage in global burden of diseases caused by various environmental factors.

Environmental factor	Sub-Saharan Africa (%)	India (%)	Asia/ Pacific (%)	China (%)	Mideast/North Africa (%)	Central and South America (%)	Former Soviet Union (%)	Industrialized nations (%)
Safe water and sanitation	10.0	9.0	8.0	3.5	8.0	5.5	1.5	1.0
Malaria	9.0	0.5	1.5	0.0	0.3	0.0	0.0	0.0
Indoor air pollution	5.5	6.0	5.0	3.5	1.7	0.5	0.0	0.0
Atmospheric pollution in cities	1.0	2.0	2.0	4.5	3.0	3.0	3.0	1.0
Agricultural and industrial wastes	1.0	1.0	1.0	1.5	1.0	2.0	2.0	2.5
Percentage in global burden of diseases	26.5	18.5	17.5	13.0	14.0	11.0	6.5	4.5

Estimates by Lvovsky (2001) based on disability-adjusted life year (DALY) estimates. DALY is a health indicator developed jointly by Harvard University, the World Bank and the World Health Organization. It is calculated as the sum of the number of years of premature mortality caused by early death as a result of each illness and the equivalent number of years lost due to injury. The values (%) in the table are for the beginning of the 1990s.

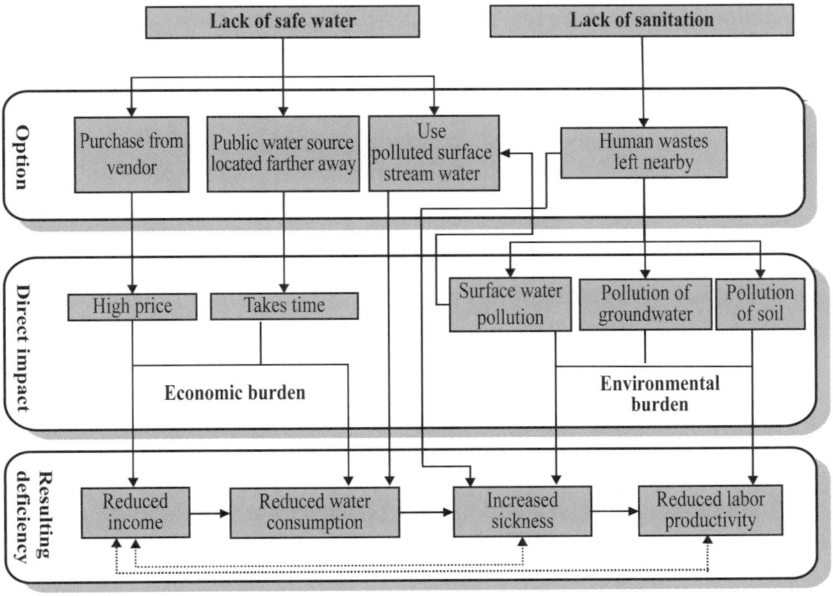

FIGURE 1-5. Impact of the lack of safe water supply and sanitation on human health and the economy.

liters per person per day). The group of developed nations with the lowest values also includes Belgium and Portugal, and there has been no increase in the values in these countries during the past 20 years.

1.1.3.2 Industry and water demand

Freshwater intake for industrial use amounts to 752 km^3/year globally (1995, Shiklomanov (1999)) and 14 km^3/year for Japan (1996, Ministry of International Trade and Industry (1998)). These Figures represent 20% and 16%, respectively, of the total intake. The water needed for industrial activities is used for cooling, cleaning, air control (regulation of temperature and humidity in the interior of factories), boilers, and as a raw material. Both freshwater and seawater are taken. In Japan, both freshwater and seawater are used; approximately 20% of the water used is seawater, representing approximately 55% of the intake quantity. 95% of this amount is used as cooling water. Cooling water is also the main use for freshwater at 72%, followed by 17% for cleaning and processing of products, 6% for air control, 1% for boiler use and 0.4% for use as a raw material. In terms of types of industry, the big consumers of water are the chemical industry, the steelmaking industry and the paper and pulp industry. These industries account for 34%, 26% and 10%, respectively, of total freshwater consumption and 25%, 25% and 11%, respectively, of total freshwater intake. The amount of industrial water use is proportional to the amount of product manufacture, but the basic unit that represents the proportionality factor differs greatly depending on the manufacturing process, the status of water reuse and other factors, and it has varied considerably throughout history. For example, in the 1930s, 60–100 m^3 of water was needed to produce one ton of steel, but now only 6 m^3 is required to do so.

The biggest reason for this reduction is the recycled use of recovered water. If the recovery factor (the ratio of recovered water to total water) is r and total water use is u, the water intake quantity is $(1-r)\,u$, and water that is taken can be used $1/(1-r)$ times (number of times circulated). Increasing the number of circulation times can reduce the intake sufficiently, but the quality of recovered water will be degraded and the consumption rate (evaporation quantity divided by use quantity; approximately 5–20% in the case of non-recycled use) will also increase, so there are limits to the degree to which water circulation can be increased.

1.1.3.3 Power generation and water demand

Both thermal power generation and nuclear power generation require large quantities of cooling water. In Japan, the amount of water used for thermal power generation is 3.5–4 m^3/s per 100 MW of power. The amount of water used for nuclear power generation is approximately one and a half times to twice this amount. In almost all cases, seawater is used. In the United States, however, 69% of the water used for power generation is taken from freshwater sources, and this represents 39% of the total intake that includes water for agriculture and daily living use and so on (Solley *et al.*, 1998). The amount of water consumed by the cooling process is 1 m^3 per MW-hour in the case of one path cooling and 2.2–2.6 m^3 in the case of a cooling tower. When these Figures are combined with the aforementioned quantities of intake, the rate of consumption comes to 1–3%.

Large quantities of water are also used in the generation of hydroelectric power. However, use for hydroelectric power generation influences the use for other purposes to a small degree, and the theoretical water power (which is expressed as the water quantity multiplied by the effective head) is more appropriate than the quantity of water itself for expressing the quantity of water resources. For these and other reasons, the quantity of water resources is normally added up using the theoretical water power rather than the quantity of water itself. There are three ways to express hydraulic power resources:

1. Theoretical water power resources, which includes all of the potential energy of flowing water
2. Technical water power resources, which constitutes the quantity that can be converted into electrical energy using existing technology, regardless of whether or not this is economically viable
3. Economic water power resources, which is the portion of technical water power resources that is economically achievable

Table 1-7 shows the estimates for each of these water power resource values. Calculating from this table, the ratio of theoretical water power resources to

TABLE 1-7. Hydro power resources in the world.

Region/country	Theoretical hydro power resources (TW•h/year)	Technical hydro power resources (TW•h/year)	Economic hydro power resources (TW•h/year)	Power generation (TW•h/year)	Development rate (%)	Runoff (km^3/year)
North America	5,817	1,509	912	697	76	7,890
Central and South America	7,533	2,868	1,199	519	43	12,030
Eurasia	22,413	7,194	3,254	945	29	16,410
Africa	3,887	2,163	1,416	291	21	4,050
Oceania	1,134	211	318	129	41	2,404
Global total	40,784	13,945	7,099	2,581	36	42,784

Hydro power resources and power generation for countries other than Japan from UNDP(2000). Runoff from Shiklomanov (1999).

runoff is 0.95 kW•h/m^3 which represents 350 m of potential energy. However, calculating from the actual water use and power generation, for example, the quantity of water used for hydroelectric power generation in the U.S. in 1995, for example, was 3,160 km^3, and the amount of power generated was 310 TW•h (Solley *et al.*, 1998), which shows 0.098 kW•h/m^3, only about 1/10 of the previous value. Moreover, according to **Table 1-7**, the development rate, which is the ratio between the amount of power generated and economic water power resources, is more than 70% in Japan and North America, but the global average is only 36%.

In most cases, hydroelectric power generation involves the use of reservoirs and regulating ponds. Currently the area of reservoirs worldwide is more than 400,000 km^2. Dividing the quantity of hydroelectric power generation by this number results in a value of 62 MW•h/(ha•year). However, there is great variation among the values for individual reservoirs. Goodland (1996) studied dam projects conducted in developing nations and reported that the value was 3.5×10^0–1.48×10^6 MW•h/(ha•year). However, 1/3 of the dams were under 20 MW•h/(ha•year) that is the range of energy production density for biomass farms, indicating that hydroelectric power generation is land-intensive.

1.1.3.4 Agriculture and water demand

Agricultural water accounts for the greatest proportion of total intake. In Japan, the proportion of total intake accounted for by agricultural water is 66% (1998, MLIT (2001)), while globally this value is 70% (1995, Shiklomanov (1999)). In recent years, the value has been decreasing.

Agricultural water is used for irrigating fields and paddies, raising livestock and so on. Irrigation enables the water needed to grow crops to be supplied by means of not only direct rainfall but also water supplied by ditches and so on. This stabilizes agricultural production and increases revenues. The amount of water needed to grow a unit weight of crops is referred to as the water use efficiency. The amount of water needed to produce 1 kg of dry matter, for example a C3 plant such as rice, is 600–1,000 liters. The amount of water needed to produce a C4 plant such as corn is 200–400 liters. In addition, the weight of a harvested product as a percentage of the total plant weight of a crop is called the harvest index. In the case of grains, the harvest index is 30–50%, so water consumption per 1 kg of harvest quantity is approximately 1,500–2,500 liters in the cast of a C3 cereal crop and 500–1,000 liters in the case of a C4 cereal crop. As meat involves many levels of the food chain, the water use efficiency for meat is even worse. **Table 1-8** shows the amount of water needed to produce 1 kg of various foodstuffs.

Based on these values, **Table 1-9** shows the estimated quantity of water consumed for food production globally by region. The value is approximately 5,500 km^3/year. Depending on the region, there is a threefold disparity in the amount of water needed per capita. At present, the cultivated acreage being irrigated throughout the world amounts to 250 million ha, 1/6 of the total area of cultivated land. In terms of production, however, this area accounts for approximately 1/3 of total production, so out of the total amount of evapotranspiration of cultivated crops, the amount for rainwater-fed agriculture is approximately 3,700 km^3 and that for irrigation agriculture is 1,800 km^3. These amounts of water are obtained from green water in the case

TABLE 1-8. Amount of water needed to produce foodstuffs.

Foodstuff	Amount of water needed for production (litter/kg)
Potatoes	500~1,500
Wheat	900~2,000
Alfalfa	900~2,000
Sorghum	1,100~1,800
Corn	1,000~1,800
Rice	1,900~5,000
Soybeans	1,100~2,000
Chicken	3,500~5,700
Beef	15,000~70,000

From Gleick (2000).

TABLE 1-9. Amount of water needed for food demand.

	Amount of water needed for present food demand	
Region	liters per person per day	km^3/year
Sahel region	1,760	329
China and other Asian nations with centrally planned economies	2,530	1,152
Eastern Europe	3,910	175
Countries of the former Soviet Union	4,300	455
Latin America	2,810	449
Mideast and North Africa	2,940	334
OECD Asia and Oceania	3,310	181
South and East Asia	2,110	1,213
Western Europe	4,690	646
North America	5,020	517
World	2,815	5,451

of rainwater-fed agriculture and from rainfall and blue water in the case of irrigation agriculture.

Irrigation water flows through such as water conduits, diversion, sprinkling and so on and undergoes transpiration through crops, or evaporates from the soil and water channels, or seeps into groundwater, etc., and the remainder is discharged. Of these amounts of water, the transpiration portion and evaporation portion are water for consumption, equivalent to 70% of the intake quantity (**Table 1-1**). Crops take the transpiration water that they need for growth from the root zone. Of the intake quantity, the proportion that reaches the root zone and is used for transpiration is called irrigation efficiency, and the range of values for irrigation efficiency is shown in **Table 1-10**. As the table shows, irrigation efficiency differs greatly depending on the method of irrigation. In arid regions in particular, water resources are precious, and excessive irrigation also results in salt damage, so microirrigation is being introduced – for example drip irrigation, in which tiny holes are made in hoses to provide water directly to the root zone. 90% of the irrigation in Israel and Cyprus and more than 50% of the irrigation in Jordan and the United Arab Emirates is conducted using microirrigation methods. However, this increases irrigation costs. Currently microirrigation and sprinkler irrigation account for approximately 0.7% and 8%, respectively, of the total irrigation area. The remainder is accounted for by surface irrigation such as furrow irrigation and border irrigation (AQUASTAT/FAO, 2002). Overall irrigation efficiency (including evaporation) is estimated to be approximately 40%. Increasing this efficiency by 10% would reduce the intake quantity by 200 km^3.

TABLE 1-10. Irrigation methods and irrigation efficiency.

Irrigation method	Irrigation efficiency (%)	Cost (USD/ha)
Unpaved earth canal and surface irrigation	40~50	
Paved channel and surface irrigation	50~60	
Pipes and surface irrigation	65~75	1,700
Irrigation using hoses	70~80	
Low- and medium-pressure sprinklers	75	2,800
Micro sprinklers, microjets, mini-sprinklers	75~85	3,950
Drip irrigation	80~90	4,000~5,000

From Phocaides, A. (2000).
Costs are initial costs. Annual maintenance costs are approximately 5% of initial costs.

Global agricultural water intake is currently increasing at a rate of 0.64% annually (1990–1995). Compared to the 3.2% per year in the 1950s, however, it has decreased dramatically. The factors determining intake quantity are as follows:

1. Population
2. Per capita crop production
3. Crop acreage per unit of crop production
4. Portion of crop acreage that is irrigated
5. Water intake per unit of irrigation area

The intake quantity is the cumulative impact of these five factors, and the combination of the rate of change for all of these factors is the rate of increase in intake quantity. **Table 1-11** shows changes of these factors during the 20th century. The slowing of the increase in crop production per crop acreage and the rate of increase in per capita crop production can be shown as reasons for the decrease in agricultural water intake.

Irrigation does more than supply the water needed to grow crops. It also washes away the salts from salt-damaged areas, prevents weathering, and is used to cause agricultural chemicals and fertilizers to be mixed in with the water in order to fertilize the crops and protect them from insect damage and sickness. Irrigation water is also for firefighting and melting snow, as well as for small-scale hydroelectric power generation and the like. In Japan, the role of irrigation water as environmental water for use in rice paddies and farm ditches in particular has been noted, and efforts are being made to evaluate agricultural water from such a broader perspective.

TABLE 1-11. Analysis of causes of changes in agricultural water intake.

Increase (%/year)	**1900–1939**	**1940–1949**	**1950–1959**	**1960–1969**	**1970–1979**	**1980–1989**	**1990–1995**
Population	0.85	1.08	1.77	1.75	2.04	1.83	1.39
Per capita crop production	0.73	0.01	1.90	1.45	0.60	0.49	−1.93
Crop acreage per unit of crop production	−0.61	0.93	−1.85	−3.15	−2.00	−2.40	−0.04
Irrigation area ratio	0.22	0.85	1.65	1.77	0.99	2.20	1.43
Water intake per unit of irrigation area	0.21	−0.97	−0.25	−0.11	0.34	−0.66	−0.17
Water intake	1.40	1.90	3.21	1.64	1.94	1.39	0.64

Crop production and crop acreage values are from FAOSTAT (2002). Values for population, irrigation area and agricultural water intake are from Shiklomanov (1999).

1.2 PRESENT AND FUTURE STATES OF WATER RESOURCES IN JAPAN

1.2.1 *The critical amount of water resources*

Japan has a population of 127,768,000 (census, 2005) in 377,846 km^2 (national municipality summary, 2005). During 30 years from 1976 to 2005, the annual precipitation average was 1,690 mm, and the annual average of the critical amount of water resources, which is defined as 'precipitation amount minus evaporation amount', was 412,700,000,000 m^3, resulting in the annual average amount of water resources per person of 3,230 m^3/(person•year). It decreased to 67% of this numerical value in dry years. The distribution of the annual averages of precipitation and the critical amount of water resources per person in each geographical division are shown in **Figure 1-6**. This indicates that in terms of water resources, waterfront areas in Kanto are very severe, and that Kinki, north Kyushu and Okinawa are hard areas.

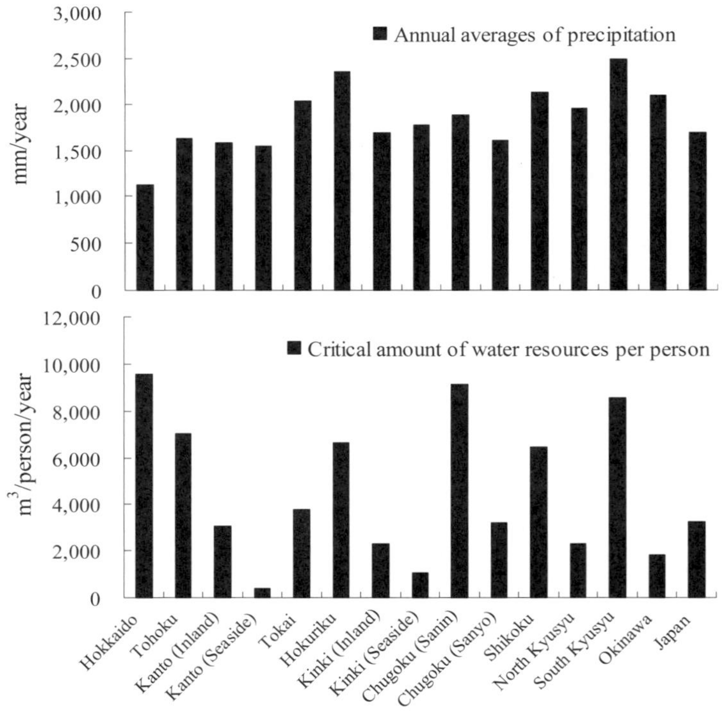

FIGURE 1-6. Regional amount of precipitation and water resources (Water Resources in Japan, 2010).

On the other hand, decreasing trend of annual precipitation and the expansion of fluctuation range in annual precipitation, and the decreasing trend of snowfall and lengthening of dry period are concerned as shown in **Figure 1-7** and **1-8**, respectively.

Figure 1-9 shows the comparison of the amount of water resources in Japan and that in other countries. Though Japan has approximately twice amount of the annual precipitation average of the world's 810 mm/year (FAQ, AQUASTAT), annual precipitation average per person, which is approximately 5,000 m^3/year, is about one third of the world's 16,400 m^3/year because of its narrow area and large population.

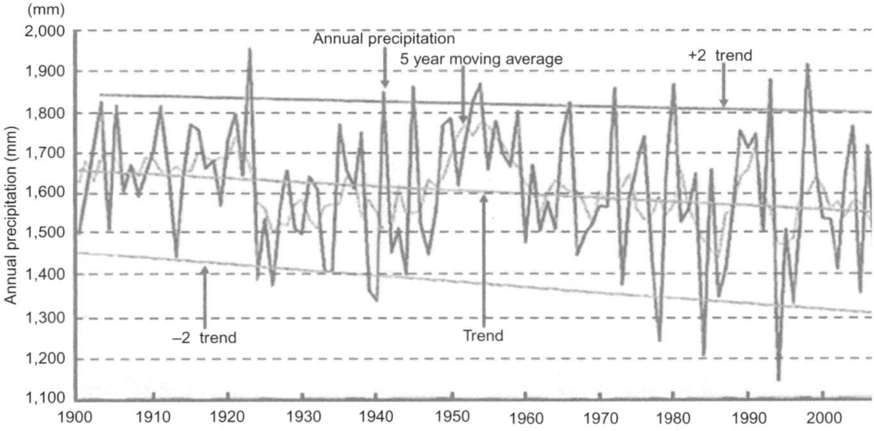

FIGURE 1-7. Temporal change of annual precipitation (Water Resources in Japan, 2010).

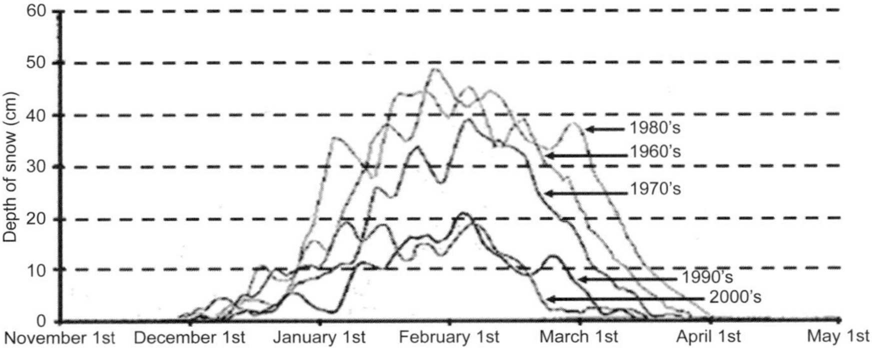

FIGURE 1-8. Change of depth of snow in Toyama (Water Resources in Japan, 2010).

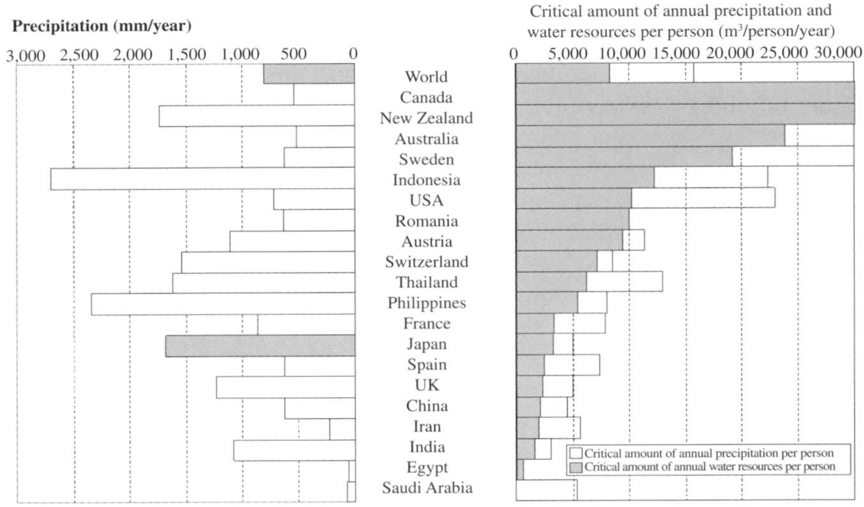

FIGURE 1-9. Precipitation etc. in countries all around the world (Water Resources in Japan, 2010).

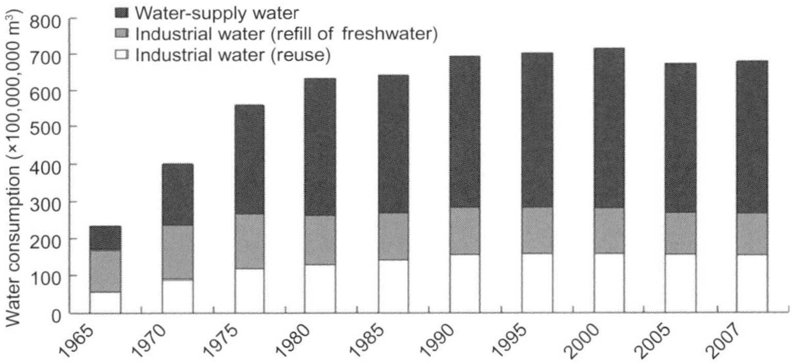

FIGURE 1-10. Change of municipal water consumption (Water Resources in Japan, 2010).

1.2.2 *Usage situation of water resources*

Water consumption in Japan in 2007 was 83,100,000,000 m^3, which consists of 15,700,000,000 m^3 for daily life water, 12,800,000,000 m^3 for industrial water, and 54,600,000,000 m^3 for agricultural water. **Figure 1-10** shows the transition of municipal water consumption which consists of industrial and water-supply water, and indicates that the municipal water consumption has almost flattened in recent years though it had increased since 1965, and that estimated amount of water reuse occupies 80% of the industrial water. **Figure 1-11** describes the

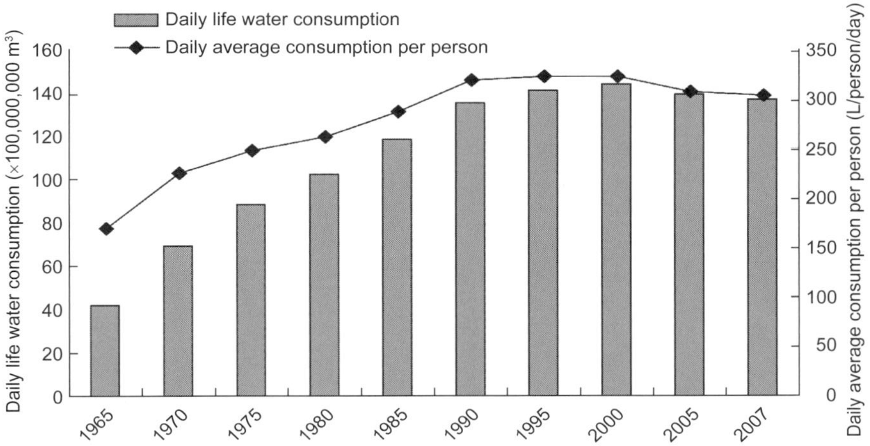

FIGURE 1-11. Change of daily life water consumption (Water Resources in Japan, 2010).

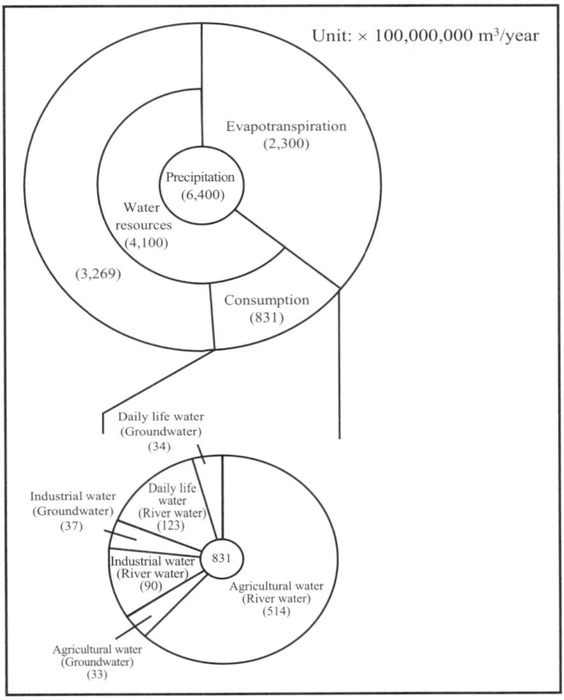

FIGURE 1-12. Water resources and consumption in Japan (Water Resources in Japan, 2010).

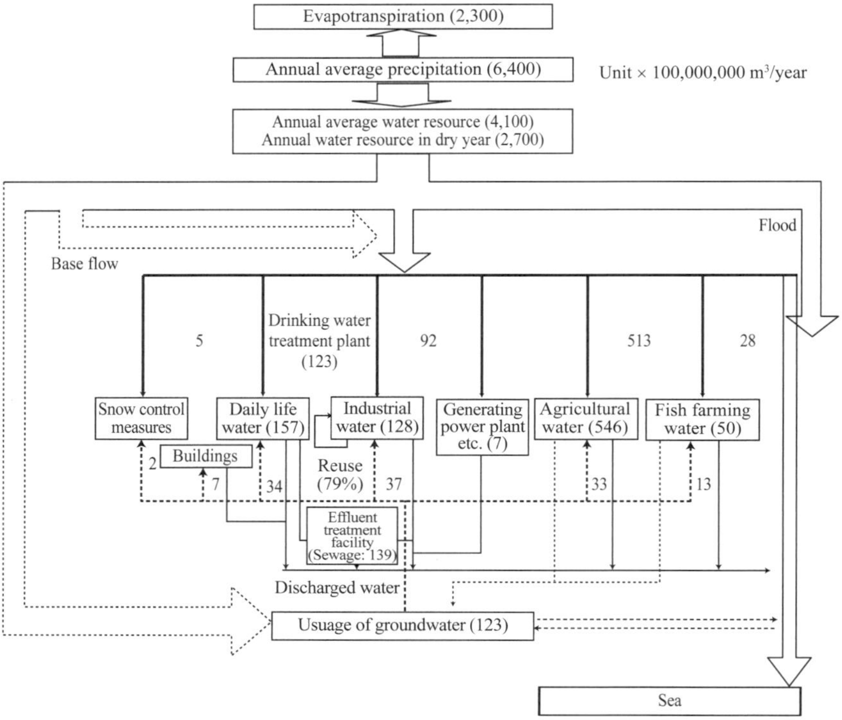

FIGURE 1-13. Water balance in Japan (Water Resources in Japan, 2010).

transition of daily life water consumption. The daily average consumption amount per person was 169 L/(person•day) in 1965. Though it had increased since 1965, it had flattened since 1990 and has been decreasing lately. And it is 303 L/(person•day) in 2007.

The relationship between the critical amount of water resources and consumption, and the balance of water in Japan are shown in **Figures 1-12** and **1-13**, respectively.

1.2.3 *Water shortage and its countermeasures*

Figure 1-14 shows the transition of the number of water shortage areas assuming that water shortage affects the areas when the supplied amount of water is reduced through water supply depressurization and temporal water shortage (water supply and industrial water), when consumers are required to save water according to the saving water rate (industrial water), and when the deterioration of river system or water intake restriction cause insufficient growth on agricultural crops (agricultural water). It shows dozens of places

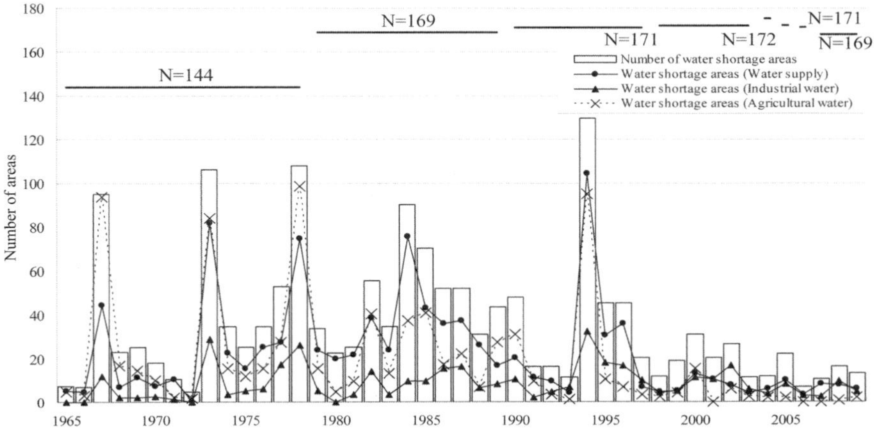

FIGURE 1-14. Number of water shortage areas of each water (Water Resources in Japan, 2010).

are affected every year. The decreasing trend of the number of water shortage areas is also shown.

About the countermeasures against water shortage, more detailed measures are considered to be important in late years in addition to maintenance of facilities for developing water resources. The conference on measures against water shortage has dealt with these actions as follows: liaison, coordination and promotion of various measures, reuse of sewage and industrial wastewater, use of rainwater, and seawater desalination (215,836 m^3 in 2010).

1.2.4 *Water resources development*

Securing the water source through water resources developing facilities such as dams is necessary to utilize precipitating water as municipal water effectively. These facilities have been building year by year (**Figure 1-15**), and the quantity of municipal water development was 182,000,000,000 m^3/year (about 12,200,000,000 m^3/year for water supply, about 6,000,000,000 m^3/year for industrial water) at the end of March, 2010, which occupies 64% of municipal water consumption. When examining how much municipal water they developed in each area (**Figure 1-16**), large quantity of water resources for water supply was developed in the inland part of Kanto, seaside part of Kanto, Tokai, inland part of Kansai, and a lot of industrial water was produced in Kanto, Sanyo and Shikoku.

1.2.5 *Water resources around Lake Biwa and the Yodo River system*

Lake Biwa which main drainage basin is Shiga prefecture is the biggest freshwater lake in Japan. It has 670.25 km^2 of water surface area and 27,500,000,000 m^3

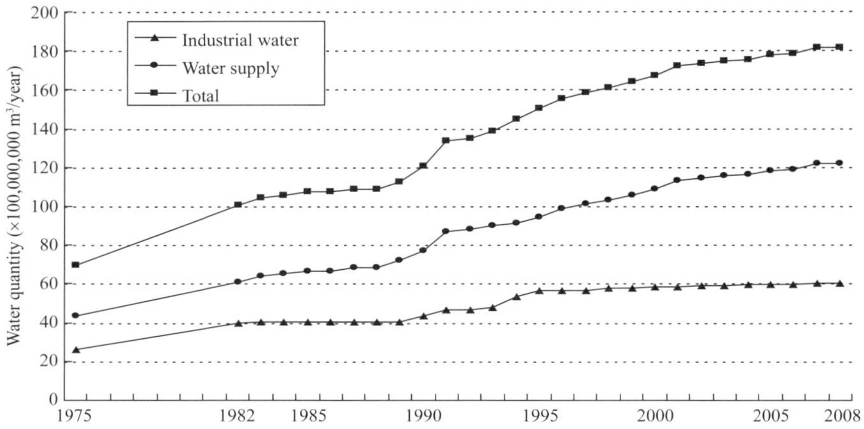

FIGURE 1-15. Quantity of municipal water development by developing facilities (Water Resources in Japan, 2010).

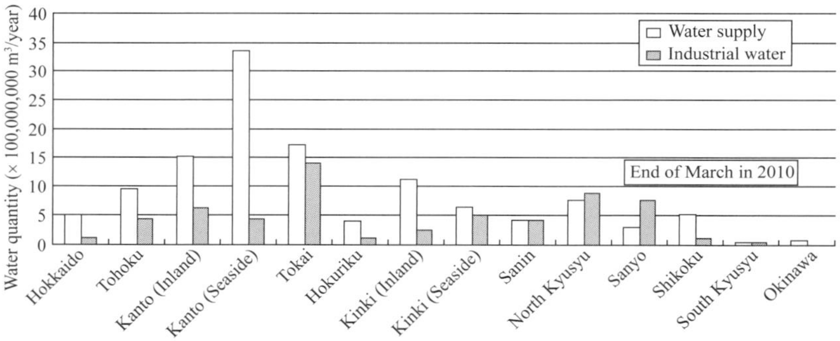

FIGURE 1-16. Quantity of municipal water development in each area (Water Resources in Japan, 2010).

of poundage. Waters along the Yodo river System spring from the lake, go through the Uji river, meet waters from small and large branch rivers such as the Kizu river and the Katsura river, flow through the Kyoto basin and the Osaka plain and inflow into Osaka bay (**Figure 1-17**). The poundage of Lake Biwa is precious water resources which corresponds with about 12 years of water consumption of 16,900,000 people in Kyoto, Osaka, and Kobe area. In the reservoirs water along it, approximately 5,428,000 m³ is used for water supply a day at 106 water purification plants, approximately 1,556,000 m³ for the industrial water a day, and approximately 117.5 m³ for the agricultural water a day.

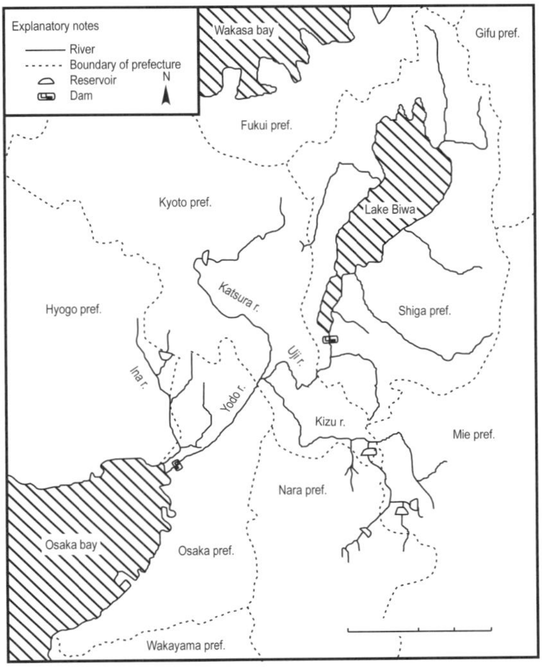

FIGURE 1-17. Lake Biwa and Yodo river system.

1.3 PRESENT AND FUTURE STATE OF WATER RESOURCES IN CHINA

With the development of industrialization and urbanization, water use and wastewater discharge in China has been increasing steadily. However, available water resources are limited. Water shortage and water pollution have become two of the most important environmental problems in China. Water pollution further exacerbates water scarcity. The Chinese government has taken great efforts to deal with these issues in order to support sustainable development (Zhu, Z. *et al.*, 2001; Liu and Diamond, 2005; Li, 2006).

1.3.1 *Available quantity of water resources*

1.3.1.1 Present State of Water Resources

Water resources are mainly derived from atmospheric precipitation and may occur in the forms of surface water and ground water, renewable annually through the hydrological cycle. In China, available quantity of water resources was about 2805.3 billion m^3 in 2005, which was ranked the 6th in the world's available water resources. However, water resources per capita in China (about 2,156 m^3 in 2005) amounted to only the fourth of the world mean (WB, 2008). The available quantity of water resources in China between 1997 and 2008 was summarized in **Table 1-12**.

According to 12 years (1997 to 2008) of statistical data assessed by the Chinese Ministry of Water Resources, total annual precipitation in China ranged from 5,687.6 billion m^3 in 2004 to 6,763.1 billion m^3 in 1998 with a mean value of 6,018.5 billion m^3. Total available water resources was between 2,413.0 billion m^3 in 2004 and 3,401.7 billion m^3 in 1998 with a mean value of 2,754.6 billion m^3. The average annual surface water and ground water showed the same variability in trend—fluctuation within a specified range with a maximum in 1998 as well as a minimum in 2004. Surface water accounted for much more than ground water. Duplication in measured volume averaged at 706.8 billion m^3. Note that the 1998 flooding of the Yangtze river may be attributed to that year's extraordinarily abundant rainfall.

1.3.1.2 Spatial and Temporal Features of Water Resources

Water availability in different parts of China varies significantly due to the monsoon climate and specific topography. China's water resources are geographically divided into six major river basins in northern China (Songhua, Liaohe, Yellow, Huaihe, Haihe, Northwest) and four major river basins in southern China (Yangtze, Pearl, Southeast, Southwest) (**Figure 1-18**). Generally speaking, water resources in southern China are much more abundant than that in northern part. In 2008, the six water resources zones in northern China

TABLE 1-12. Available quantity of water resources in China between 1997 and 2008*.

Year	Precipitation	Water resources	Surface water	Ground water	Duplication in measured volume
1997	58,169	27,855	26,835	6,942	5,923
1998	67,631	34,017	32,726	9,400	8,109
1999	59,702	28,196	27,204	8,387	7,395
2000	60,092	27,701	26,562	8,502	7,363
2001	58,122	26,868	25,933	8,390	7,455
2002	62,610	28,255	27,243	8,697	7,685
2003	60,416	27,460	26,251	8,299	7,090
2004	56,876	24,130	23,126	7,436	6,432
2005	61,010	28,053	26,982	8,091	7,020
2006	57,840	25,330	24,358	7,643	6,671
2007	57,763	25,255	24,242	7,617	6,604
2008	62,000	27,434	26,377	8,122	7,065

*Data source: MWR (1997–2010); unit: 100 million m^3.

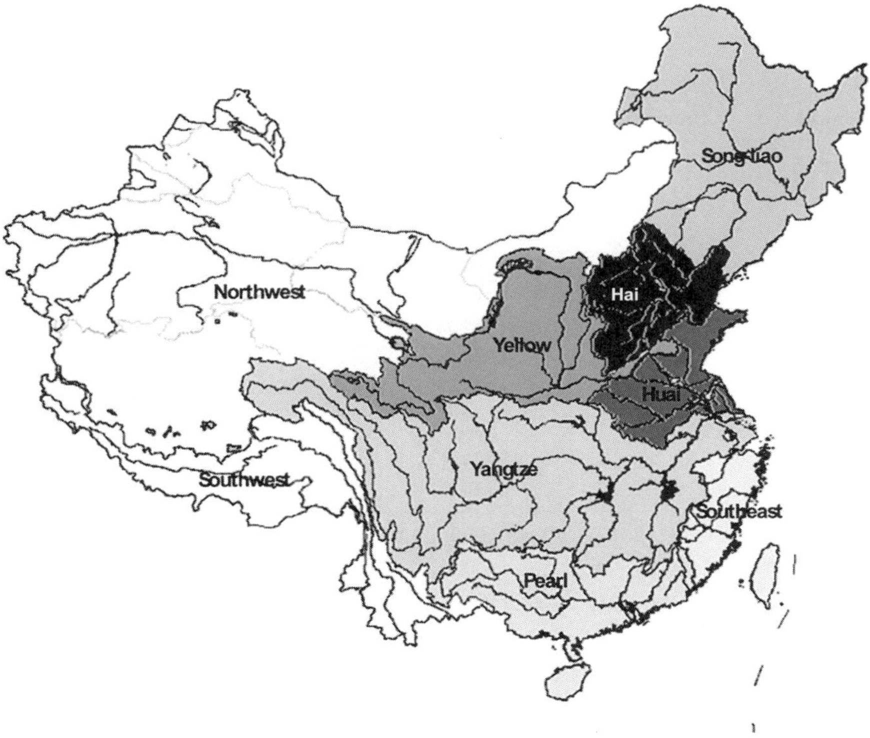

FIGURE 1-18. Map of major river basins in China (Jiang, Y., 2009). The increasing darkness corresponds to the decreasing annual per capita water availability.

had a total of 460.1 billion m^3 with available water resources per capita value of about 776.7 m^3. The four water resources zones in southern China had a total of 2,283.4 billion m^3 with a available water resources per capita of about 3,286.9 m^3 (**Table 1-13**).

Due to strong monsoonal climates, the annual rainfall is mainly concentrated in the summer season and gradually declines in a spatial gradient from more than 2,000 mm at the southeastern coastline to usually less than 200 mm at the northwestern hinterlands (MWR, 2004). China suffers from highly variable rainfall in different seasons, giving rise to frequent floods and droughts, often simultaneously in various regions. The volume ratio of observed maximum annual runoff to minimum annual runoff for regions north of the Yangtze river is more than 10, while the ratio is less than 5 for rivers in the southern part of China. As for the seasonal distribution, the runoff in northern China is more concentrated than that in the southern China at times. For rivers north of the Yangtze river, the total runoff during four consecutive

TABLE 1-13. Available quantity of water resources in major river basins in 2008*.

Regions	Precipitation	Water resources	Surface water	Ground water	Duplicated in measured volume
Total	62,000.3	27,434.3	26,377	8,122	1,057.3
North	19,534.8	4,600.7	3,681.6	2,455.2	919.1
Songhua	4,353.2	982.7	788.5	426.1	194.2
Liaohe	1,586.8	393.9	305.1	171.7	88.8
Haihe	1,729.5	294.5	126.9	242.1	167.6
Huanghe	3,443.1	559.0	454.2	344.7	104.9
Huaihe	2,902.1	1,047.2	782.1	430.6	265.1
Northewestern	5,520.1	1,323.4	1,224.8	840	98.6
South	42,465.5	22,833.6	22,695.4	5,666.8	138.2
Yangtze	19,120.6	9,457.2	9,344.3	2,416.3	113
Pearl	10,438	5,696.8	5,682.3	1,293.7	14.6
Southeastern	3,269.5	1,735.2	1,724.4	454.4	10.7
Southwestern	9,637.5	5,944.4	5,944.4	1,502.4	0.0

*Data source: MWR (2010) for year 2008; unit: 100 million m^3.

months of high flow usually accounts for 80% of the total amount of annual runoff, and even 90% for the Haihe river. In contrast, rivers south of the Yangtze river make up about 60% of the annual runoff (Zhang, H., 2005).

1.3.1.3 Water Pollution

Database comprising water quality for 150,000 kilometers of national monitoring rivers showed that river length with grade I water quality accounted for 3.5% of the total river length, grade II 31.8%, grade III 25.9%, grade IV 11.4%, grade V 6.8% and inferior to grade V 20.6% in 2008 (**Figure 1-19**) (Refer to the APPENDIX for details on the Environmental Quality Standards for Surface Water.). Water quality for northwest rivers, southwest rivers, southeast rivers, the Yangtze river, and the Pearl river was relatively good, with grade III and superior water amounting from 64%~95%. Water quality for the Haihe river, the Yellow river, the Huaihe river, the Liaohe river, and the Songhua river was relatively poor, with grade III and superior quality water amounting from 35%~47%.

Water quality survey of 44 lakes in 2008 indicated that lake areas with water quality of grade III and superior amounted to 44.2%, grade IV and grade V at 32.5%, and inferior to grade V at 23.3% (**Figure 1-20**). Trophic state evaluation of 44 lakes showed that it had 1 lake with oligotrophy, 22 lakes with mesotrophy, 10 lakes with mild eutrophy and 11 lakes with moderate eutrophy.

Among 378 reservoirs assessed by the Ministry of Water Resources in 2008, there were 303 reservoirs with grade III and superior water quality, 59 reservoirs with grade IV and V water quality, and 16 reservoirs with inferior to grade V water quality (**Figure 1-21**). Trophic state evaluation of reservoirs

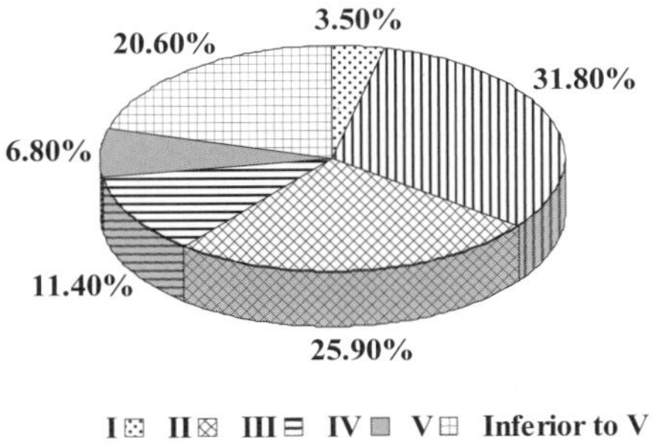

FIGURE 1-19. Water quality of rivers in China in 2008 (MWR, 2010).

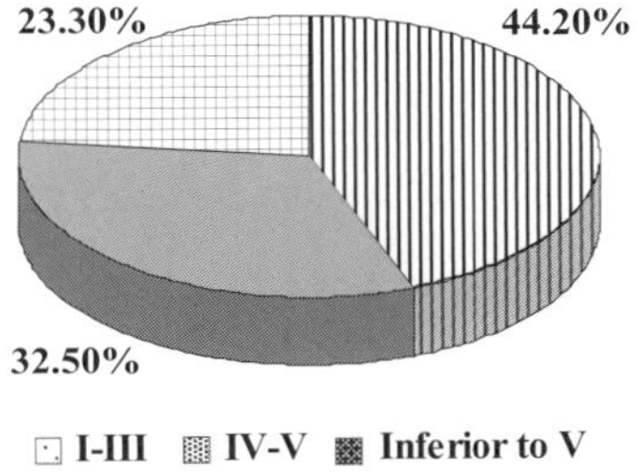

FIGURE 1-20. Water quality of lakes in China in 2008 (MWR, 2010).

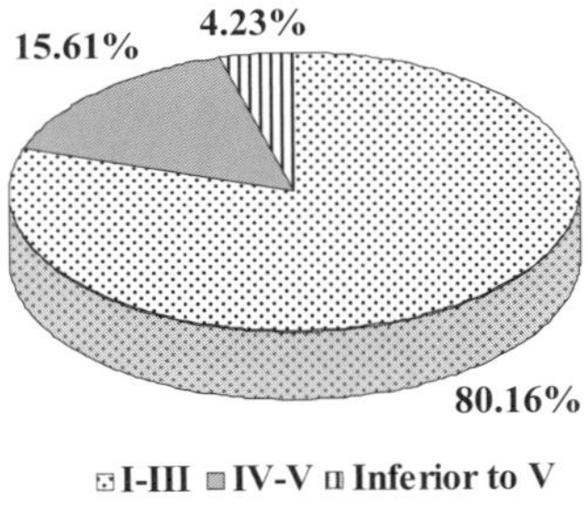

FIGURE 1-21. Water quality of reservoirs in China in 2008 (MWR, 2010).

showed 241 reservoirs with mesotrophy, 86 reservoirs with mild eutrophy, 18 reservoirs with moderate eutrophy and 2 reservoirs with heavy eutrophy.

In 2008, monitoring data for 641 wells showed 2.3% wells with grade I-II quality (applicable for all purposes), 23.9% wells with grade III quality (suitable for drinking and domestic, industrial and agricultural use), and 73.8 wells

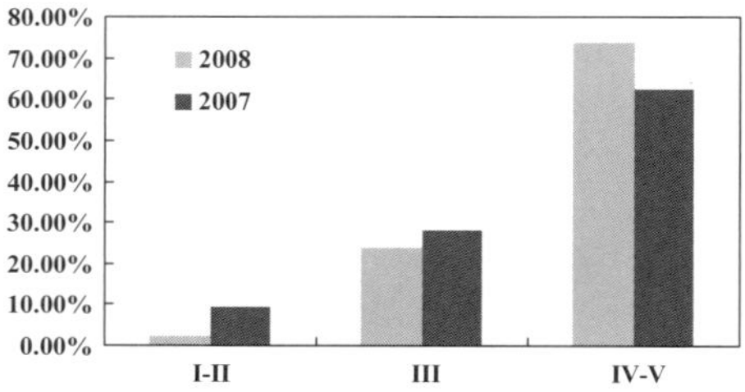

FIGURE 1-22. Comparative study of water quality of wells in China in 2007 and 2008 (MWR, 2010).

with grade IV–V quality (suitable for purposes other than drinking), suggesting that water quality of ground water in 2008 deteriorated compared with that in 2007 (**Figure 1-22**) (Refer to the appendix for details on the Quality Standard for Ground Water).

1.3.2 *Utilization state of water resources*

During the year of 1997~2002, total water supply and water consumption fluctuated around 555.0 billion m^3 and 552.6 billion m^3, respectively (**Table 1-14**). For water supply, surface and ground water averaged 446.5 billion m^3 (80.5%) and 106.2 billion m^3 (19.1%), respectively. For water consumption, agricultural, industrial and domestic use averaged 381.7 billion m^3 (69.1%), 113.8 billion m^3 (20.6%) and 69.1 billion m^3 (10.3%), respectively. Most of the water used in China was for agriculture. There was no record of water used in ecology during the 1997~2002 period. However, national policy reforms during the 2003~2008 period resulted in a gradual increase of water usage in ecology from 8.0 billion m^3 to 12.0 billion m^3. Total water supply progressively increased by 11.1%, 86.1% of which was from surface water. Concurrently, water used in the agricultural, industrial and domestic sectors increased by 23.3 billion m^3, 22.1 billion m^3 and 9.6 billion m^3, respectively. Note that agricultural water usage declined during the 2003~2008 period relative to the 1997~2002 period. This reflects change in the national economic structure regarding implementation of highly-efficient irrigation technology and comprehensive water-saving methodology. Correspondently, water usage in industrial and, in particularily, the domestic sectors increased rapidly. Year 2008 industrial water usage increased by 24.6% compared with that in 1997. Moreover, the year 2008 domestic water usage increased by 38.9% compared with that in 1997. To some extent, these figures indicate prosperous socio-economical growth along with better living conditions in China.

The socio-economic indices of water use during the 1997~2008 period were tabulated in **Table 1-15**. The annual per capita water use fluctuated around 436 m^3. The water use per 10,000 RMB gross domestic product (GDP) decreased significantly from 726 m^3 in 1997 to 193 m^3 in 2008, indicating improvement in China's economic structure. The water use per 10,000 RMB added value of industry declined from 288 m^3 in 2000 to 108 m^3 in 2008, showing remarkable enhancement in efficient industrial water use. The water irrigated per mu (1 mu = 1/15 hectare) of farmland gradually decreased from 492 m^3 in 1997 to 435 m^3 in 2008. Water use per capita in urban and rural areas was around 216 L per day and 79 L per day, respectively.

Total water supply reached 591.0 billion m^3 in 2008, of which surface water, groundwater and other water sources were 479.6 billion m^3 (81.2%), 108.5 billion m^3 (18.3%) and 2.9 billion m^3 (0.5%), respectively (**Table 1-16**). For surface water, 33.8% derived from storage projects, 38.6% from diversion projects, 24.9% from lifting projects and 2.7% from grade-I river basins. For ground water, 80.1% originated from shallow ground, 19.4% from deeply defined water and 0.5% from slightly saline water.

The water supply of northern and southern China were 262.2 billion m^3 and 328.8 billion m^3, accounting for 44.4% and 55.6% of the national total water supply, respectively. Water supply in the southern provincial administrative regions derived mostly from surface water, accounting for more than 90%; meanwhile water supply in the northern regions came mostly from ground water, especially in Hebei, Henan, Shanxi provinces and Beijing municipality, accounting for more than 50%.

Based on statistical data assessed by Ministry of Water Resource in China, water consumption totaled at 591.0 billion m^3 in 2008, of which domestic, industrial, agricultural and ecological water were 72.9 billion m^3 (12.3%), 139.7 billion m^3 (23.7%), 366.3 billion m^3 (62.0%) and 12.0 billion m^3 (2.0%), respectively (**Table 1-16**). Compared with that in 2007, water consumption increased by 9.1 billion m^3, of which domestic water increased by 1.9 billion m^3, industrial water decreased by 0.7 billion m^3, agricultural water increased by 6.5 billion m^3 and ecological water increased by 1.4 billion m^3.

Utilization of water resources varied significantly amongst the different provincial administrative regions. Water consumption in the Jiangsu, Guangdong provinces and the Xinjiang autonomous region were more than 40.0 billion m^3. Water consumption in the Qinghai, Hainan provinces, the Tibet autonomous region and Tianjin, Beijing municipalities were less than 5.0 billion m^3. Agricultural water in the Gansu, Hainan provinces and the Xinjiang, Ningxia, Tibet, Inner Mongolia autonomous regions accounted for more than 75% of their total water consumption. In addition, industrial

TABLE 1-14. Supply and use of water resources in China.

Year	Water supply	Surface water	Ground water	Other	Water use	Agriculture	Industry	House-hold and service	Ecology
1997	5,623	4,566	1,031	26	5,566	3,920	1,121	525	–
1998	5,470	4,420	1,028	22	5,435	3,766	1,125	544	–
1999	5,613	4,514	1,075	24	5,591	3,869	1,159	563	–
2000	5,531	4,440	1,069	21	5,498	3,784	1,139	575	–
2001	5,567	4,448	1,097	22	5,567	3,825	1,141	601	–
2002	5,497	4,403	1,072	22	5,497	3,738	1,143	616	–
2003	5,320	4,288	1,016	16	5,320	3,431	1,176	633	80
2004	5,548	4,505	1,026	17	5,548	3,584	1,232	649	83
2005	5,633	4,574	1,036	23	5,633	3,583	1,284	676	90
2006	5,795	4,706	1,066	23	5,795	3,662	1,344	695	93
2007	5,819	4,725	1,071	23	5,819	3,602	1,402	710	105
2008	5,910	4,796	1,085	29	5,910	3,664	1,397	729	120

Data source: MWR (1997–2010); unit: 100 million m^3; –: no record.

TABLE 1-15. Socio-economic indices of water use.

Year	Per capita water use (m^3/person)	Water use/ GDP (m^3/ 10,000 yuan)	Water use/ Value added of Industry (m^3/10,000 yuan)	Water use/ Mu of farmland irrigation (m^3/Mu)	Per capita water use (Urban area) (L/d)	Per capita water use (Rural area) (L/d)
1997	458	726	–	492	220	84
1998	435	683	–	488	222	87
1999	440	680	–	484	227	89
2000	430	610	288	479	219	89
2001	436	580	268	479	218	92
2002	428	537	241	465	219	94
2003	412	448	222	430	212	68
2004	427	399	196	450	212	68
2005	432	304	166	448	211	68
2006	442	272	154	449	212	69
2007	442	229	131	434	211	71
2008	446	193	108	435	212	72

Data source: MWR (1997–2010); –: no record; 1 Mu = 1/15 hectare.

water in the Fujian, Jiangsu, Hubei, Guizhou, Anhui provinces and Shanghai, Chongqing municipalities accounted for more than 30%, while domestic water in Beijing, Tianjin and Chongqing municipalities accounted for more than 20%.

TABLE 1-16. Supply and use of water resources in various river basins in 2008.

	Water suply				Water use				
Regions	Surface water	Ground water	Other	Total	Household and service	Industry	Agriculture	Ecology	Total
North	1,655.9	949.2	16.8	2,621.9	261.5	340.8	1,959.2	60.4	2,621.9
Songhua	236.1	174.9	0	411.1	33.4	79.21	294.2	4.3	411.1
Liaohe	88.3	111.7	2.6	202.6	31.7	30.4	137.1	3.3	202.6
Haihe	123.3	240.6	7.7	371.5	57.1	51.3	254	9.1	371.5
Yellow	253.9	128.1	2.2	384.2	39.8	60.8	277.2	6.5	384.2
Huaihe	432.4	175.8	3	611.2	81.9	98.6	421.7	9	611.2
Northwestern	521.82	118.1	1.4	641.3	17.6	20.43	575.1	28.2	641.3
South	3,140.5	135.6	11.9	3288	467.7	1,056.3	1,704.2	59.8	3,288
Yangtze	1,861.7	83.2	6.6	1,951.5	250.5	718	948.1	34.9	1,951.5
Pearl	837	40.1	4.1	881.2	155.7	210.3	502.3	12.9	881.2
Southeastern	333.4	9.1	1	343.6	49.5	119.8	162.6	11.6	343.6
Southwestern	108.4	3.2	0.2	111.8	11.9	8.3	91.2	0.4	111.8
Total	4796.4	1,084.8	28.7	5,909.9	729.2	1,397.1	3,663.4	120.2	5,909.9

Data source: MWR (2010) for year 2008; unit: 100 million m^3.

1.3.3 *Water shortage and countermeasures*

Despite abundant natural water resources, China still faces severe water shortages caused by an increasing population and other natural factors. Water pollution also adds to the problem. Therefore, the government has developed countermeasures in order to obtain sustainable development for China.

1.3.3.1 Water Shortage

(1) *Spatial Feature of Water Shortage*

China's annual per capita water resources were around 2,000 m^3 during the 2000~2008 period. Precipitation varies yearly, as reflected by the fluctuation in annual water resources (**Figure 1-23**). In 2005, annual per capita water resources were about 2,156 m^3 which were far below the world average (6,794 m^3) at that time (WB, 2008). Due to a large population base, the per capita water resources stayed low despite abundant water resources. According to international standards, the per capita water resources of less than 3,000 m^3 is regarded as mildly dry; less than 2,000 m^3 per capita is regarded as moderately dry; less than 1,000 m^3 per capita is regarded as serious water shortage; per capita water resources of less than 500 m^3 is regarded as water scarcity. Overall, China was considered to be a mildly arid country, though some years exhibited less than 2,000 m^3 per capita water resources (**Figure 1-23**).

Water resources in southern China are much more abundant than that in northern regions. The spatial and temporal disparities of water resources do not match the distribution of China's population, industry and agriculture. Seasonal variation in precipitation, especially during dry spells, worsens the situation of water shortage in certain regions.

In 2008, water resources in six core areas in northern China (Songhua, Liaohe, Haihe, Yellow, Huaihe and Northwest rivers) totaled at 460.07 billion m^3, accounting for 16.8% of the national amount. The other 83.2% came from four water source zones (Yangtze, Pearl, Southwest and Southeast rivers) in southern China (**Table 1-13**). The few northern basins often faced serious water shortages. In the Huang-Huai-Hai river basins, 35% of China's population had access to only 7.7% of China's naturally available water resources with averages of 500 m^3 per capita or less than 6,000 m^3 per hectare (China's sustainable development of water resources project team of Chinese Academy of Engineering, 2002).

According to 2007 statistics (**Table 1-17**), North China accounted for only 2.2% of the nation's naturally available water resources, alongside 11.9% of its population, 14.9% of its cultivated land, and 14.5% of its GDP. Per capita water resources in the region were only 360 m^3. Southwest China accounted for 41.9% of its available water resources, alongside 15.1% of its population, 15.7% of its cultivated land, and 8.1% of its GDP. According to international standards, eight municipalities, provinces or autonomous regions, including

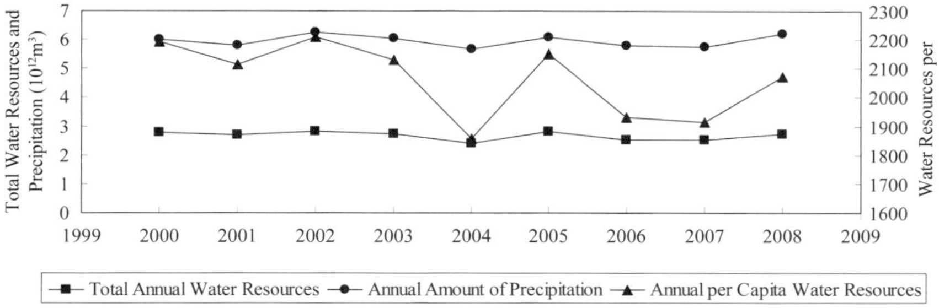

FIGURE 1-23. Variation of Annual Water Resources and Precipitation.
Data source: MWR (2000–2008) for total amount of annual water resources or precipitation; NBS (2008) for annual per capita water resources; The annual per capita water resources in 2008 are the division of the total amount of water resources by the average population in 2008.

TABLE 1-17. Spatial distribution of China's water resources and other social variables in 2007.

Regions	Total amount of water resources (10^8 m^3, %)	Per capita water resources (m^3)	*Population (millions, %)	Area of cultivated land (10^3 hectares, %)	Gross regional product (10^8 yuan, %)
*North China	554.2 (2.2)	360	154.08 (11.9)	18,191 (14.9)	39,938 (14.5)
Northeast China	1,099.6 (4.4)	1,015	108.35 (8.4)	21,459 (17.6)	23,373 (8.5)
East China	4,707.8 (18.6)	1,250	376.61 (29.1)	24,336 (20.0)	1,04,790 (38.0)
Central and South China	6,157.7 (24.4)	1,697	362.86 (28.1)	24,168 (19.9)	71,706 (26.0)
Southwest China	10,594.3 (41.9)	5,434	195.00 (15.1)	19,110 (15.7)	22,453 (8.1)
Northwest China	2,141.5 (8.5)	2,244	95.43 (7.4)	14,472 (11.9)	13,364 (4.9)

*North China includes Beijing municipality, Tianjin municipality, Hebei province, Shanxi province and Inner Mongolia autonomous region.
*The regional population was the average of the year-end statistical data of 2006 and 2007. The data were not adjusted on the basis of sampling errors and survey errors.
Data source: NBS (2008).

Beijing, were undergoing water scarcity. Per capita water resources in Beijing were only 148.2 m^3. Only six municipalities, provinces or autonomous regions were not short of water resources (**Figure 1-24**).

However, it should be pointed out that the regions classified 'not short of water resources' actually face water shortages. In accordance with another

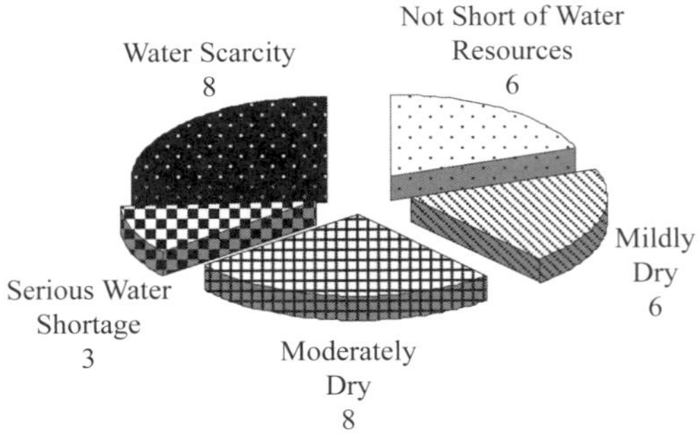

FIGURE 1-24. Situation of Water Resources in 31 Municipalities, Provinces or Autonomous Regions in 2007.
Data source: NBS (2008); Hong Kong, Macao and Taiwan were not included.

standard, if runoff depth equivalent to the amount of water resources is less than 150 mm, the region is facing ecological imbalance. Therefore, it is unsuitable for further human development. Although per capita water resource for 2007 in the Qinghai province and the Xinjiang autonomous region were 12,029.5 m^3 and 4,167.8 m^3, respectively (NBS, 2008), the two regions were still undergoing serious water shortages due to ecological factors.

(2) *Temporal Feature of Water Shortage*

In addition to the deficiency of water resources per capita in various regions, the temporal distribution of water resources in China is also extremely uneven. The annual precipitation varies greatly from year to year and from season to season. Water shortages become apparent during the dry seasons or even dry years.

The amount of precipitation in North and Northwest China was about one third of that in other regions. Over the past 100 years, precipitation differences in inter-regional increased, with rainfall gradually declining in North China at rates of 20–40 mm/decade, and rising in South China at rates of 20–60 mm/decade (Jian, X. *et al.*, 2009). The gap between northern and southern resources continues to widen.

Global atmospheric temperature continues to rise due to the greenhouse effect. Rise in temperature will intensify evaporation of surface water, further reducing the quantity of water resources and worsening the situation of water shortages. It is predicted that the average nationwide temperatures will increase by 2.3 to 3.3°C by 2050 as compared to 2000 (Jian, X. *et al.*, 2009).

(3) *Depletion of Groundwater Resources*

Groundwater is an important source of water for China's urban, industrial and agricultural sectors. It is especially essential for the northern regions due to lack of surface water. Groundwater generally accounts for more than 50% of the region's total water resources. Due to inefficient development and utilization of groundwater, this Figure continues to drop year by year. Furthermore, excessive pollution exacerbates the situation.

In late 1950s, the exploitation of groundwater in the Beijing plain was around 0.4 billion m^3 per year and gradually increased during the 1960s. Theis value scaled up during the 1970s, resulting in over-exploitation. In 1971, the total exploitation of ground water was 1.38 billion m^3, and then to 2.56 billion m^3 in 1978. Due to a series of dry years in the 1980s, groundwater extraction rose even higher at around 2.6 billion m^3 to 2.8 billion m^3. This value began to stabilize during the 1990s. Entering the 21st century, groundwater utilization remains steady ranging from 2.4 billion m^3 to 2.6 billion m^3 per year (Zhang, Z. *et al.*, 2009).

In the North China Plain, the amount of shallow and deep groundwater resources exploited was about 2.64 billion m^3 and 1.24 billion m^3 per year, respectively. In contrary to the water tables in the late 1950s, the water tables for shallow groundwater dropped by 10 to 30 meters in the piedmont plain area, 5 to 10 meters in the central and east plain area, and 0 to 5 meters in the coastal plain area by 2001 (Zhang, Z. *et al.*, 2009).

According to an incomplete survey of 21 provincial-level administrative regions in 2008, depression cones for shallow groundwater and complex cones for deep groundwater within an area of 70,000 km^2 were defined (MWR, 2008). According to some other data, about 24.8% of all groundwater sources were polluted. The polluted area measured at 49.6 km^2 of which 24.9% was grade IV and V in groundwater quality standards (Zeng, H., 2004).

(4) *Effects of Serious Water Pollution on Water Shortage*

During the early foundation stages of the People's Republic of China, the quantity and quality of water resources were in good condition. Water pollution arose as a result of rapid industrialization and urbanization, in conjunction with a growing population increased water shortages. Extent of water pollution was wide, effecting whole river basins in critical regions. Ever since the 1980s, the government has taken steps in dealing with water pollution. However, conditions remain serious for most parts. In addition to surface water, offshore water and groundwater have also been contaminated, further contributing to water shortages.

With the development of China, industrial and domestic wastewater emissions increased annually from 1998 to 2008 in pace with water demand (**Figure 1-25**). Total industrial and domestic wastewater discharges have steadily risen to 57.17 billion tons, with COD discharges and NH_3–N discharges amounting to 13.21 million tons and 1.27 million tons, respectively (MEP, 2008). Industrial COD discharges had been brought under control due to an increasing share of treated industrial wastewater. Domestic wastewater emission increased gradually, most of which were untreated. Since 1999,

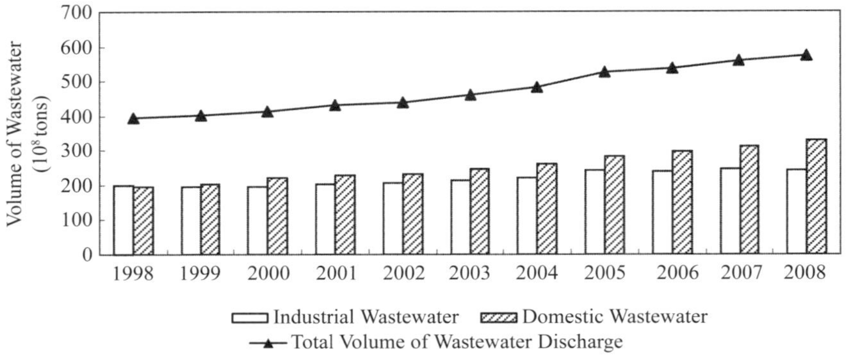

FIGURE 1-25. Variation of Wastewater Discharge.
Data source: MEP (1998–2008).

domestic wastewater discharges has surpassed industrial discharge and has become the dominant pollution source. As of 2005, 52% of organic pollutants (Biochemical Oxygen Demand) were attributable to domestic sources (Jian, X. *et al.*, 2009). About one third of industrial wastewater and more than 90% of domestic wastewater were discharged into rivers or lakes without any forms of treatment. No less than 100 million tons of wastewater was discharged directly every day in cities and towns over the country (Gan, P., 2006). In addition, non-point source pollution from agricultural and rural regions has become the leading source of water pollution for some severely polluted waters.

Although some regions are abundant in surface water, available water resources are limited due to severe water pollution. Since the early 1990s, overall water quality in China does not seem to have changed much, but regional trends were quite different. There were improvements in the South, contrasting with deterioration in the North (Jian, X. *et al.*, 2009). According to year 2008 data on all monitored river sections mentioned above (For details, see **Figure 1-19** to **1-22**), 38.8% of the evaluated river length failed to meet grade II. Certain areas in river basins had been seriously polluted. The proportion of poor quality water (worse than grade V) of the Haihe river was as high as 51.9% in 2008 (MWR, 2008), which intensified the shortage of utilizable water in North China. On the contrary, water quality in southern China was relatively better. Water pollution greatly influenced the available amount of water resources.

1.3.3.2 Countermeasures

Despite high speed economical growth, water shortages will continue to hinder further development in China. China's population is expected to continue growing to 1.6 billion by 2050. Water shortage will only continue to worsen if no effective steps are put into practice. More efforts must be made by the government in order to keep sustainability of water resources. The public also play an important role in the process.

(1) *Legislate for Effectual Water Management*

A perfect legal framework is important in guaranteeing efficient water management. Therefore, the Chinese government has made strenuous efforts in improving the relevant legal system in order to achieve such efficiency. The existing legal framework can be classified into four levels: Constitution, national laws and their implementation, national and sectoral administrative regulations, and local laws and regulations (Jian, X. *et al.*, 2009). Supervision and inspection at all governmental levels must be enforced. However, additional efforts must be made to complete the law system.

(2) *Save Water and Increase Water Use Efficiency*

Water-saving practice is not only a fundamental state policy of this country that would be adhered to for a long time, but also the best way to reduce cost of water supply and wastewater discharge. In 1984, the State Council issued a notice on promotion of water saving in urban areas. Afterward, a series of regulations and laws on water saving were drawn up by relevant departments. Since the beginning of the 7th Five-Year Plan, the government designated water-saving as a major scientific research program and provided a lot of funding towards the development.

A number of urban and rural areas were selected to put into practice the water-saving demonstration projects at the same time, which promoted the development of water-saving in agriculture, industries, cities and society. Forty cities had been designated as water-saving cities by 2010. China is a large agricultural country, resulting in a large proportion of water used in the fields. Most of the agricultural water is used to irrigate. In 2008, water for agricultural use accounted for 62% of the total water use of the whole country (**Table 1-14**). Therefore, it is vital to develop water-saving irrigation methods. Micro-irrigation, drip irrigation, and low pressure pipe irrigation are just some of the water-saving technologies popularized all over the country.

For a long time, price of water in China had been unreasonably low, contradicting the policy of water-saving, pollution control and unified water resources management. Since the 1990s, China has gradually increased the price of water in both urban and rural areas. For example in Beijing, residential water price from 1991 to 2004 increased from 0.12 to 3.70 yuan/m^3 (Jian, X. *et al.*, 2009), and settled at 4.0 yuan/m^3 by the end of 2009. The residential water prices are adjusted to be made up of three parts: water supply fees, water resources fees and wastewater treatment fees. Price reforms hope to increase the efficiency of water utilization and arouse public awareness in saving water. However, the average cost of wastewater is still less than the actual cost of treatment. Statistics from The World Bank indicate water price should account for 4% of the household's income; China averages at about 2%.

Efficient use of water is essential to addressing water shortages. The water conservation movement and Water Price Reform are expected to strengthen public awareness of using water responsibly.

(3) *Accelerate Construction of Water Diversion Projects*

Optimized reallocation of water resources is a main measure in countering the uneven distribution of water resources. A number of inter-basin water transfer projects have been constructed since the 1960s, in order to cope with water shortages in some regions. The eastern route and central route of the South-to-North water diversion project were launched in 2002 and 2003, respectively. Construction of the western route of the South-to-North water diversion project, which involves working on the Qinghai-Tibet plateau, is not scheduled to begin until 2010 and will involve overcoming some major engineering and climatic challenges. When finished, the work will link China's four main rivers – the Yangtze, Yellow river, Huaihe river and Haihe river. It will eventually divert 44.8 billion m^3 of water annually to the population centre of the drier north. Water diversion from south to north is a feasible approach to cope with the spatial disparity of water resources, and it is also a way to resolve the crisis of groundwater over-exploitation in northern China. Besides the South-to-North water diversion project, a number of water resources allocation projects are under construction or will be constructed in the future; water from the Yellow river will be diverted to Tianjin to help alleviate the city's water shortage, and the Wudu Water Diversion Project is under construction in order to provide water for agricultural use in arid areas in Sichuan province. The inter-basin water transfer projects are going to ease the imbalance between supply and demand of water resources in some special regions.

(4) *Prevention and Control of Water Pollution*

Industrial wastewater control began during the 1970s. By the year 2007, investment in industrial wastewater control rose to 19.61 billion yuan, and 78,210 wastewater treatment facilities had been built. Proportion of industrial wastewater meeting discharge standards rose from 76.9% to 92.4% from 2000 to 2008 (MEP, 2000–2008). Municipal wastewater control was put in practice in the mid-1980s. Investment on urban environmental infrastructures was 146.7 billion yuan in 2007. Treatment rate of municipal wastewater rose from 34.3% to 62.9% from 2000 to 2007 (NBS, 2008). Since the mid-1990s, pollution control in key river basins was paid more and more attention. The government carried out a series of decisions, such as pollution control plans for the Three Rivers and Three Lakes. The total discharge of major pollutants decreased steadily, and water quality of some lakes had been improved.

In 2003, groundwater protection activities were launched for rational development in groundwater utilization. Environmental and geological problems resulting from over-exploitation received more attention.

To strengthen mitigation of water pollution remains a major task for the government in the future. Improving water quality, reducing COD discharge and controlling pollution in major river basins were all important aspects of environmental protection discussed in the National 11th Five-Year Plan (**Table 1-18**).

The Five-Year Plan emphasized on accelerate construction projects for wastewater treatment and recycling. It was planned that sewage treatment will be implemented in every city until 2010, raising treatment rate of domestic sewage above 70% with a daily domestic sewage treatment capacity of 100 million tons nationwide. The government invested huge sums of money towards the construction projects for water pollution control every year. Available data from the Ministry of Water Resources indicated that the State had completed 112.92 billion yuan worth of investment on the control of major rivers in 2008 (MWR, 2008).

In 2011, the next Five-Year Plan will be drawn up and then put into practice. Officials of the Ministry of Environmental Protection and Water Resources said that wastewater treatment will be strengthened in cities and towns. Improved wastewater treatment facilities will gradually be built and wastewater treatment in every city and town will be brought into effect. Water pollution control for five rivers in the north and the Three Lakes is still a major concern for the government. The Three Gorges Reservoir area and areas along the South-to-North Water Diversion Project cannot be ignored. The plan involves 23 Municipalities, Provinces or Autonomous Regions, 2.75 million km^2 of key river basins, 788 million of population, and up to 812.87 billion m^3 of water resources (China Water Net, 2009). However, it is hard to achieve targets without support from the general public during the process of industrialization and urbanization.

TABLE 1-18. Indicators about water quality in the National 11th Five-Year plan.

Indicators	2005 (Beginning)	2010 (End)
Total amount of COD discharge (10^4 tons)	1,414	1,270
Proportion of monitoring sections of surface water worse than Grade V (%)	26.1	Less than 22
Proportion of monitoring sections of the seven key river systems better than Grade III (%)	41	Greater than 43

Date source: MEP (2007).

In order to limit pollutant discharge, the State Council issued the Interim Measures on the Collection of Pollution Discharge Fee in 1982, which marked the establishment of a emission fee collection system. From then on, the system had made gradual improvements, allowing local governments at all levels to adjust the system according to local conditions. It was emphasized in the Eleventh Five-Year Plan that China will make more efforts in collecting and enforcing emission fees and further improving the emission fee collection system.

1.3.4 *Development of water resources*

Besides traditional water sources such as rivers, lakes and groundwater, developments in untraditional water sources has been under the scope. Wastewater reclamation, seawater desalination and rainwater collection are three of the most important technologies that have been developed.

Treated wastewater can be reclaimed as low-quality water source for industrial, agricultural, residential and ecological use. China also realized the importance of wastewater reuse as a result of the growing crisis in water shortage. In the early 1980s, China started on relevant research and construction of wastewater reuse projects. In 2008, the proportion of recycled water used in industries reached 83.8% (MEP, 2008). In some big cities like Beijing and Dalian, wastewater reuse demonstration projects have been established, and the treated wastewater has been used for public toilet flushing, car washing, greenbelt irrigation, street cleaning and river water replenishment. The quantity of municipal waste water reused has reached 1.6 billion tons by 2007 (NBS, 2000–2007). However, only a small portion of the treated wastewater has been reused at present.

The use of desalinized water in the more water-intensive industries, including thermal power, petroleum and chemicals, is a promising solution to China's thirst for water. As the technology matures, daily production had reached 31,000 m^3 and the cost of desalinization dropped to 5 yuan per ton in 2003. However, many problems still exist, such as low daily production, high costs, and no written regulations. The government provided tax incentives and subsidies to boost the production and use of desalinized water in the eastern coastal regions. It was expected to ease the water shortage in the future. Based on the plan, daily production of desalinated seawater will reach 0.8 to 1.0 million m^3 in 2010 and 2.5 to 3.0 million m^3 in 2020 (**Table 1-19**).

Rainwater is another available water source, but utilization is still limited. Rainwater collection system was built in some cities and rural areas. Pilot-projects were first established in Beijing in 2000, and the collection of rainwater reached 2 million m^3 by 2007. The potential of available urban rainwater at the national scale was estimated to be 4 billion m^3 by 2010, equivalent to 20% of the total water use in cites all over the country. Estimated utilizable quantity of rainwater will come up to 6 billion m^3 by 2030 (Qiao, X. and Li, W., 2007).

TABLE 1-19. Development goals of seawater utilization.

Year	Amount of seawater desalinated		Amount of seawater directly used	Contribution to the coastal water-deficient regions (%)
	10^4 m^3/day	10^4 m^3/year	10^8 m^3/year	
2010	80–100	2.6–3.3	550	16–24
2020	250–300	8.3–9.9	1,000	26–37

Data source: NDRC (2005).

APPENDIX

The classification of surface water bodies by functions (Environmental quality standard for surface water, GB3838-2002, China).

This standard is applicable to the surface water bodies of rivers, lakes and reservoirs within the territory of the People's Republic of China.

Grade I is mainly applicable to the water from sources, and the national nature reserves.

Grade II is mainly applicable to first class of protected areas for centralized sources of drinking water, the protected areas for rare fishes, and the spawning fields of fishes and shrimps.

Grade III is mainly applicable to second class of protected areas for centralized sources of drinking water, protected areas for the common fishes and swimming areas.

Grade IV is mainly applicable to the water areas for industrial use and entertainment which is not directly touched by human bodies.

Grade V is mainly applicable to the water bodies for agricultural use and landscape requirement.

Groundwater quality classification (Quality Standard for Ground Water, GB/T 14848-93, China)

Grade I: Low concentration of natural background levels of unpolluted groundwater which can be used for any purposes directly.

Grade II: The natural background levels of groundwater which can be used for any purposes.

Grade III: Based on long term of public health monitoring and Drinking-Water Quality Standard. Water of this Grade is mainly used for drinking water source as well as industry, irrigation water supply.

Grade IV: Based on the water-quality demand of industry and agriculture uses. It is mainly used as industry and agriculture water supply. After treated, it also can be use for drinking.

Grade V: Not fit for drinking. But can be used for other purposes after simple treated based on designated uses.

1.4 SUSTAINABLE DEVELOPMENT AND WATER RESOURCES

1.4.1 *Sound water cycle*

Water is essential for life support. We live by having to do with various kinds of water, not only drinking water. Water is used for the fishery, recreation such as bathing, and relaxation in addition to the municipal water (daily life water, industrial water) and agricultural water as mentioned above. We are also concerned with it through culture including religious rituals and as an important factor for the landscape. We are related to plants and creatures which live in the water through the ecosystem. This means that maintaining the sound aqueous environment, and establishing the sound water cycle including the waterside which have close relationship with our lives in the social environment are essential in terms of healthy human living environment. The sound water cycle is defined as the state 'the function of water which contribute to human activities and preservation of the environment is ensured under the appropriate balance in the process of a sequence of flow around the drainage basin'.

Conserving the function of and enjoyment of water are indispensable not only for survival, but also the maintenance of daily life. It means the following three factors are important: the quantity of water, the quality of water, and the places where water exists. In addition to restoring the soundness of resources, the viewpoint of not only flow but also stock and store around in water district are notable. When they are realized, plenty and clean water will be ensured in the close water area, rich ecological system will be secured including at the waterside, peaceful and relaxing places where people chat will be constructed, and succession of culture and creation of new culture can be expected.

1.4.2 *New viewpoints and frameworks on water resources*

The first step toward the conservation of water resources is the preservation and recovery of water source areas. Although water resource planning has been started from the flowing water until now, we should change to plan water resources projects with considering forests for conservation of water resources. Secondly, we have to take into account the sound water cycle and all function of water, not only the currently targeted water use when thinking about water resources plans. Furthermore, detailed correspondence such as the recognition how important water resources of small rivers are and realization of the reuse of the sewerage as a new and stable water resource at cities will be needed.

REFERENCES

AQUASTAT/FAO (2002) http://fao/org/landandwater/aglw/aquastatnew/main/index.stm

Blanke, A. *et al.*, (2007) Water saving technology and saving water in China, *Agricultural Water Management*, 87, pp. 139–150.

BYQ Report on Water Environment (2008) Lake Biwa-Yodo river Water Quality Preservation Organization.

China Water Net (2009) http://news.h2o-china.com/html/2009/12/241259888970_1.shtml

China's sustainable development of water resources project team of Chinese Academy of Engineering (2002) Comprehensive Report of Strategy on Water Resources for China's Sustainable Development. Beijing: China Waterpower Press.

Esrey, S.A. *et al.*, (1986) Epidemiological evidence for health benefits from improved waste and sanitation in developing countries, *Epidemiological Reviews*, 8, pp. 117–128.

Falkenmark, M. (2000) Plants: the real water consumers, Newsflow, 1/00, *Global Water Partnership* pp. 2–3.

Falkenmark, M. and Rockström, J. (2006) The New Blue and Green Water Paradigm: Breaking New Ground for Water Resources Planning and Management, *Journal of Water Resource Planning and Management*, ASCE, pp. 129.

FAOSTAT (2002) http://faostat.fao.org/site/339/default.aspx

Gan, P. (2006) Analysis of Water Pollution and Treatment Technology in China, *Jiangxi Chemical Industry* 4: pp. 82–83.

Gleick, P.H. (1996) Basic water requirements for human activities: meeting basic needs, *Water International*, pp. 21, pp. 83–92.

Houghton, J.T. *et al.*, (2001) Climate Change: The Scientific Basis, Cambridge University Press.

IWMI (2000) World water supply and demand in 2025, in F.R. Rijsberman (ed), World Water Scenarios: Analysis (ed.), World Water Council, International Water Management Institute.

Jian, X. *et al.*, (2009) Addressing China's water scarcity: recommendations for selected water resource management issues, Washington, D.C.: World Bank.

Jiang, Y. (2009) China's water scarcity, *Journal of Environmental Management*, 90, pp. 3185–3196.

Kalbermatten, J.M. *et al.*, (1982) Appropriate sanitation alternatives: a technical and economic appraisal, (in) World Bank Studies in Water Supply and Sanitation I, Johns Hopkins University Press.

Li, S.T. (2006) Urban water pollution issues. China Youth, http://news.xinhuanet.com/environment/2006–09/13/content_5084123.htm accessed in July, 2007.

Liu, J. and Diamond, J. (2005) China's environment in a globalizing world. *Nature*, 435, pp. 1179–1186.

L'vovich, M.L. *et al.*, (1995) Use and Transformation of Terrestrial Water Systems.

Lvovsky, K. (2001) Health and Environment, *Environment Strategy Papers*, 1, World Bank.

Margat, J. (1996) Contribution to comprehensive assessment of the freshwater resources of the world, assessment of water resources and water availability in the world.

MEP (Ministry of Environmental Protection, P.R. China) (1998–2008) National Environmental Statistical Bulletin 1998–2008. *Ministry of Environmental Protection*, Beijing, China.

MEP (Ministry of Environmental Protection, P.R. China) (2000–2008) National Environmental Statistical Bulletin 2000–2008. *Ministry of Environmental Protection*, Beijing, China.

MEP (Ministry of Environmental Protection, P.R. China) (2007) The National eleventh Five-Year Plan for Environmental Protection. *Ministry of Environmental Protection*, Beijing, China.

Ministry of Land, Infrastructure and Transport (2001) Water Resources in Japan (White Paper on Water Resources), Land and Water Bureau, Department of Water Resources, Ministry of Land, Infrastructure and Transport, Printing Bureau, Ministry of Finance.

MWR (Ministry of Water Resources, P.R. China) (1997–2010) Water Resources Bulletin 1998–2008. *Ministry of Water Resources*, Beijing, China.

MWR (Ministry of Water Resources, P.R. China) (2004) Water Resources in China. *Ministry of Water Resources*, Beijing, China. http://www.mwr.gov.cn/english1/20040802/38161.asp retrieved in July 2007.

MWR (Ministry of Water Resources, P.R. China) (2008) Water Resources Bulletin, Water Resources Development Statistical Bulletin & Water Quality Report 2008. *Ministry of Water Resources*, Beijing, China.

MWR (Ministry of Water Resources, P.R. China) (2010) Water Resources Bulletin 2008. *Ministry of Water Resources*, Beijing, China.

NBS (National Bureau of Statistics, P.R. China) (2000–2007) Environmental Statistical Data 2000–2007, *National Bureau of Statistics*, Beijing, China.

NBS (National Bureau of Statistics, P.R. China) (2008) China Statistical Yearbook 2008. *National Bureau of Statistics*, Beijing, China.

NDRC (National Development and Reform Commission, P.R. China) (2005) Plan for Seawater Utilization. *National Development and Reform Commission*, Beijing, China.

Phocaides, A. (2000) Technical Handbook on Pressured Irrigation Techniques , FAO, Rome.

Postel, S. (1999) Last Oasis, The World Watch Environment Alert Series, The World Watch Institute.

Qiao, X. and Li, W. (2007) Discussion about Rainwater Utilization in China Comparing with the utilization state in other countries. Collected Papers at the Forum of Build a resource-saving and environment-friendly society.

Shiklomanov, I.A. (1999) World water resources and their use: assessment and outlook for 2025, (in) F.R. Rijsberman (ed.), World Water Scenarios: Analysis, World Water Council.

Solley *et al.*, (1998) Estimated use of water in the United States in 1995, U.S. Geological Survey Circular 1200.

UN (United Nations) (1977) Report of the United Nations Water Conference, Mar del Plata, March, pp. 14–25, No.E.77.II.A.12, United Nations Publications.

UN (United Nations) (1989) Convention on the rights of the child, *General Assembly resolution* 44/25 of 20 Nov. 1989.

UN (United Nations) (1997) Convention on the law of the non-navigational uses of international watercourses, UN General Assembly Doc. A/51/869(11, April).

UNDP (2000) World Energy Assessment, energy and the challenge of sustainability.

Vorosmarty, C.J. *et al.*, (1997) The storage and aging of continental runoff in large reservoir systems of the world, *Ambio*, 26(4), 210–219.
Water Resources in Japan (2009) Water Resources Department under Land and Water Bureau in the Ministry of Land, Infrastructure, Transport and Tourism.
Water Resources in Japan (2010) Water Resources Department under Land and Water Bureau in the Ministry of Land, Infrastructure, Transport and Tourism.
WB (World Bank) (2008) World Development Indicators 2008. *World Bank*, Washington DC, USA.
WHO/UNICEF/WSSCC (2000) Global water supply and sanitation assessment 2000 report.
Yolles, P. (1993) Water quantity requirements in various sanitation alternatives, Pacific Institute for Studies in Development, Environment, and Security, Oakland.
Zeng, H. (2004) Groundwater quality in China is getting worse. China Mining News, pp. 12–16 (T00).
Zhang, H. (2005) Strategic study for water management in China. Nanjing: Southeast University Press.
Zhang, Z. *et al.*, (2009) Investigation and Assessment of Sustainable Utilization of Groundwater Resources in the North China Plain. Beijing: Geological Publishing House.
Zhu, Z.Y. *et al.*, (2001) Water shortage: A serious problem in sustainable development of China, *International Journal of Sustainable Development and World Ecology*, 8, pp. 233–237.

2

Water Pollution

2.1 MAJOR WATER POLLUTION ISSUES

2.1.1 *Characteristics and roles of water*

Water is essential for the survival and growth of all living things on Earth owing to its basic characteristics. The melting and boiling points of fresh water under standard condition are 0 and 100°C (around –1.9 and 103°C for sea water), respectively, which are very high compared to the hydrogen compounds of chemical elements near oxygen on the periodic table. In addition, the range between these two points is quite narrow. Therefore, we can see all three phases of water, gaseous (vapor), liquid (water), and solid (ice), under general atmospheric conditions.

Surface water will evaporate even in liquid phase. The evaporated amount increases under less humidity and higher temperature. This characteristic of water enables its global circulation, providing its renewal and purification.

The heat of vaporization of water is 582.8 and 539.8 cal/g at 25 and 100°C, respectively, and its heat of fusion is 79.7 cal/g. These values are high compared with other substances. Water has a high heat capacity as shown by its high specific heat of 1.0 cal/g at 15°C, which is twice of that of general fluid and 3.4-times that of wood. These characteristics of water, together with the occurrence of its three phases, greatly contribute to our survival and the temperature control of the environment. The fact that a diversity of life exists, and many living organisms undergo early development in water, is proof of its importance.

The volume of water is minimized at 4°C under standard atmospheric pressure, and increases by 10% when frozen. This unique property enables aquatic organisms to survive in deep water bodies even in severe cold areas by the occurrence of stratification, as shown in **Figure 2-1**, because water in the hypolimnion is prevented from freezing owing to the low heat conductivity of water of 1.4×10^{-4} cal/(cm•sec•°C).

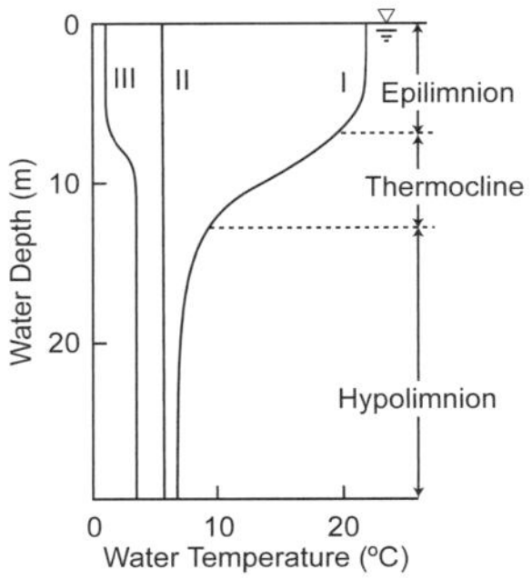

FIGURE 2-1. Thermal stratification (I: Summer stratification II: Circulation III: Winter stratification).

Water has a very high surface tension of 72.75 dyne/cm, which is three-times that of ethyl alcohol, and enables soil to sustain water in a soil pore through a capillary phenomenon (capillary water potential), which allows plants to receive water and grow. Water also has a high solubilization of materials that are ingested by organisms and are easy to transport and transform through physical actions and biochemical reactions.

2.1.2 *Major water pollution issues*

The word 'water pollution' is used to express a state in which water is loaded with impurities, and the quality of water and sediment negatively affect the health and environment of living organisms. Some of the main water pollution issues are shown below.

2.1.2.1 Organic substances and the depletion of dissolved oxygen

Organic biodegradable substances cause the most common and primitive type of water pollution, which is measured and evaluated based on the biochemical oxygen demand (BOD) and chemical oxygen demand (COD). These affect various species of aquatic organisms, particularly fish and shellfish, cultivation of which has shifted from commercially expensive types to cheaper types with an increase in the extent of pollution. Water containing such substances is not suitable as drinking water.

These substances consume dissolved oxygen (DO) through their aerobic degradation. The depletion of DO has an adverse effect and deteriorates aquatic environments. Aquatic organisms cannot survive, and people suffer from offensive odors released from the sediment and water.

The changing trend in DO and BOD loaded into a river as it flows downward are classically expressed using a Streeter-Phelps equation as follows:

$$dC/dt = -k_1 \bullet C \qquad (2\text{-}1)$$

$$dO_2/dt = k_2 \bullet (O_2^* - O_2) - a \bullet k_1 \bullet C \quad (2\text{-}2)$$

where C is BOD concentration, t is downward flow time, k_1 is the degradation rate constant of BOD, O_2 is the DO concentration, O_2^* is the saturation concentration of DO and is in the range of 7 to 9 mg/L depending on the water temperature and salinity, k_2 is the reaeration rate constant, and a is the ratio of DO consumption to degraded BOD.

The typical results obtained from these equations are shown in **Figure 2-2**. The minimum DO concentration and place where it happens should be noted. To maintain a good aquatic ecosystem, the DO concentration needs to be higher than 50% of the saturation DO concentration.

2.1.2.2 Suspended solids

Suspended solids (SS) are defined as particles with a size of more than 1 μm, which decreases the light penetration into water and increases the color and turbidity. SS have a negative effect on aquatic plants and fish, the purification of drinking water, aesthetic quality, agricultural and recreational uses, and so on. If SS consist of volatile suspended solids (VSS), i.e., organic suspended solids, they settle and accumulate on the sediment and consume DO, resulting in DO depletion at bottom water. This problem is enhanced through stratification in water bodies (see **Figure 2-3**). A serious problem may occur in water bodies where the VSS content in sediment exceeds 10% at dry weight. VSS problems usually occur concurrently with BOD and COD problems.

2.1.2.3 Waterborne diseases

Waterborne diseases are caused by water-mediated pathogenic microorganisms. Enteric bacteria, viruses and helminths are waterborne pathogenic

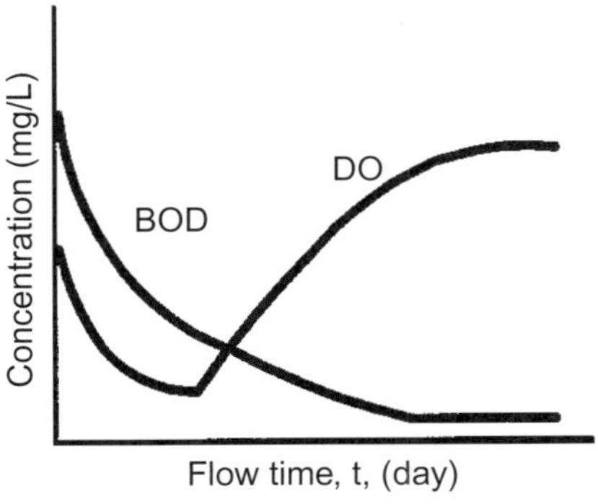

FIGURE 2-2. BOD and DO change with the flow direction of the river.

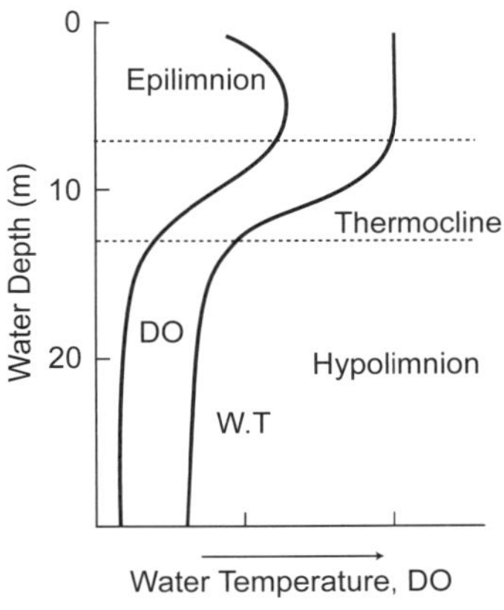

FIGURE 2-3. DO under stratification.

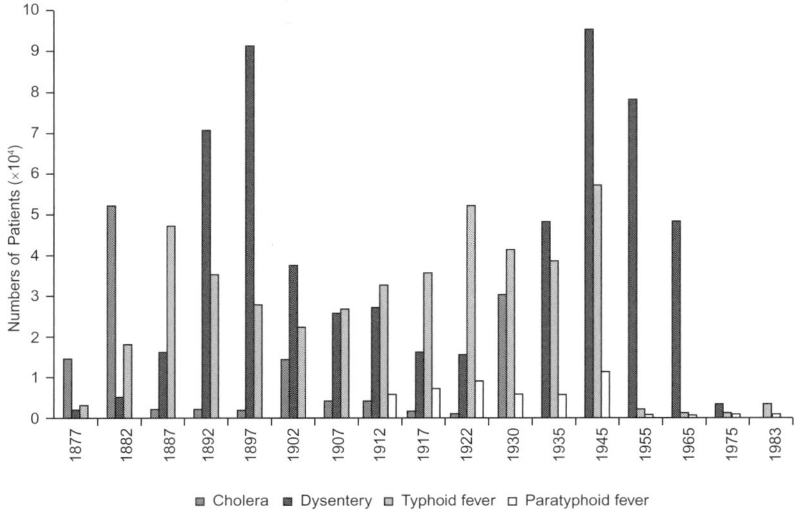

FIGURE 2-4. Changes in numbers of patients throughout the years. (Data: Ministry of Health and Welfare in Japan)

microorganisms that can cause diverse public health problems. The changes in patient numbers in Japan over the years are shown in **Figure 2-4**. The numbers decreased dramatically from 1965 to 1975. The construction of water and sewage works and solid waste management systems, as well as education on good sanitation practices, have made a significant contribution to this decrease.

However, the prevention of pathogenic protozoa such as *Giardia lamblia* and *Cryptosporidium parvum*, and pathogenic viruses tolerant to chorination disinfection, has recently garnered significant attention.

2.1.2.4 pH

Higher (alkaline) and lower (acidic) pH than 7.0 destroys the living environments of aquatic organisms, and limits the use of water for drinking, agriculture, industry, recreation, and other purposes.

2.1.2.5 Eutrophication

Phytoplankton grows as shown in the following Richards equation, and when grown excessively, causes several problems in enclosed water bodies with a high concentration of phosphorus and nitrogen. This state is called eutrophication or enrichment.

$$106HCO_3^- + 106H_2O + 16NH_3 + H_3PO_4 + Light$$
$$(CH_2)\ 106(NH_3)\ 16H_3PO_4 + 106O_2 + 106OH \qquad (2\text{-}3)$$

The term 'enclosed water bodies' indicates lakes and dams with a hydraulic retention time of more than 10 days, or an enclosed sea area with a 'Geographical Enclosed Index (GEI)' greater than 1.0. GEI is defined in the following equation:

$$GEI = S^{(1/2)} \cdot D_1/(W \cdot D_2) \qquad (2\text{-}4)$$

where S is the surface area, W is the width of the section of mouth (inlet and outlet) open to the sea, and D_1 and D_2 are the maximum water depth of the enclosed sea area and the section of the mouth, respectively.

The GEI values of Tokyo bay, Ise bay, and Osaka bay, where eutrophication is a severe problem, are shown in **Table 2-1**. Both phosphorus and nitrogen loads into an enclosed water body are also necessary conditions for eutrophication.

TABLE 2-1. Values of geographical enclosed index (GEI) in enclosed coastal seas in Japan.

Name	E	I1/W	I(km)	W(km)	D_1(m)	D_2(m)	S(km^2)
Tokyo bay	4.52	7.14	66	17	66	66	1,000
Ise bay	3.63	6.09	70	12	100	100	1,738
Mikawa bay	1.91	2.33	30	13	35	35	604
Osaka bay	4.35	0.63	5.7	9.0	113	113	1,529
Mutsu bay	3.88	5.14	54	11	94	94	1,660
Sagami bay	0.66	0.72	57	79	1,650	1,650	2,700
Suruga bay	0.86	1.16	65	56	2,245	2,245	2,300
Ariake	12.89	21.33	96	4.5	165	117	1,700
Hakata bay	1.73	2.08	18.3	8.8	22	17	138
Omura bay	55.13	80.0	26	0.3	54	54	321
Kagoshima	1.48	3.30	177.2	23.4	230	230	1,129

Seto-Inland-Sea E = 3.4~3.8

(Data: Environmental Protection Agency in Japan)

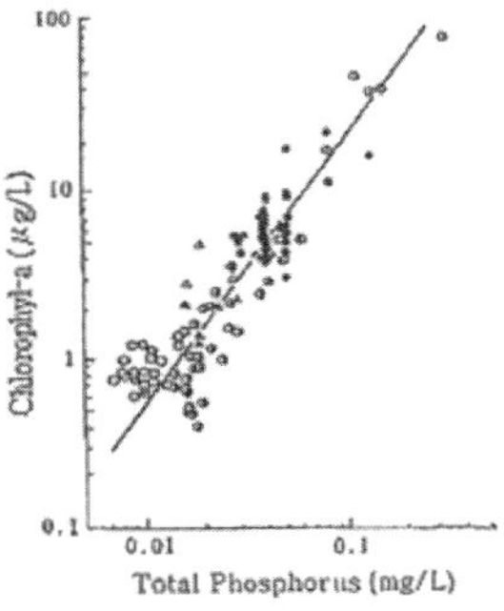

FIGURE 2-5. Relationship between TP and Chlorophyl-a. (Tsuno, 1991).

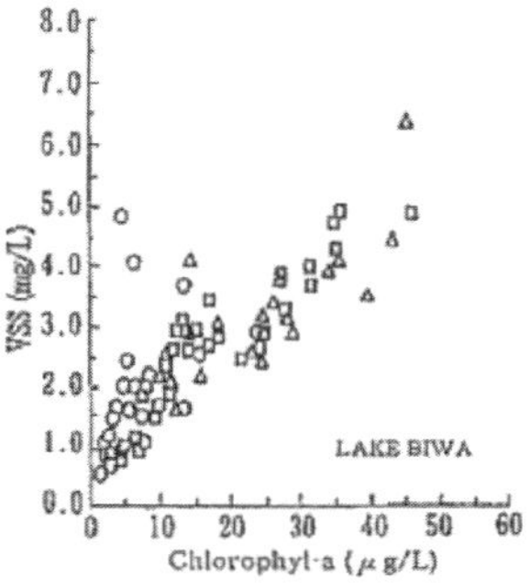

FIGURE 2-6. Relationship between Chlorophyl-a and VSS. (Tsuno, 1991).

As shown in Equation (2-3) and **Figures 2-5** and **2-6**, if phytoplankton grows by ingesting 1 and 7.2 g of phosphorus and nitrogen, respectively, 115 g of SS and 109 g of COD are correspondingly produced. The problems caused by COD and VSS therefore occur through the growth of phytoplankton. Red-tide problems also occur owing to eutrophication, resulting in the closure of swimming locations, the killing of fish, and a release of offending odors from the water. Eutrophication effects are enhanced through thermal and chemical stratification, as shown in **Figure 2-7**.

Certain unpleasant substances, some of which are toxic and odorous, are also produced with the growth of phytoplankton. In addition, shells are occasionally unable to be harvested owing to an accumulation of toxic substances produced by phytoplankton through filter feeding. Many people also experience poor tasting and odorous drinking water, which is caused by substances such as geosmin and 2MIB at concentrations of as low as 20 to 30 ng/L, which are produced by blue-green algae in lakes. Major water quality issues caused by phytoplankton are listed in **Table 2-2**.

2.1.2.6 Oil

Oil pollution limits the use of water for drinking, agriculture, industry, fishing, recreation, and other purposes. Oil also causes odors in fish for consumption at concentrations of as low as 0.002 to 0.1 mg/L.

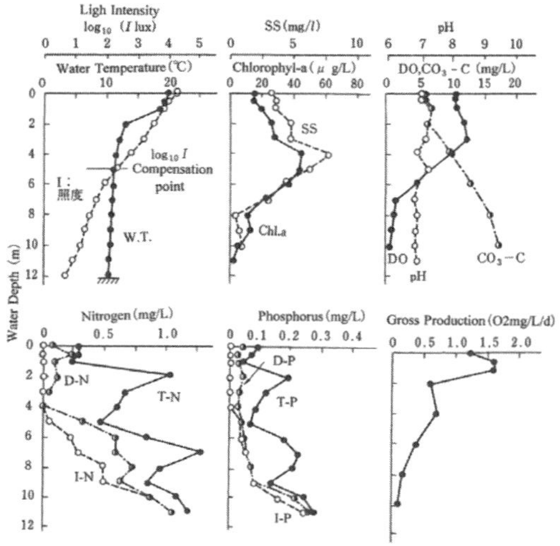

FIGURE 2-7. Example of thermal and chemical stratification during the summer. (Yuno lake on August 7, 1978) (Tsuno, 1979).

TABLE 2-2. Examples of water problems caused by phytoplankton.

Water problem	Associated phytoplankton
Fresh water red tide	Diatom: *Asterionella* Dinoflagellates: *Gymnodinium, Peridinium* Chrysophyceae: *Uroglena*
Water bloom	Blue-green algae: *Anabaena, Micmcystis, Aphanizomenon*
Red tide	Diatom: *Skeletomvma* Dinoflagellates: *Gymnodinium, Noctiluca, Heterosigma, Gonyaulax* Raphydo algae: *Chattonwlla*
Taste & odor	Blue-green algae: *Oscillatoria, Phormidium, Anabaena*
Toxic substance	Blue-green algae: *Micmcystis viridis* (microcystin) *Anabaenaflos-aquae* (anatoxin) Dinoflagellates: *Alexandrium, Dinophysis* *Gonyaulax* (saxitoxin)

2.1.2.7 Heavy metals and recalcitrant substances

Heavy metals and recalcitrant organic substances (persistent organic pollutants) are impossible or difficult to degrade chemically or biologically, and thus are accumulated in the environment, becoming concentrated in aquatic living organisms through the food chain. They are mainly accumulated in the lipids of the body. These kinds of pollutants have serious adverse effects on living organisms including human beings, such as acute toxicity; destruction of the nervous system, internal organs, and/or the skeletal system; genotoxicity; and carcinogenecity.

Organic mercury and cadmium cause Minamata disease and Itai-itai disease, respectively, which are typical examples of damage from hazardous heavy metals. It must be noted that even inorganic mercury can cause a serious disease such as Minamata disease because it is easily transformed to organic mercury by microorganisms existing in sediment. Cadmium has strange characteristics compared with other heavy metals; it does not have an obvious adverse effect on plant growth, but is easily ingested by plants and accumulated in their fruit. For this reason, cadmium concentrated in rice causes Itai-itai disease.

PCBs are typical recalcitrant organic substances that are widely dispersed through the food chain and have even been detected in penguins in Antarctica. Examples of the accumulation of mercury through the food chain and the concentration of PCBs in mussels are shown in **Figures 2-8** and **2-9**, respectively. It was shown that the mercury content increases along with a higher level of the food chain, and that PCB isomers are concentrated within a range of 1,000 to 100,000 times in mussels depending on their Kow values.

These types of pollutants are easily dispersed into global areas through a bypass of the food chain. The discharge of these types of substances into the environment should be severely inhibited.

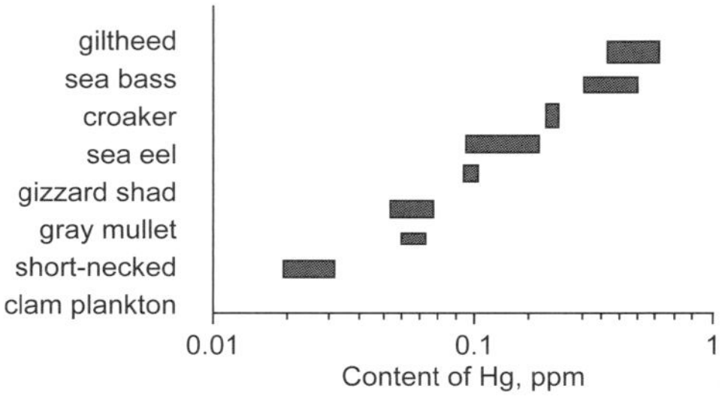

FIGURE 2-8. Hg content in fish in Tokuyama Bay (J.EPA, 1973). The concentration of Hg in water is less than 0.001 mg/L. The content of Hg in sediment is 0.2–0.4 ppm.

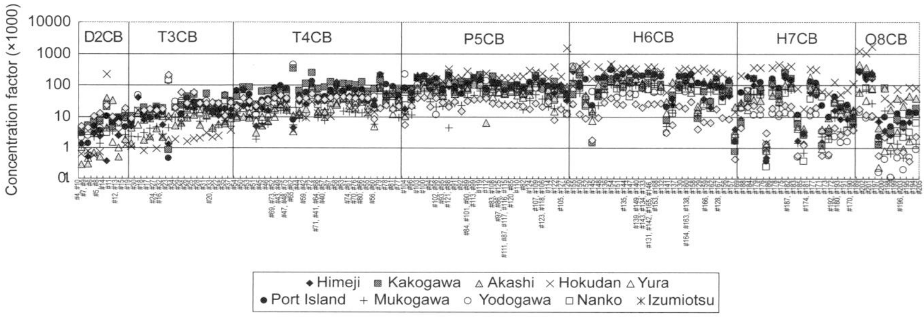

FIGURE 2-9. Concentration factor of each isomer of PCBs at each sampling point (Tsuno *et al.*, 2005).

2.1.2.8 Micropollutants

Some organic materials may affect human health even at a slight concentration in the environment. These kinds of materials, called micropollutants, have acute toxicity, genotoxicity, and cancer risk. Pesticides and organic halogenated chemicals are included in this category and have recently gained the attention of researchers. Endocrine disrupting chemicals (EDCs) and pharmaceutical and personal care products (PPCPs) are also attracting attention, and are considered as emerging contaminants.

2.2 HISTORY OF WATER POLLUTION AND COUNTERMEASURES IN JAPAN AND CHINA

2.2.1 *History of water pollution and countermeasures in Japan*

Japan has experienced several serious water pollution problems caused by a rapid economic development, urbanization, and changes in life style since the 1870s. An outline of the history of water pollution and countermeasures are shown in **Table 2-3**. Pollution from heavy metals, arsenic, and acids

TABLE 2-3. Outline of the history of water pollution (•) and countermeasures (○) in Japan.

1978–1881	
Heavy metals	• Heavy metal pollution in Watarase river basin by wastewater from mining plots
1900–1950	
Organics	• Wastewater from paper mills, textile and food industries, and so on
Heavy metals	• A 'strange disease' occurred in Jintsu river basin (Itai-itai disease)
1950s	
Heavy metals	• Oral presentation by Dr. Ogino and Kohno on the cause of "Itai-itai disease (cadmium)" in annual medical conference
	• Establishment of "Minamata disease" research group in Kumamoto University.
	• Organic mercury was presented as the cause of Minamata disease by the this group
Organics	• Affilication of fishermen by paper mill wastewater in Edo river
Oil	• Affilication of fishermen by petrochemical wastewater in Yokkaichi
General	o Factory Effluent Control Law
	o Water Quality Conservation Law
1960s	
Oil	• Oil odor problems in fish in Ise and Mizushima bays
	o Pollution Control Law for Sea Areas to prevent oil pollution from ships
Heavy metals	• Occurrence of embryonic Minamata disease
	• The second Minamata disease occurred in the Agano river basin.
Organics	• Cadmium from a mining plot was determined as the cause of Itai-itai disease
	• H_2S problem caused by polluted sediment in Tagonoura
General	• Fresh trout deaths in Kiso river
	o Basic Law for Environmental Pollution Control (1967)
	o Sea Water Pollution Protection Law
1970s	
General	o Water pollution Control Law
	o Environmental standards on water quality
Heavy metal	o Safety standard of cadmium-polluted rice.
	o Mercury standards for food fish and removal from sediment
	o Strengthening of mercury standard on wastewater and the environment

(Continued)

TABLE 2-3. Continued.

	• Hexa positive valency chromium pollution
Eutrophication	• Severe deterioration of water quality and the death of a great number of cultivated fish in Seto Inland Sea
	o Establishment of Special Protection Law for Seto Inland Sea
	• Taste and odor problems for drinking water from Tone river and Yodo river basin
Organics	o Control of COD loading
Oil	• Oil contamination of seaweed
Recalcitrant	o Addition of n-exane extraction index into environmental standard
	• Human and fish contamination by PCBs
Chemicals	o Addition of PCBs into environmental and effluent standards
Worm effluent	o PCB standards for the removal of contaminated sediment
	o Restriction Law for the Production and Judgment of Chemicals
	• Worm sea water problem caused by effluent from industry and power generation plants
1980s	
Eutrophication	• Taste and odor problems in drinking water
	• Fresh water red-tide problems
	o Addition of phosphorus and nitrogen into environmental and effluent standards for lakes
	o Special Law for Lake Water Conservation
Ground water	o Addition of grand water pollution control into Water Pollution Control Law
Acid rain	o Comprehensive survey by Environmental Protection Agency
1990s	
General	o Environmental Basic Law (1993)
Grey water	o Addition of household grey water control into the Water Pollution Control Law
Eutrophication	o Addition of phosphorus and nitrogen control into the environmental and effluent standards for enclosed coastal sea areas
Ground water	o Addition of ground water pollution control into Water Pollution Control Law
	o Environmental standards associated with ground water
Chemicals	o Amendment of environmental standards
	o Pollutant Release and Transfer Register Law
	o Special Control Law for Dioxins
EDCs	o Comprehensive survey by national ministries

caused by effluent from mining plots was generated between 1880 and 1950, and typical examples include the heavy metal pollution of the Watarase river basin, and Itai-itai disease caused by cadmium in the Jintsu river basin. Water pollution from organic substances was also caused by wastewater discharged from paper mills and textile and food factories owing to the rapid development of such industries during this time. Serious adverse effects to fish and shellfish occurred by vigorous sediment pollution, the depletion of DO, and other factors. In the 1950s, Minamata disease was caused by organic mercury, which was transformed by microorganisms in the sediment from inorganic mercury included in the industrial wastewater discharged in Minamata Bay. In addition, wastewater including oil from tankers and petrochemical industries has polluted sea areas, resulting in oily odors in fish populations. To solve the serious water pollution problems that have occurred in many locations around Japan, two laws were established, i.e., the Factory Effluent Control Law and the Water Quality Conservation Law, but they have not been effective owing to their applicability in only few designated areas.

In the 1960s, these problems worsened. To overcome this, the Basic Law for Environmental Pollution Control was established in 1967, and in 1970 was amended to remove article on harmonization with economy development and introduce responsibility of the company manager for the caused public nuisance, even if no misfeasance verified. Based on this, environmental standards and a Water Pollution Control Law were established, and effective strategies against water pollution began to be implemented. The Water Pollution Control Law was implemented throughout Japan, and an article regarding more severe effluent standards than the national level depending on the state of the water quality in each area was addressed.

In the 1970s, a survey on how to overcome public nuisance was conducted, and actual countermeasures based on the survey results were established. Cadmium, which causes Itai-itai disease, was also found to be included in the wastewater discharged from electric industries, and safety standards for rice, which contributes greatly to preventing cadmium exposure to human beings, was presented. In addition, countermeasures against organic mercury, which causes Minamata disease, started to be implemented. These countermeasures also included inorganic mercury owing to its easy transformation to organic mercury by microorganisms in nature, the presentation of safety standards for fish and shellfish as food items, the removal standard for polluted sediment, and a strengthening of environmental and effluent standards. For protection against oil pollution, environmental and effluent standards were established and discharge from ship prohibited. On the other hand, hexa positive vallent cromium contamination by cromium slag appeared, and oil-disease and the contamination of fish and shellfish by PCBs broadly used for machinery, heat exchangers, and paint occurred. To prevent adverse effects by PCBs, environmental, effluent, and sediment removal standards were established, and the Chemical Assessment and Production Control Law, which addressed chemicals use, was created. Eutrophication became a problem in enclosed coastal seas, representatively, Seto Inland Sea. A Special Law for the Environmental Protection of Seto Inland Sea was established, and a pollutant-loading control strategy was introduced into the Water Pollution Control Law.

After the 1980s, new water pollution problems appeared. In addition to enclosed coastal seas, eutrophication also became a problem in lake waters. Environmental and effluent standards for phosphorus and nitrogen, and the Special Law for Lake Water Quality Preservation were established. In addition, countermeasures against grey wastewater from residential areas were required to decrease the pollution load, and were introduced into the Water Pollution Control Law. Additionally, the chemical pollution of ground water occurred, and the environmental standards were therefore amended to deal with chemical pollution. On top of that, environmental standards for ground water were newly established and ground water pollution control strategies introduced into the Water Pollution Control Law. The Chemical Assessment and Production Control Law was also amended and a Pollutant Release and Transfer Register (PRTR) Law was established for recording the discharge amount and management of certain chemicals. Dioxine was found to be produced and discharged into the environment through the incineration of solid waste. The Special Law for Dioxine Control and environmental standards based on this law were established. Recently, endocrine disrupting chemicals have attracted attention, and a comprehensive survey on this subject was carried out. Pharmaceutical and personal care products are now attracting attention as environmental pollutants. Waterborne diseases are caused by chlorination-tolerant microorganisms such as pathogenic spore-forming bacteria, viruses, and protozoa.

Japan has learned several quite important lessons through these serious and sad experiences with water pollution:

1. It is very important to consistently watch and monitor human health and the state of the environment, and to respond immediately to determine the causes of and solve any problems that may occur. To implement these matters, national and domestic research institutes for environmental studies have been established and public nuisances prevented before becoming serious. The dioxine issue is one such success story.
2. Economic development cannot be accomplished without water quality conservation. Water pollution damages human health and the living environment, and prevents economic development. Additional money and time are required to recover a damaged environment. Further, as many people may suffer from such damage, economic can be developed only on the base of a sound environment.
3. Environmental conservation does not have an adverse effect on economic development, and as such, the development of a new field of economics with new technologies for conserving the environment, will be developed. In addition, not only preventing water pollution but also saving energy and maintaining material recycling processes are needed.

2.2.2 *History of water pollution and control in China*

Since 1840, China has suffered unremittingly from war. Cities with domestic services including drinking water supply and wastewater treatment were almost completely destroyed, particularly during World War II and the succeeding Civil War periods. After 1949, the New Republic devoted all of its energy and financial resources into infrastructure construction, during which basic safety guarantees of drinking water supply and sanitation services were gradually recovered. Unfortunately, however, during the 1960s and 1970s, the political movement replaced economic development, becoming the main task of the Chinese government. National construction was abandoned, and the development of drinking water supply and water pollution control stopped. During these periods, almost no archives regarding environmental protection can be found to indicate the status of the water environment for that time. Medical records show that several epidemic diseases occurred, which may have been caused by low-quality drinking water and poor sanitary services.

However, several significant events regarding water services did take place during this period, one of which was the construction of the Miyun reservoir. This so-called largest artificial reservoir in Asia is located in the north-eastern suburban area of Beijing, and was built between 1958 and 1960, with an annually average storage capacity of 4 billion m^3. The reservoir has solved the water supply problem for the Beijing metropolitan area, providing about two-thirds of the drinking water for the city's population. A good example was set, and local governments began to pay more attention to their own water conservancy projects to provide a sufficient water supply.

Another event that should be mentioned is the First National Conference on Environmental Protection, which was held in 1973. After attending the United Nations Conference on Human Environment at Stockholm in 1972, the central government decided to initiate environmental protection standards in China. At this conference, a micro-concept of environmental protection, called Three-Simultaneously, was brought forward, which stated that environmental protection technologies and processes should be designed, constructed, and operated simultaneously, as the main part of each project. This concept is still applicable, and is an essential indicator in environmental impact assessments for certain projects. As a result of the national environmental protection conference, a regulation called Provisions Concerning Environmental Protection and Improvement became the first legislative regulation in China, and was symbol of the real beginning of a Chinese march toward environmental protection. Unfortunately, towing to the political and economic conditions of that period, the regulation was put on a shelf. This kind of low-efficiency management lasted until the end of the Cultural Revolution in the late 1970s, during which the environmental protection of Chinese waters entered a new era. An outline of the history of water pollution and countermeasures in China is shown in **Table 2-4**.

TABLE 2-4. Outline of the history of water pollution (•) and countermeasures (○) in China.

Prior–1840	○ Ancient sewer systems and simple septic tanks
1840–1949	○ China's first modern sewer network in Tsingdao city
	○ China's first modern wastewater treatment plant in Shanghai city (1923)
1949–1979	○ Construction of China's first modern reservoir, the Miyun reservoir, in Beijing city
	○ First National Conference on Environmental Protection (1973)
1980s	○ Water Pollution Prevention and Control Law (1984)
	○ Sanitary Standards for Drinking Water (1985 Edition)
	○ Environmental Protection Law (1989)
1990s	• Heavy metal pollution in Pearl river
	• Heavy metal pollution in Xiangjiang river
	• Taihu lake pollution and its control action, Zero Hours
	• Eutrophication crisis in Dianchi lake
	• Eutrophication crisis in Chaohu lake
	• Water pollution in Huaihe river
	○ Development Plan for Urban Water Supply Technology for the year 2000 (1992)
	○ Regulation of Abstraction and Protection of Urban Groundwater (1993)
	○ Implementing Measures for the Water Licensing System (1993)
	○ Regulation of Urban Water Supply (1994)
	○ Administrative Measures for the Sanitary Supervision of Drinking Water (1996)
	○ Administrative Measures for the Supervision of the Water Licensing System (1997)
	○ Administrative Regulation of Urban Water Supply Quality (1999)
2000-now	○ National Water Law (2000)
	• Songhuajiang river crisis caused by a nearby chemical factory explosion, and drinking water supply failures in and around cities (2005)
	○ Sanitary Standards for Drinking Water (2006 Edition)
	• Algal bloom and eutrophication crisis in Taihu lake, and drinking water supply failure in Wuxi city (2007)
	○ Construction of Three Gorges Project
	• Cd pollution in Beijiang river, Guangxi province
	○ Upgrade of discharge from Class I-B to Class I-A by Pollutants Discharge Standards for Municipal Wastewater Treatment Plants around Taihu lake (2008)
	○ Initiation of Water Special Project of National Mid- and Long-Term (2006–2020) Science and Technology Plan (2008)
	• Dalian sea pollution crisis caused by a nearby oil storage tank explosion (2010)

The reformation of the Chinese political and economic systems began in 1978. As the entire industrial system of the nation, especially heavy industry, was taking shape, different environmental issues were taking place. The idea of environmental protection was gradually given significant attention by the government and the public. The content of the Provisions Concerning Environmental Protection and Improvement began to be reviewed and actualized. In 1983, the first edition of the Environmental Quality Standards of Surface Water was released, in which surface waters were categorized into five classes based on their different usages and water qualities. Different quality standards for water, air, drinking water, wastewater discharge, and so on were then issued, and have subsequently been revised into different editions about every ten years or so. In 1988, the State Environmental Protection Administration of China was founded, and the branch agency in each province was granted special powers aside from the local government to deal with matters of economic development when environmental issues arise. Finally, in 1989, after a significant amount of effort by millions of citizens on the progress of Chinese environmental protection, the Environmental Protection Law of China was approved, after which environmental protection became not only a technical concern, but also a political, economic, and social one.

After the United Nations Conference on the Environment and Development in Rio de Janeiro in 1992, China accepted the concept of sustainable development and made it a basic national policy for future development. Still, along with the rapid growth of the Chinese economy, more and more environmental problems have occurred within a very short time period. Such problems include eutrophication, heavy metals, persistent organic pollutants (POPs), pharmaceutical and personal care products (PPCPs), frog infestations, sandstorms, and red tides. There is a popular saying in the academic world that the environmental issues faced by western countries over the past decades have now occurred in China with in just a couple of years. To deal with this, more intensive reformations in terms of administration and legislation have been carried out. In 2006, the former State Environmental Protection Administration was reorganized into the Ministry of Environmental Protection, which was granted additional power and freedom for related affairs. Meanwhile, an Environment Law, which may take the place of the current Environmental Protection Law, is now being drafted to form a complete legislation system for environmental affairs.

Although substantial work has been carried out to develop Water Supply and Sanitation Service since the creation of the People's Republic of China in 1949, much remains to be done, particularly in rural areas where the bulk of the population is still concentrated.

Some of the main metropolitan areas and larger cities have been very successful in expanding their wastewater treatment coverage, especially within their core urban areas. However, even as late as 2007, only slightly more than half of the wastewater produced has received any treatment, while around 200 Chinese cities still have no coverage for sanitation services. This situation represents a formidable challenge for the future development of China.

According to data for March 2008, there are a total of 1,321 municipal WWTPs in mainland China, with a designed treatment capacity of 80.43 million m^3 per day. The sewerage network is 156,000 km long. In 2006, 16.31 billion m^3 of municipal sewage and industrial wastewater were treated nationwide, 13.04 billion m^3 of which was municipal wastewater. The average municipal sewage treatment rate was 43.8%, and among the provincial administrative regions, the highest rate (about 90%) was achieved in Beijing, followed by Shanghai (about 78%), while the lowest rate of treatment (10%) was recorded in Guangxi province.

Along with the rapid growth of the Chinese economy, different kinds of water pollution issues have emerged in a short period of time, and we are facing tremendous problems that western countries never even imaged. More advanced technologies in engineering fields and a more intensive reformation of the political arena are urgently needed to help China find a new solution to its water pollution problem. Under these conditions and circumstances, in 2006, China's State Council adopted the National Mid- and Long-Term Science and Technology Plan (2006–2020), which aims at achieving the proper water pollution control and efficient management of water bodies over the next fifteen years. The water component of the Science and Technology Plan has given Chinese scholars and engineers renewed energy to concentrate their work on improving water environments. We believe that these efforts, focused on such tasks as protecting water sources, improving waste and storm-water treatment, and recovering energy from sludge digestion, will come to fruition in the near future. Although China still has a long way to catch up with developed countries in these regards, the country is set to achieve significant advances in the fields of wastewater treatment and water environmental protection in the near future, as well as in other important aspects related to the nation's rapid development.

2.3 PRESENT STATE OF WATER QUALITY IN JAPAN AND CHINA

2.3.1 *Environmental standards for water quality in Japan*

2.3.1.1 Introduction

The legal system for water environmental protection in Japan is as follows. The Basic Environment Law shows the principle and philosophy of environmental conservation, and various laws have been enacted to cope with the actual circumstances. Environmental standards are indicated in the government's policy objectives in Article 16 of the Basic Environmental Law, and the government has set desirable standards for both human health protection and preservation of the living environment. Reviews reflecting the current scientific knowledge are continuously conducted. In recent years, 1,4-dioxane was added to the standards for human health protection on November 30, 2009, and zinc was set as the first environmental standard item for the preservation of aquatic life in 2003. The contaminant levels are unacceptable, and it is necessary to make the environment as clean as possible.

2.3.1.2 Environmental standards for the protection of human health

The items and maximum contaminant levels are announced by the Ministry of Environment. Current items and its maximum contaminant level are listed in **Table 2-5**. The environmental standards related to human health protection are applied to all public water areas, and items such as heavy metals, inorganics, organo chlorines, solvents, and pesticides are listed. About twenty-six items have also been nominated for required monitoring and discussed as items of concern in the environmental standards.

2.3.1.3 Environmental standards for the preservation of living environments

The environmental standards related to the preservation of living environments are specific to each type of water area. The type of water area is determined according to the adaptability of the area usage and the living conditions of the aquatic organisms with regard to the natural environment, water supply, fisheries, industry, agriculture, and so on. The types of water areas are categorized into rivers, lakes, and seas, as shown **Table 2-6**. In addition, the standards are classified for each type of water area. The pH, DO, total coliform, and total zinc are the standard items for rivers, lakes, and seas. BOD is for only river areas, whereas COD_{Mn} is a standard item for both lake and sea areas. Total nitrogen and total phosphorus are items for the prevention of eutrophication in designated lakes and enclosed coastal sea areas. Total zinc is indicated for aquatic organisms. Common items for all water areas, and the original items for each area, were established.

2.3.1.4 Environmental standard for ground water

Groundwater is a widely used resource, and about 30% of the water supply demand in Japan is groundwater. Environmental quality standards for groundwater were announced based on Article 16 of the Basic Environment Law in March 1997. The objectives are promoting the comprehensive protection of groundwater quality. The standards are set for twenty-eight items, which are almost the same as the health-related standards for public waters. The differences from the health-related standards for public waters are as follows. A vinyl chloride monomer is listed in the groundwater standard at a value of 0.002 mg/L, which is not regulated in the health-related standards for public waters. In addition, *trans*-1,2-dichloroethylene should be taken into account, whereas *cis*-1,2-dichloroethylene is listed in the health-related standards for public waters. These

TABLE 2-5. Environmental water quality standards for protecting human health, and effluent standards (hazardous substances).

Environmental standards related to human health protection		**Effluent standards (hazardous substances)*[1]**	
Item	**Maximum contaminant level**	**Item**	**The maximum permissible limit**
cadmium	0.01 mg/L	cadmium and its compounds	0.1 mg/L
total cyanide	Not detectable	cyanide	1 mg/L
		organophosphorous pesticides	1 mg/L
lead	0.01 mg/L	lead and its compounds	0.1 mg/L
hexavalent chromium	0.05 mg/L	hexavalent chromium compound	0.5 mg/L
arsenic	0.01 mg/L	arsenic and its compounds	0.1 mg/L
total mercury	0.0005 mg/L	mercury, alkyl mercury and other mercury compounds	0.005 mg/L
alkyl mercury	Not detectable	alkyl mercury compound	Not detectable
polychlorinated biphenyl(PCBs)	Not detectable	polychlorinated biphenyl	0.003 mg/L
dichloromethane	0.02 mg/L	dichloromethane	0.2 mg/L
carbon tetrachloride	0.002 mg/L	carbon tetrachloride	0.02 mg/L
1,2-dichloroethane	0.004 mg/L	1,2-dichloroethane	0.04 mg/L
1,1-dichloroethylene	0.1 mg/L	1,1-dichloroethylene	0.2 mg/L
cis-1,2-dichloroethylene	0.04 mg/L	*cis*-1,2-dichloroethylene	0.4 mg/L
1,1,1-trichloroethane	1 mg/L	1,1,1-trichloroethane	3 mg/L
1,1,2-trichloroethane	0.006 mg/L	1,1,2-trichloroethane	0.06 mg/L
trichloroethylene	0.03 mg/L	trichloroethylene	0.3 mg/L
tetrachloroethylene	0.01 mg/L	tetrachloroethylene	0.1 mg/L
1,3-dichloropropene	0.002 mg/L	1,3-dichloropropene	0.02 mg/L
thiuram	0.006 mg/L	thiuram	0.06 mg/L
simazine	0.003 mg/L	simazine	0.03 mg/L
thiobencarb	0.02 mg/L	thiobencarb	0.2 mg/L

(*Continued*)

TABLE 2-5. Continued.

Environmental standards related to human health protection		**Effluent standards (hazardous substances)*1**	
Item	**Maximum contaminant level**	**Item**	**The maximum permissible limit**
benzene	0.01 mg/L	benzene	0.1 mg/L
selenium	0.01 mg/L	Selenium and its compounds	0. 1 mg/L
Nitrate and nitrite nitrogen	10 mg/L	Ammonia, Nitrate and nitrite nitrogen	100 mg/L*2
fluorine	0.8 mg/L	fluorine	8 mg/L (land area) 15 mg/L (sea area)
Boron	1 mg/L	Boron	10 mg/L (land area) 230 mg/L (sea area)
1,4-dioxane	0.05 mg/L		
Dioxins	Water quality 1 pg-TEQ/L Bottom quality 150 pg-TEQ/g	Dioxins	10 pg-TEQ/L

Remarks:
1. Standard values are used for annual average values. However, the value for total cyanide is the maximum value.
2. 'Not detectable' indicates that when the substance is measured through a specific method, the amount is less than the quantitative limit defined by that method.
3. The standard values for boron and fluoride are not applied to coastal waters.
4. More severe standards can be applied by prefectural law based on The Water Pollution Control Act*1.
5. Total amount of ammonia-N multiplied by 0.4, nitrite-N, and nitrate-N*2.

standards are applied to all groundwater because the twenty-eight chemicals listed pose a risk to human health. As with other environmental standards, these standards should be promptly achieved and maintained.

2.3.1.5 Effluent standard

Effluent standards were established as a regulatory action for clear environmental standards regarding water quality. The effluent standards are based on the Water Quality Pollution Control Act. The items included in the effluent

TABLE 2-6(a). Environmental water quality standards for preserving living environments.
1. Rivers (excluding lakes).

A

Parameter class	Water use	Environmental standard				
		pH	BOD	SS	DO	Total coliform
AA	Water supply class 1, conservation of natural environments, and uses listed in A-E	6.5–8.5	1 mg/L or less	25 mg/L or less	7.5 mg/L or more	50 MPN/ 100 mL or less
A	Water supply class 2, fishery class 1, bathing, and uses listed in B-E	6.5–8.5	2 mg/L or less	25 mg/L or less	7.5 mg/L or more	1,000 MPN/ 100 mL or less
B	Water supply class 3, fishery class 2, and uses listed in C-E	6.5–8.5	3 mg/L or less	25 mg/L or less	5 mg/L or more	5,000 MPN/ 100 mL or less
C	Fishery class 3, industrial water class 1, and uses listed in D-E	6.5–8.5	5 mg/L or less	50 mg/L or less	5 mg/L or more	–
D	Industrial water class 2, agricultural water, and uses listed in E	6.0–8.5	8 mg/L or less	100 mg/L or less	2 mg/L or more	–
E	Industry water class 3 and conservation of the environment	6.0–8.5	10 mg/L or less	Floating materials such as garbage should not be observable.	2 mg/L or more	–

Remarks:
1. Standard values are used for daily averages (the same applies to lakes/reservoirs and seas/coastal areas).
2. For irrigation (agricultural) water, the hydrogen ion concentration must be from 6.0 to 7.5, and the dissolved oxygen level must be greater than 5 mg/L (the same applies to lakes and reservoirs).

Notes:

1. Nature conservation: Conservation of sightseeing and other environments.
2. Water supply class 1: Purify water using filters and other simple means Water supply class 2: Purify water using sedimentation filters and other ordinary means Water supply class 3: Purify water using pre-treatment and other advanced methods.
3. Fishery class 1: For oligosaprobic members of Salmonidae (salmon/trout) species such as Salmomasoumasou, and Salvelinusleucomaenisu, and marine products for fishery classes 2 and 3. Fishery class 2: For alpha-oligosaprobic marine products such as Salmonidae (salmon/trout) species, sweetfish, and marine products for fishery class 3. Fishery class 3: For beta-oligosaprobic marine products such as koi and crucian carp.
4. Industrial water class 1: Water purified using sedimentation and other ordinary means Industrial water class 2: Purify water using chemical additives and other advanced means Industrial water class 3: Purify water using special means.
5. Environmental conservation: Limit of no disruption in the day-to-day lives of the population (including walks along the beach).

B

Parameter class	Adaptability to aquatic life habitat conditions	Standard values Total zinc
Aquatic life A	Water bodies inhabited by aquatic organisms such as char, salmon, and trout, and their prey, which favor relatively low-temperature ranges.	0.03 mg/L or less
Special aquatic life A	Water bodies categorized as 'Aquatic life A' need to be conserved, particularly as breeding or nursery grounds for aquatic life categorized therein.	0.03 mg/L or less
Aquatic life B	Water bodies inhabited by aquatic organisms such as carp and crucian, and their prey, which favor relatively high-temperature ranges.	0.03 mg/L or less
Special aquatic life B	Water bodies categorized as 'Aquatic life B' need to be conserved, particularly as breeding or nursery grounds for aquatic life categorized therein.	0.03 mg/L or less

TABLE 2-6(b). Environmental water quality standards for preserving living environments.
2. Lakes (natural lakes and artificial reservoirs having 10 million m^3 of water or more).

A

Parameter class	Water use	Standard values				
		pH	**BOD**	**SS**	**DO**	**Total coliform**
AA	Water supply class 1, fishery class 1, conservation of the natural environment, and uses listed in A-C	6.5–8.5	1 mg/L or less	1 mg/L or less	7.5 mg/L or more	50 MPN/ 100 mL or less
A	Water supply classes 2 and 3, fishery class 2, bathing, and uses listed in B-C	6.5–8.5	3 mg/L or less	5 mg /L or less	7.5 mg/L or more	1,000 MPN/ 100 mL or less
B	Fishery class 3, industrial water class 1, agricultural water, and uses listed in C	6.5–8.5	5 mg/L or less	15 mg /L or less	5 mg/L or more	—
C	Industrial water class 2 and conservation of the environment	6.5–8.5	8 mg/L or less	Floating matter such as garbage should not be observed	2 mg /L or more	—

Notes:

1. Conservation of the natural environment: Conservation of sightseeing and other environmental concerns.
2. Water supply class 1: Purify water using filters and other simple means Water supply classes 2 and 3: Purify water using sedimentation filters and other ordinary means, and pre-treatment and other advanced methods.
3. Fishery class 1: For marine products inhabiting oligotrophic lakes such as sockeye salmon, and marine products for fishery class 2 and 3. Fishery class 2: For marine products inhabiting oligotrophic lakes such as Salmonidae (salmon/trout) species, sweet fish, and marine products for fishery class 3. Fishery class 3: For marine products inhabiting oligotrophic lakes such as koi and crucian carp.
4. Industrial water class 1: Water purified using sedimentation and other ordinary means Industrial water class 2: Purify water using such advanced means as chemical additives and special purification means.
5. Conservation of the environment: Limit of no disruption in the day-to-day lives of the population (including walks along the beach).

B

Parameter class	Water use	Standard values	
		Total nitrogen	Total phosphorus
I	Conservation of the natural environment and uses listed in II–V	0.1 mg/L or less	0.005 mg/L or less
II	Water supply classes 1, 2, and 3 (except special types), fishery class 1, bathing, and uses listed in III–V	0.2 mg/L or less	0.01 mg/L or less
III	Water supply class 3 (special types) and uses listed in IV–V	0.4 mg/L or less	0.03 mg/L or less
IV	Fishery class 2 and uses listed in V	0.6 mg/L or less	0.05 mg/L or less
V	Fishery class 3, industrial water, agricultural water, and conservation of the environment	1 mg/L or less	0.1 mg/L or less

Remarks:

1. Standard values are used for annual average values.
2. Specified aquatic areas are lakes at risk for lake phytoplankton blooms. The standard for total nitrogen applies to lakes where the total nitrogen is a factor for lake-phytoplankton blooms.
3. The standards for total phosphorus do not apply to irrigation water.

Notes:

1. Conservation of the natural environment: Conservation for sightseeing and other environmental concerns.
2. Water supply class 1: Purify water using filters and other simple means Water supply class 2: Purify water using sedimentation filters and other ordinary means Water supply class 3: Purify water using pre-treatment and other advanced methods (a 'special item' is a special purification means capable of removing odor-producing substances).
3. Fishery class 1: For marine products such as the Salmonidae (salmon/trout) species, sweet fish, and marine products for fishery classes 2 and 3. Fishery class 2: For marine products such as smelt and marine products for fishery class 3. Fishery class 3: Marine products such as koi and crucian carp.
4. Conservation of the environment: Limit of no disruption in the day-to-day lives of the population (including walks along the beach).

(*Continued*)

TABLE 2-6(b). Continued.

C

Parameter class	Adaptability to aquatic living habitat conditions	Standard values Total zinc
Aquatic life A	Water bodies inhabited by aquatic organisms such as char, salmon, and trout, and their prey, which favor relatively low-temperature ranges.	0.03 mg/L or less
Special aquatic life A	Water bodies categorized as 'Aquatic life A' need to be conserved, particularly as breeding or nursery grounds for the aquatic life categorized therein.	0.03 mg/L or less
Aquatic life B	Water bodies inhabited by aquatic organisms such as carp and crucian, and their prey, which favor relatively high-temperature ranges.	0.03 mg/L or less
Special aquatic life B	Water bodies categorized as 'Aquatic life B' need to be conserved, particularly as breeding or nursery grounds for the aquatic life categorized therein.	0.03 mg/L or less

TABLE 2-6(c). Environmental water quality standards for preservingliving environments.
3. Seas and coastal areas.

A

Parameter class	Water use	Standard values				
		pH	**BOD**	**DO**	**Total coliform**	**N-hexane extracts (oil content etc.)**
A	Fishery class 1, bathing, conservation of the natural environment, and uses listed in B-C	7.8–8.3	2 mg/L or less	7.5 mg/L or more	1,000 MPN/ 100 mL or less	Not detectable
B	Fishery class 2, industrial water and the uses listed in C	7.8–8.3	3 mg/L or less	5 mg/L or more	—	Not detectable
C	Conservation of the environment	7.0–8.3	8 mg/L or less	2 mg/L or more	—	—

Remarks:

1. For Fishery class 1, the standard for the total coliform amount in waters used to raise oysters for raw consumption is 70 MPN/100 mL or less.

Notes:

1. Conservation of the natural environment: Conservation for sightseeing and other environmental concerns.
2. Fishery class 1: For marine products such as red sea breams, yellowtails, and seaweed, and marine products for fishery class
   2. Fishery class 2: Marine products such as mullets and dried seaweed.
3. Conservation of the environment: Limit of no disruption in the day-to-day lives of the population (including walks along the beach).

(*Continued*)

TABLE 2-6(c). Continued.

B

Parameter class	Water use	Standard values	
		Total nitrogen	Total phosphorus
I	Conservation of the natural environment and uses listed in II-IV (except fishery classes 2 and 3)	0.2 mg/L or less	0.02 mg/L or less
II	Fishery class 1, bathing, and the uses listed in III-IV (except fishery classes 2 and 3)	0.3 mg/L or less	0.03 mg/L or less
III	Fishery class 2 and the uses listed in IV (except fishery class 3)	0.6 mg/L or less	0.05 mg/L or less
IV	Fishery class 3, industrial water, and conservation of habitable environments for marine biota	1 mg/L or less	0.09 mg/L or less

Remarks:

1. Standard values are used for annual average values.
2. Specified aquatic areas are at risk for marine phytoplankton blooms.

Notes:

1. Conservation of the natural environment: Conservation for sightseeing and other environmental concerns.
2. Fishery class 1: A large variety of fish, including benthic fish and shellfish, are taken in good balance and stability.
   Fishery class 2: Marine products (mainly fish) are taken, with the exception of some benthic fish and shellfish.
   Fishery class 3: Specific types of marine products highly resistant to pollution are mainly taken.
3. Conservation of habitable environments for marine biota: Level at which bottom-dwelling organisms can survive year-round.

C

Parameter class	Adaptability to aquatic habitat conditions	Standard values
		Total zinc
Class A organisms	Water areas inhabited with aquatic life	0.02 mg/L or less
Special class A organisms A	Of the water areas inhabited by Class A organisms, those that should be conserved as spawning/rearing areas of aquatic life	0.01 mg/L or less

standards are categorized into hazardous and other substances. The hazardous substances and their maximum permissible limit are shown in **Table 2-5**. The effluent standards for living environments are shown in **Table 2-7**. These standards are applied to wastewater from specific factories, which are defined as factories and/or business institutions that have specific facilities designated by a Cabinet order. The items in the effluent standards for hazardous substances are almost equivalent to those in the water quality standards for

TABLE 2-7. Effluent standard (living environment items).

Living environment items	Permissible limit
Hydrogen ion activity (pH)	Non-Coastal Regions 5.8–8.6 Coastal Regions 5.0–9.0
Biochemical oxygen demand (BOD)	160 mg/L (Daily Average 120 mg/L)
Chemical oxygen demand (COD)	160 mg/L (Daily Average 120 mg/L)
Suspended solids (SS)	200 mg/L (Daily Average 150 mg/L)
n-hexane extracts (mineral oil)	5 mg/L
n-hexane extracts (animal and vegetable fats)	30 mg/L
Phenols	5 mg/L
Copper	3 mg/L
Zinc	5 mg/L
Dissolved iron	10 mg/L
Dissolved manganese	10 mg/L
Chromium	2 mg/L
Number of coliform groups	Daily Average 3,000/cm^3
Nitrogen	120 mg/L (Daily Average 60 mg/L)
Phosphorus	16 mg/L (Daily Average 8 mg/L)

Remarks:

1. The effluent standards listed in this table apply to the effluents of factories or commercial facilities with an effluent discharge of 50 m^3 or more per day on average.
2. The effluent standards for biochemical oxygen demand (BOD) apply exclusively to the effluents discharged in public waters other than seas and lakes; the effluent standards for chemical oxygen demand (COD) apply exclusively to the effluents discharged in sea and lake areas.
3. The effluent standards for nitrogen content only apply to lakes and reservoirs designated by the Ministry of the Environment as being susceptible to lake plankton blooms, sea areas designated by the Ministry of the Environment as being susceptible to marine phytoplankton blooms, and effluents discharged into public water areas.
4. The effluent standards for phosphorus content only apply to lakes and reservoirs designated by the Ministry of the Environment as being susceptible to lake plankton blooms, sea areas designated by the Ministry of the Environment as being susceptible to marine phytoplankton blooms, and effluents discharged into public water areas.

human health protection, but also include organic phosphorus pesticides. It is assumed that an effluent is diluted more than time fold at the discharge area, and therefore most of the items in the effluent standards are set, and their maximum permissible limits are 10 times the environmental standard limits for human health protection. In addition to hazardous materials, fifteen other items are set as living environment items, which are applied to the wastewater from specific factories discharging 50 m^3 of water per day or more on average. The registered number of specific factories is about 270,000, and the number of factories discharging wastewater at 50 m^3/day or more was about 3,400 as of March 2010. The ratio factories discharging wastewaters at 50 m^3/day or more to the total number of factories is only about 13%. This means that most of the small specific factories are not affected by these items. Therefore, local governments can reduce the number of water restrictions to improve the situation. In addition to the uniform effluent standards, prefectures and major cities can establish stricter effluent standards or add new regulatory items through ordinances based on the Water Quality Pollution Control Act. For example, more stringent prefectural effluent standards for nitrogen and phosphorus have been applied in Shiga prefecture, where Lake Biwa is located.

2.3.2 *Environmental standards for water quality in China*

2.3.2.1 Introduction

China has established a system of environmental protection standards at both the national and local levels. Environmental protection standards at the national level include environmental quality standards, pollutant discharge (control) standards, and standards for environmental samples. Environmental protection standards at the local level include environmental quality and pollutant discharge standards. By the end of 2005, the state had promulgated over 800 national environmental protection standards. The municipalities of Beijing and Shanghai, and the provinces of Shandong and Henan, have promulgated over thirty local environmental protection standards.

The first water environmental standard in China was promulgated in 1983. Thereafter, a series of water environmental standards, including five quality standards, thirty-five discharge standards, 154 method standards, and seventeen reference material standards, have been issued or promulgated.

To date, a total of five water quality standards have been established, and after several revisions, are currently executed in China. The standard code, execution date, and number of pollutant items restricted in the standard are listed in **Table 2-8**.

2.3.2.2 Environmental quality standards for surface water

The first water quality standards (also the first water environmental standard) were on surface water quality, and were promulgated in 1983 and revised

TABLE 2-8. The current water quality standards in China.

Standard name	Standard code	Execution date	Standard items
Environmental quality standards for surface water	GB 3838-2002	20020601	109
Quality standards for groundwater	GB/T14848-93	19941001	39
Sea water quality standards	GB3097-1997	19980701	35
Water quality standards for fisheries	GB11607-89	19900301	33
Standards for irrigation water quality	GB5084-92	19921001	29

three times in 1988, 1999, and 2002. The last version (GB 3838-2002) was promulgated and executed in 2002. The standards list the detailed indexes to be controlled, and the limits for water quality, the analysis methods, and the enforcement and supervision of the standards are based on the different environmental functions of the surface water and the targets for its protection. These standards apply to all useable surface waters including rivers, lakes, canals, channels, and water reservoirs in China. Waters with specific functions shall apply special water quality standards accordingly.

The latest version of these performance standards is GB 3838-2002, which defines five grades of water quality according to the different environmental functions, and lists the quality standards for every grade (**Table 2-9**).

- Grade I
  - Headwaters and national reservation zones
- Grade II
  - First class of preservation areas for drinking water sources
  - Habitats of rare and precious aquatic organisms
  - Spawning zones for fish and shrimp
- Grade III
  - Second level of reservation areas for drinking water sources
  - Reservation zones of general fish populations
  - Swimming areas
- Grade IV
  - General industrial uses
  - Recreation water bodies for indirect human contact
- Grade V
  - Agricultural uses
  - General landscaping requirement

TABLE 2-9. Environmental quality standards for basic surface water items (mg/L).

		Standard limits (mg/L)				
No.	**Item**	**Grade I**	**Grade II**	**Grade III**	**Grade IV**	**Grade V**
1	Temperature	Weekly average, increasing less than 1 and decreasing less than 2 degrees				
2	pH	6–9				
3	Dissolved oxygen	7.5	6	5	3	2
4	Index of permanganate	2	4	6	10	15
5	COD_{Cr}	15	15	20	30	40
6	BOD_5	3	3	4	6	10
7	Ammonia-nitrogen (NH_4)	0.15	0.5	1.0	1.5	2.0
8	Total phosphorus (measured as P)	0.02 (lakes and reservoirs, 0.01)	0.1 (lakes and reservoirs, 0.025)	0.2 (lakes and reservoirs, 0.05)	0.3 (lakes and reservoirs, 0.05)	0.4 (lakes and reservoirs, 0.05)
9	Total nitrogen (measured as N)	0.2	0.5	1.0	1.5	2.0
10	Copper	0.01	1.0	1.0	1.0	1.0
11	Zinc	0.05	1.0	1.0	2.0	2.0
12	Fluoride (as F^-)	1.0	1.0	1.0	1.5	1.5
13	Selenium	0.01	0.01	0.01	0.02	0.02
14	Arsenic	0.05	0.05	0.05	0.1	0.1
15	Mercury	0.00005	0.00005	0.0001	0.001	0.001
16	Cadmium	0.001	0.005	0.005	0.005	0.01
17	Chromium (VI)	0.01	0.05	0.05	0.05	0.1
18	Lead	0.01	0.01	0.05	0.05	0.1
19	Cyanide	0.005	0.05	0.2	0.2	0.2
20	Volatile phenol	0.002	0.002	0.005	0.01	0.1
21	Oil	0.05	0.05	0.05	0.5	1.0
22	Anion surfactant	0.2	0.2	0.2	0.3	0.3
23	Sulfide	0.05	0.1	0.2	0.5	1.0
24	Total bacilli	200	2,000	10,000	20,000	40,000

2.3.2.3 Quality standards for ground water

These standards were formulated for the protection and sustainable development of ground water resources, the prevention and control of ground water pollution, the protection of ground water environments, and human health. They serve as the basis for the prospecting and assessment of ground water, as well as its development, use, supervision, and management. These standards provide the classification of ground water quality, the methods for monitoring and assessing ground water quality, and water quality protection. These apply to ordinary ground water other than geothermal water, mineral water, and brine.

The quality standards for ground water (GB/T 14848-93) were issued in 1993 and executed in 1994. They define five grades and regulate the limits for thirty-five items.

- Classification of ground water quality and its index
    - Grade I
        - Low natural background content for the chemical component of ground water
        - For all uses
    - Grade II
        - Natural background content for the chemical component of ground water
        - For all purpose of uses
    - Grade III
        - Based on health criteria
        - For drinking water sources and agricultural and industrial uses
    - Grade IV
        - Based on the requirements of agricultural and industrial uses, and suitable for drinking after proper treatment
    - Grade V
        - Not suitable for drinking

The items and their standards for every grade of ground water are listed in **Table 2-10**.

2.3.2.4 Seawater quality standard

The standard provides water quality requirements for sea areas with different functions. It applies to sea areas within the jurisdiction of the People's Republic of China. The first standards for sea water quality were issued and executed in 1982, and replaced by the revised version in 1998 (GB 3097-1997). The sea water quality standards (GB 3097-1997) define four seawater quality grades for different environmental functions.

TABLE 2-10. Classification indexes for ground water quality.

		Standard limit (mg/L)				
No.	**Item**	**Grade I**	**Grade II**	**Grade III**	**Grade IV**	**Grade V**
1	pH		6.5–8.5		5.5–6.5; 8.5–9	<5.5; >9
2	Total hardness	150	300	450	550	>550
3	Solvable solid	300	500	1,000	5,000	>2,000
4	Sulfate	50	150	250	350	>350
5	Chloride	50	150	250	350	>350
6	Iron	0.1	0.2	0.3	1.5	>1.5
7	Manganese	0.05	0.05	0.1	1.0	>1.0
8	Copper	0.01	0.05	1.0	1.5	>1.5
9	Zinc	0.05	0.5	1.0	5.0	>5.0
10	Molybdenum	0.001	0.01	0.1	0.5	>0.5
11	Cobalt	0.005	0.05	0.05	1.0	>1.0
12	Volatile phenol	0.001	0.001	0.002	0.01	>0.01
13	Anion synthesis detergent	Zero	0.1	0.3	0.3	>0.3
14	Index of permanganate	1.0	2.0	3.0	10	>10
15	Nitrate (as N)	2.0	5.0	20	30	>30
16	Nitrite (as N)	0.001	0.01	0.02	0.1	>0.1
17	Ammonia-nitrogen	0.02	0.02	0.2	0.5	>0.5
18	Fluoride	1.0	1.0	1.0	2.0	>2.0
19	Iodide	0.1	0.1	0.2	1.0	>1.0
20	Cyanide	0.001	0.01	0.05	0.1	>0.1
21	Mercury	0.00005	0.0005	0.001	0.001	>0.001
22	Arsenic	0.005	0.01	0.05	0.05	>0.05
23	Selenium	0.01	0.01	0.01	0.1	>0.1
24	Cadmium	0.0001	0.001	0.01	0.01	>0.01
25	Chromium (VI)	0.005	0.01	0.05	0.1	>0.1
26	Lead	0.005	0.01	0.05	0.1	>0.1
27	Beryllium	0.00002	0.0001	0.0002	0.001	>0.001
28	Barium	0.01	0.1	1.0	4.0	>4.0

(*Continued*)

TABLE 2-10. Continued.

No.	Item	Standard limit (mg/L)				
		Grade I	Grade II	Grade III	Grade IV	Grade V
29	Nickel	0.005	0.05	0.05	0.1	>0.1
30	DDT (mg/L)	Zero	0.005	1.0	1.0	>1.0
31	HCHs (mg/L)	0.005	0.05	5.0	5.0	>5.0
32	Total colon bacillus (/L)	3.0	3.0	3.0	100	>100
33	Total bacilli (/mL)	100	100	100	1,000	>1,000
34	Total a radio-activity (Bq/L)	0.1	0.1	0.1	0.1	>0.1
35	Total b radio-activity (Bq/L)	0.1	1.0	1.0	1.0	>1.0

- Classification of seawater quality
  - Grade I
    - Water bodies for sea fisheries
    - Maritime nature conservation areas
    - Rare and precise marine organism reserves
  - Grade II
    - Water bodies for aquaculture
    - Bathing beaches
    - Maritime sports or recreation with direct skin contact
  - Grade III
    - Water bodies for general industrial uses
    - Scenic and tourist areas bordering on the sea
  - Grade IV
    - Coastal port water bodies
    - Ocean exploitation and operation areas

The standards for every grade of seawater are listed in **Table 2-11**.

2.3.2.5 Standards for irrigation water quality

These standards were formulated for the purpose of preventing the pollution of soil, ground water, and farm produce, and protecting the environment and health of people and livestock. It provides standard methods for the application, sampling, and monitoring of irrigation water quality. The standards

TABLE 2-11. Seawater quality standards.

		Standard limit (mg/L)			
No.	**Item**	**Grade I**	**Grade II**	**Grade III**	**Grade IV**
1	Floating substances	No oil film, foam, or other floating substances on the surface			No obvious oil film, floating foam, or other substances on the surface
2	Color, odor, and taste	No unusual colors, odors, or taste			
3	SS	Man-made increase ≤10		Man-made increase ≤100	Man-made increase ≤150
		10,000			
4	Total coliforms (/L)	Seawater for shellfish cultivation being eaten raw ≤700			–
5	Fecal coliforms	2,000 Seawater for shellfish cultivation being eaten raw ≤700			–
6	Pathogens	No pathogen exists for shellfish cultivation being eaten raw			–
7	Temperature	Man-made increase not exceeding 1°C of the local area during summer and 2°C in the other seasons		Man-made increase not exceeding 4°C of the local area	
8	pH	7.8–8.5, and not exceeding 0.2 pH of the normal change range		6.8–8.8, and not exceeding 0.5 pH of the normal change range	
9	Dissolved oxygen	6	6	4	3
10	COD	2	3	4	5
11	BOD_5	1	3	4	5
12	Inorganic nitrogen (measured as N)	0.20	0.30	0.40	0.50
13	Non-ionic ammonia (measured as N)		0.020		

(Continued)

TABLE 2-11. Continued.

<table>
<tr><th></th><th></th><th colspan="4">Standard limit (mg/L)</th></tr>
<tr><th>No.</th><th>Item</th><th>Grade I</th><th>Grade II</th><th>Grade III</th><th>Grade IV</th></tr>
<tr><td>14</td><td>Active phosphorate (measured as P)</td><td>0.015</td><td colspan="2">0.030</td><td>0.045</td></tr>
<tr><td>15</td><td>Mercury</td><td>0.00005</td><td colspan="2">0.0002</td><td>0.0005</td></tr>
<tr><td>16</td><td>Cadmium</td><td>0.001</td><td>0.005</td><td colspan="2">0.010</td></tr>
<tr><td>17</td><td>Lead</td><td>0.005</td><td>0.010</td><td>0.020</td><td>0.050</td></tr>
<tr><td>18</td><td>Chromium (VI)</td><td>0.020</td><td>0.030</td><td colspan="2">0.050</td></tr>
<tr><td>19</td><td>Total chromium</td><td>0.005</td><td>0.010</td><td colspan="2">0.050</td></tr>
<tr><td>20</td><td>Arsenic</td><td>0.020</td><td>0.030</td><td colspan="2">0.050</td></tr>
<tr><td>21</td><td>Copper</td><td>0.005</td><td>0.010</td><td colspan="2">0.050</td></tr>
<tr><td>22</td><td>Zinc</td><td>0.020</td><td>0.050</td><td>0.10</td><td>0.50</td></tr>
<tr><td>23</td><td>Selenium</td><td>0.010</td><td colspan="2">0.020</td><td>0.050</td></tr>
<tr><td>24</td><td>Nickel</td><td>0.005</td><td>0.010</td><td>0.020</td><td>0.050</td></tr>
<tr><td>25</td><td>Cyanide</td><td colspan="2">0.005</td><td>0.10</td><td>0.20</td></tr>
<tr><td>26</td><td>Sulfide (measured as S)</td><td>0.02</td><td>0.05</td><td>0.10</td><td>0.25</td></tr>
<tr><td>27</td><td>Volatile phenol</td><td colspan="2">0.005</td><td>0.010</td><td>0.050</td></tr>
<tr><td>28</td><td>Oil</td><td colspan="2">0.05</td><td>0.30</td><td>0.50</td></tr>
<tr><td>29</td><td>HCHs</td><td>0.001</td><td>0.002</td><td>0.003</td><td>0.005</td></tr>
<tr><td>30</td><td>DDT</td><td>0.00005</td><td></td><td>0.0001</td><td></td></tr>
<tr><td>31</td><td>Malathion</td><td>0.0005</td><td></td><td>0.001</td><td></td></tr>
<tr><td>32</td><td>Methyl parathion</td><td>0.0005</td><td></td><td>0.001</td><td></td></tr>
<tr><td>33</td><td>Benzo[a]pyrene</td><td></td><td colspan="2">0.0025</td><td></td></tr>
<tr><td>34</td><td>Anion surfactant (measured as LAS)</td><td>0.03</td><td></td><td>0.10</td><td></td></tr>
<tr><td>35</td><td>Radio-nuclides (Bq/L)</td><td>^{60}Co</td><td colspan="2">0.03</td><td></td></tr>
<tr><td></td><td></td><td>^{90}Sr</td><td colspan="2">4</td><td></td></tr>
<tr><td></td><td></td><td>^{106}Rn</td><td colspan="2">0.2</td><td></td></tr>
<tr><td></td><td></td><td>^{134}Cs</td><td colspan="2">0.6</td><td></td></tr>
<tr><td></td><td></td><td>^{137}Cs</td><td colspan="2">0.7</td><td></td></tr>
</table>

apply to irrigation water drained from surface water, groundwater, treated municipal sewage, and industrial effluent with similar water quality as municipal sewage. It is not suitable for treated wastewater from such sectors as pharmaceuticals, bio-products, chemical agents, pesticides, oil refineries, coking, and organic chemicals.

The first standards for irrigation water quality were issued and executed in 1985, and replaced by the revised version in 1992 (GB 5084-92). GB 5084-92 defines three grades of water standards according to different functions for irrigation on upland crops, paddy crops, and vegetables. The standards are shown in **Table 2-12**.

2.3.2.6 Water quality standards for fisheries

These standards were formulated to prevent and control water pollution in fishery waters, and ensuring the normal growth and breeding of fish, shrimp, shellfish, and alga, as well as the quality of aquatic products. These standards apply to seawater and freshwater fishery areas where fish and shrimp spawn, feed, live throughout the winter, migrate, and breed. The quality standards for fisheries (GB 11607-89) were issued in 1989 and executed in 1990 (**Table 2-13**).

2.3.3 *Present state and issues of water quality in Japan*

In Japan, national and local governments have monitored the water quality of public waters in accordance with the monitoring plan prepared by the Water Pollution Control Act, and the results are announced by the Ministry of Environment each year.

The compliance ratio of environmental standards for twenty-seven items in the field of health protection in the fiscal year of 2009 was 99.1% of the monitoring points, and environmental quality standards have been achieved in most locations. The yearly compliance trend of BOD and COD for the living environment items in the environmental standards are shown in **Figure 2-10**. The overall ratio is 87.6% while looking at each area, 92.3% for rivers, 50.0% for lakes, and 79.2% for coastal seas. Improvements in river areas are progressing, but the ratio in lakes remains low.

Tokyo bay, Ise bay, and Seto Inland Sea are typical enclosed coastal seas, and special counter measures such as a restriction of the total loading rate of COD, total nitrogen, and total phosphorus are applied. **Figure 2-11** shows the changes in mean concentration of BOD and COD. The COD concentration in coastal seas has been continuing at the same level as during the fiscal year of 1979, at 1.5 to 2.0 mg/L. The BOD concentration in rivers twenty years ago was 2.5 mg/L, and has declined every year, reaching 1.3 mg/L in 2008. For lakes, the COD concentration ten years ago was 3.5 mg/L, and has remained stable.

TABLE 2-12. Standards for irrigation water quality.

		Standard limits (mg/L)		
No.	**Item**	**Upland crops**	**Paddy crops**	**Vegetables**
1	BOD_5	80	150	80
2	COD_{Cr}	200	300	150
3	Suspended substances	150	200	100
4	Anion surfactant (LAS)	5	8	5
5	Kjeldahl nitrogen	12	30	30
6	Total phosphorous (measured as P)	5	10	10
7	Temperature		35	
8	pH		5.5–8.5	
9	Total salinity		1,000 (non-saline-alkaline soil) 2,000 (saline-alkaline soil)	
10	Chloride		250	
11	Sulfide		1	
12	Total mercury		0.001	
13	Total cadmium		0.005	
14	Total arsenic	0.05	0.1	
15	Chromium (VI)		0.1	
16	Total lead		0.1	
17	Total copper		1	
18	Total zinc		2	
19	Total selenium		0.02	
20	Fluoride		2.0–3.0	
21	Cyanide		0.5	
22	Oil	5	10	1
23	Volatile phenol		1	
24	Benzene		2.5	
25	Trichloroaldehyde	1	0.5	0.5
26	Acrolein		0.5	
27	Boron		1.0–3.0	
28	Total bacilli (/L)		10,000	
29	Ascarid ovum		2	

TABLE 2-13. Water quality standard for fisheries.

No.	Item	Standard limit (mg/L)
1	Color, odor, and taste	No effects on fish, shrimp, shellfish, or algae
2	Floating substances	No obvious oil film or foam on the surface
3	Suspended substance	Man-made increase ≤10, harmless to fish, shrimp, and shellfish
4	pH	Fresh water 6.5–8.5 and seawater 7.0–8.5
5	Dissolved oxygen	>5 during 16 h in a 24 h period
6	BOD_5	≤5
7	Total coliforms	≤5,000/L (≤500/L for shellfish culture)
8	Mercury	≤0.0005
9	Cadmium	≤0.005
10	Lead	≤0.05
11	Chromium	≤0.1
12	Copper	≤0.01
13	Zinc	≤0.1
14	Nickel	≤0.05
15	Arsenic	≤0.05
16	Cyanide	≤0.005
17	Sulfide	≤0.2
18	Fluoride (measured as F^-)	≤1
19	Non-ionic ammonia	≤0.02
20	Kjeldahl nitrogen	≤0.05
21	Volatile phenol	≤0.005
22	Yellow phosphorus	≤0.001
23	Oil	≤0.05
24	Acrylonitrile	≤0.5
25	Acrolein	≤0.02
26	HCHs	≤0.002
27	DDT	≤0.001
28	Marathion	≤0.005
29	Pentachlorophenol	≤0.01
30	Dimethoate	≤0.1
31	Methamidophos	≤1
32	Methylparathion	≤0.0005
33	Furadan	≤0.01

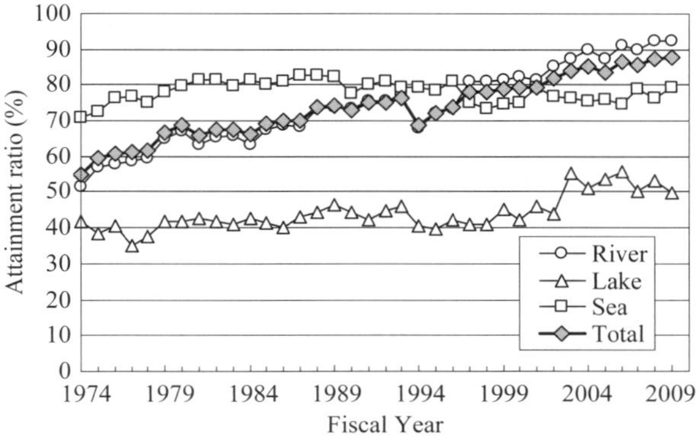

FIGURE 2-10. Changes in environmental quality standards (EQS) achieved.

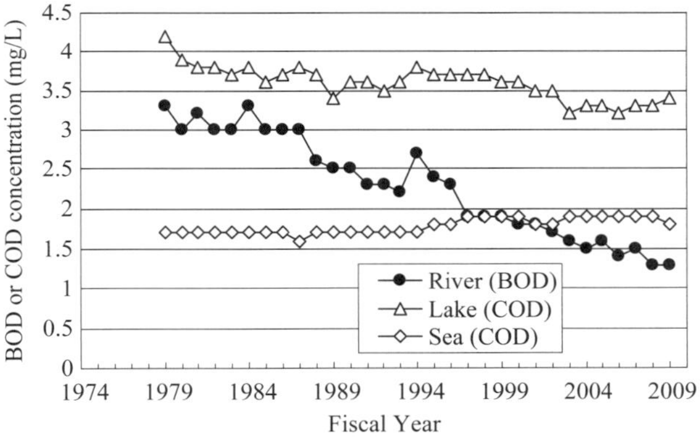

FIGURE 2-11. Changes in water quality for each water area.

The environmental standards for total nitrogen and total phosphorus in lakes were set in 1982, and the benefits of these standards have been monitored since 1984. While a total of 115 water areas were classified for these standards, only 60 have successfully achieved them, which is an achievement ratio of 52.5%. The achievement rate of total nitrogen is 15.4%, whereas that of total phosphorus is 58.3%. The changes in the total nitrogen concentration for each type of lake are shown in **Figure 2-12**, whereas those for total phosphorus concentration are shown in **Figure 2-13**. The recent nutrient concentrations in lakes are almost at the same level.

The achievement ratio of the environmental standards for total phosphorus and total nitrogen amounts in sea waters is as follows. A total of 151 water areas were classified for these standards, and the number of the areas achieving

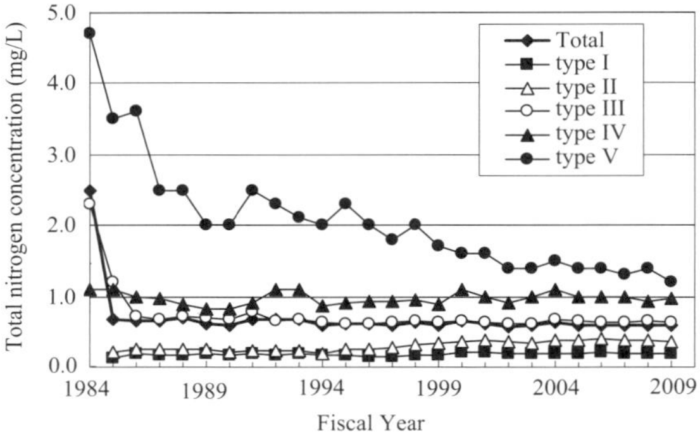

FIGURE 2-12. Changes in total nitrogen concentration in each lake type.

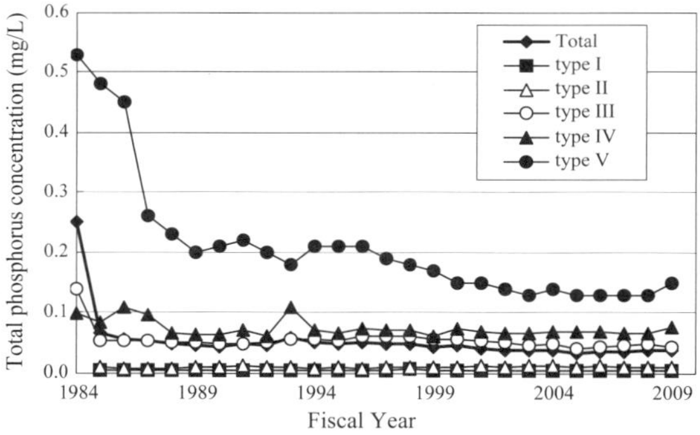

FIGURE 2-13. Changes in total phosphorus concentration in each lake type.

the standards is 123. Therefore, the achievement ratio is 81.5%. Changes in the total nitrogen concentration for each type of sea are shown in **Figure 2-14**, and those of total phosphorus concentration are shown in **Figure 2-15**. The nutrient concentrations in the sea areas were improved during the late 1990s, and became almost stable after 2000.

The total zinc amount was set as the only item of the water quality standards for aquatic organism preservation, and the concentrations in water environments were measured at 4,110 different points. The standard value of total zinc in river waters is 0.03 mg/L, and the achievement rate is 96.2%. The standard value for total zinc in lakes is also 0.03 mg/L, with a 99.6% achievement rate. The standard value of total zinc in normal sea areas is 0.02 mg/L, and the percentage of points where the zinc concentration is under this level

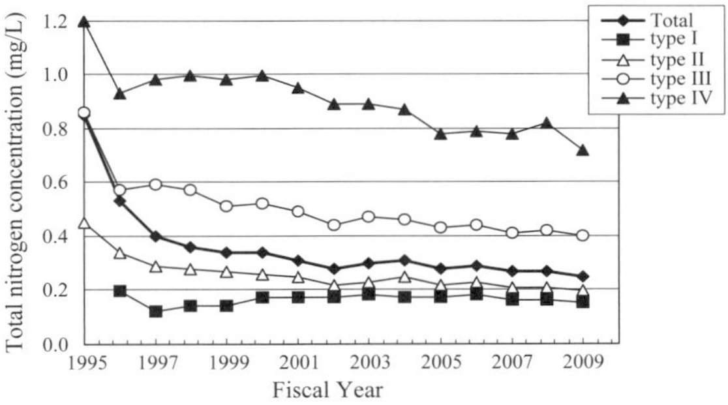

FIGURE 2-14. Changes in total nitrogen concentration for each sea type.

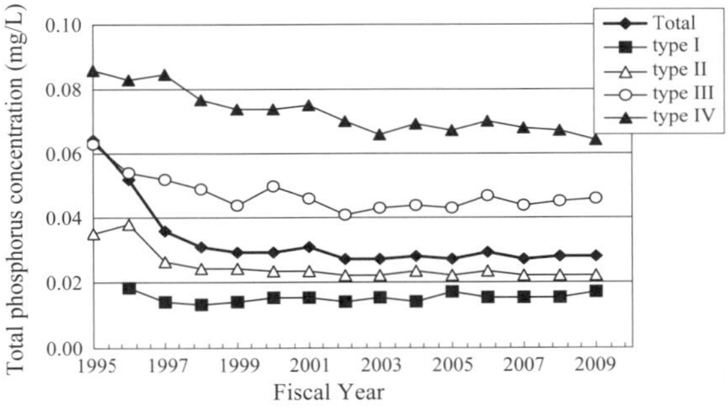

FIGURE 2-15. Changes in total phosphorus concentration for each sea type.

is 99.1%, whereas the standard value in special sea areas is 0.01 mg/L, and the rate was 92.2%. The types of environmental quality standards were specified in June 2006 for nine river and lake water areas in four water systems, the Kitakami river, Tama river, Yamato river, and Yoshino river. At all measurement points in these rivers and lakes, the total zinc concentration was no more than 0.03 mg/L, which means the environmental quality standards of total zinc were achieved in all currently regulated water areas.

Although the achievement ratio of the water quality standards continues to improve, red tides have still occurred. Ninety-nine such events happened in Seto Island Sea, and forty-one in Ariake sea, in 2007. Blue tides have also been observed in Tokyo and Mikawa bays in Aichi prefecture. A total of 6.9% of the groundwater monitoring points did not achieve the standards, and 4.4% of the monitoring points were contaminated by nitrate or nitrite nitrogen owing to fertilization or domestic wastewater. Groundwater points contaminated by trichlorethylene are still being discovered. Continuous countermeasures are required to maintain a desirable environment.

2.3.4 *Present state and issues of water quality in China*

2.3.4.1 Introduction

China is a country short of water resources, with a per capita amount of only 2,220 m^3, which is about a quarter of the world's average. Since the late 1970s, China's economy has developed rapidly and continuously, resulting in an increasing demand for water resources. Many cities and regions are facing serious water shortages. At the same time, the water quality has deteriorated seriously in China in recent decades. Serious pollution has occurred in a wide range of rivers, lakes, reservoirs, groundwater areas, and coastal waters. Water pollution, water shortages, aquatic ecosystem degradation, and landscape destruction are the main water environmental problems taking place in China, and water pollution in particular has become one of the key factors hampering the nation's sustainable development.

2.3.4.2 Status of water environments in China

(1) *Status of river waters*

The water quality of ten first-order basins was assessed (**Table 2-15**). Overall, the surface waters in China are still seriously polluted. According to the Environmental Quality Standards for Surface Water (GB 3838-2002), during the flood season, non-flood season, and the entire year of 2008, the river length belonging to Grades I–III accounted for 60.1%, 61.6%, and 61.2%, respectively, and slightly increased from 2007 (**Table 2-14**). The main pollution indicators were COD_{Mn}, COD_{Cr}, ammonia, BOD_5, and volatile phenol.

Among the ten first-order basins and key rivers assessed, the rivers located in southwest and northwest China had the highest percentages of clean water

TABLE 2-14. Water quality status of the rivers in 2007 and 2008.

			Percentage of river length (%)					
Year	Period	Total river length assessed (km)	Grade I	Grade II	Grade III	Grade IV	Grade V	Worse than Grade V
2008	Whole year	147,727.5	3.5	31.8	25.9	11.4	6.8	20.6
	Flood season	146,824.0	2.7	28.9	28.5	14.2	7.4	18.3
	Non-flood season	146,650.4	6.0	31.6	24.0	10.3	6.0	22.1
2007	Whole year	143,604.4	4.1	28.2	27.2	13.5	5.3	21.7
	Flood season	142,732.9	2.5	28.2	27.3	16.8	6.6	18.6
	Non-flood season	142,282.2	6.0	28.3	25.2	12.8	5.5	22.2

Note: Cited from 2008 annual report on water source quality in China.

(Grades I–III), while those in the Huaihe, Yellow and Haihe river areas had the highest percentages of water worse than Grade V (**Table 2-15**). From 2007, the water quality was found to have improved in the rivers in southwest and northwest China, the Liaohe and Haihe river areas, but remained unchanged.

TABLE 2-15. Water quality status of ten first-order basins and key rivers in 2008.

		Percentage of river length (%)					
First-order water basins	**Total river length (km)**	**Grade I**	**Grade II**	**Grade III**	**Grade IV**	**Grade V**	**Worse than Grade V**
All of China	147,727.5	3.5	31.8	25.9	11.4	6.8	20.6
Songhua river basin	13,562.4	0.8	17.0	29.2	25.2	6.3	21.5
Songhuajiang river	10,945.7	1.0	17.9	33.3	19.6	5.6	22.6
Liaohe river basin	5,496.7	1.5	27.1	17.4	10.8	13.1	30.1
Liaohe river	2,552.6	0.0	4.6	27.8	9.2	24.8	33.6
Haihe river basin	12,996.2	2.4	19.6	13.2	10.7	2.2	51.9
Haihe river	10,309.5	2.4	16.1	6.8	11.9	2.0	60.8
Yellow river basin	13,847.7	5.2	12.7	21.3	13.5	10.5	36.8
Huaihe river basin	14,130.5	0.5	15.6	23.3	18.1	11.3	31.2
Huaihe river	12,025.6	0.5	15.6	22.3	20.2	12.8	28.6
Yangtze river basin	41,176.6	3.7	36.2	29.2	9.0	7.5	14.4
Southeast river basin	5,035.2	5.2	38.3	20.7	12.1	8.9	14.8
Pearl river basin	18,541.5	0.0	38.8	29.8	11.0	6.8	13.6
Pearl river	13,886.4	0.0	32.4	35.2	11.1	7.0	14.3
Southwest river basin	13,406.7	0.2	48.0	46.1	2.8	0.2	2.7
Northwest river basin	9,534.0	21.2	65.8	7.0	3.5	1.7	0.8

Note: Cited from 2008 annual report of water source quality of China.

Overall, China's seven major river systems, i.e., the Yangtze, Yellow, Pearl, Songhuajiang, Huaihe, Haihe, and Liaohe rivers, were moderately polluted as of 2008 (**Figure 2-16**). Among the 409 monitoring areas located in 200 rivers, those with Grades I–III, Grades IV–V, and worse than Grade V water quality accounted for 55.0%, 24.2%, and 20.8% respectively. The main pollutants were COD_{Mn}, oils, ammonium, and BOD_5.

Among the seven main rivers, the Yangtze and Pearl rivers, located in southern China, were overall comparatively clean owing to their abundant water flow and high dilution capability. Songhuajiang river was slightly polluted, while the Yellow, Huaihe, and Liaohe rivers were moderately polluted, and Haihe river was heavily polluted. The sections corresponding to Grades I–III and worse than Grade V accounted for 28.6% and 50.8% respectively, among the sixty-three monitoring sections of the Haihe river. The water quality of the rivers shown in **Figure 2-16** was characterized using only conventional indicators, such as COD, BOD_5, and ammonia. The situation is even worse when additional hazardous materials such as endocrine disrupting organic substances are also taken into consideration.

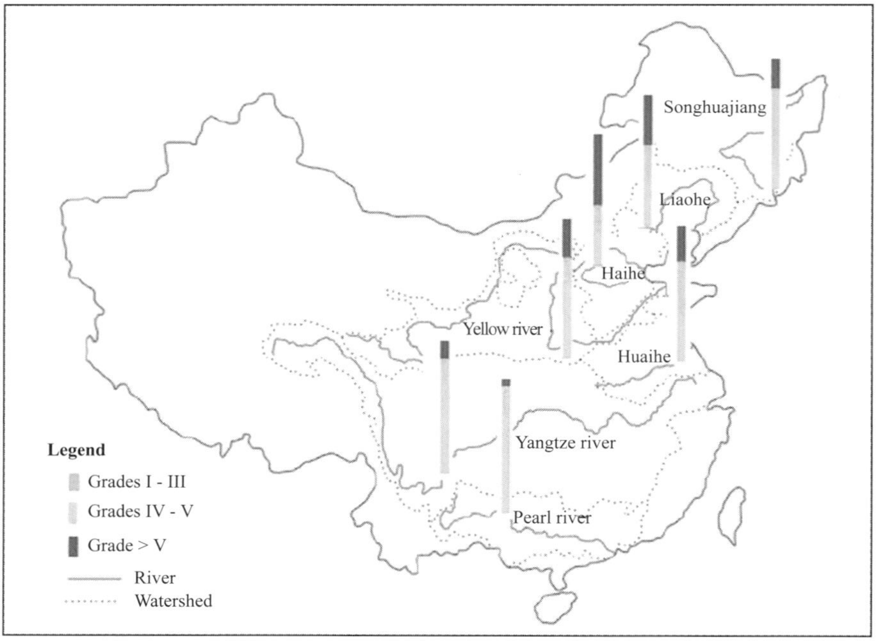

FIGURE 2-16. The current water quality of seven major rivers in China. The length of the bars are normalized to 1; the lengths of the bars represent the percentages of each river section with water quality between Grades I–III, Grades IV–V, and worse than Grade V, respectively.
Note: Cited from the 2008 Report on the State of the Environment in China.

The rivers in Zhejiang-Fujian provinces are slightly polluted. The monitoring areas meeting Grade II or III, and those meeting Grade IV, accounted for 71.9% and 28.1% of the total monitoring areas, respectively. The rivers located in southwest and northwest China are comparatively clean overall, and the sections corresponding to Grades I–III accounted for 88.2% and 92.8%, respectively.

The water quality classification of the monitoring areas since 2001 in the seven main rivers is shown in **Figure 2-17**. From 2001 to 2008, the overall river water quality improved slightly: the percentage of areas with Grades I–III quality increased, while the percentage of areas worse than Grade V decreased. However, the overall river pollution has not been significantly improved in recent years.

(2) *Water environmental status of lakes and reservoirs*

According to the 2008 Report on the State of the Environment in China, 39.3% of the key lakes and reservoirs had a water quality worse than Grade V, and only 21.4% belonged to Grades I–III, which are grades fit for drinking

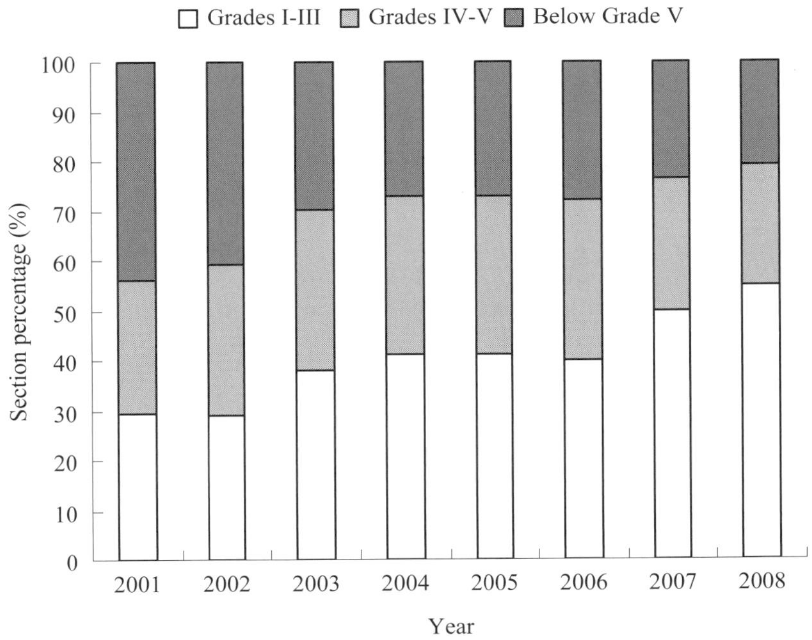

FIGURE 2-17. The water quality classification of the seven main rivers in 2002–2008.
Note: Cited from the 2000–2008 Report on the State of the Environment in China.

water and human contact (**Table 2-16**). The main pollutants of these lakes and reservoirs are total nitrogen (TN) and total phosphorous (TP), indicating that the lakes in China are facing serious eutrophication problems.

(3) *Three lakes*

Taihu lake, the third largest freshwater lake in China with a surface area of approximately 2,338 km^2, is located downstream from the Yangtze river. The Taihu lake watershed is an economically developed district with convenient transportation. It has only 3% of the total population in China, yet accounts for 12% of China's total GDP. Taihu lake has been seriously polluted in recent years, and the water was mesoeutrophic in 2008 (**Figures 2-18** and **2-19**). Although great effort has been made regarding water pollution control during the last decades, the water quality of Taihu lake has not been improved significantly, and the water quality of monitoring sites worse than Grade V accounted for 61.9% in 2008. Water blooms have occurred every year and have greatly deteriorated the water quality and water supply around the lake.

Dianchi lake, located at the Yunnan-Guizhou plateau, is the sixth-largest fresh water highland lake in China. It has a surface area of about 300 km^2 and maximum and mean depths of 10 and 4.4 m, respectively. In recent years, a cyanobacterial bloom in occurs throughout the entire lake annually. Dianchi lake has been seriously polluted by TN, ammonia, and TP, and has become hypereutrophic in the Caohai region of the lake (**Figures 2-18** and **2-20**). In 2008,

TABLE 2-16. Water quality classification of key lakes and reservoirs in 2008.

Lakes	Number	Grade I	Grade II	Grade III	Grade IV	Grade V	Worse than Grade V
Three lakes[a]	3	0	0	0	0	1	2
Big fresh-water lakes[b]	10	0	2	1	3	1	3
Urban lakes[c]	5	0	0	0	1	0	4
Reservoirs[d]	10	0	2	1	2	3	2
Sum	28	0	4	2	6	5	11
Percentage (%)		0	14.3	7.1	21.4	17.9	39.3

Note:

a. Dianchi lake, Taihu lake, and Chaohu lake.

b. Dalai, Baiyangdian, Hongze, Nansi, Bositeng, Poyang, Dongting, Jingbo, Erhai, and Xingkai.

c. Kunming lake (Beijing), Xuanwu lake (Nanjing), West lake (Hangzhou), East lake (Wuhan), and Daming lake (Ji'nan).

d. Shimen, Qiandaohu, Danjiangkou, Miyun, Dongpu, Yuqiao, Songhua, Dahuofang, Menlou, and Laoshan.

e. Cited from the 2008 Report on the State of the Environment in China.

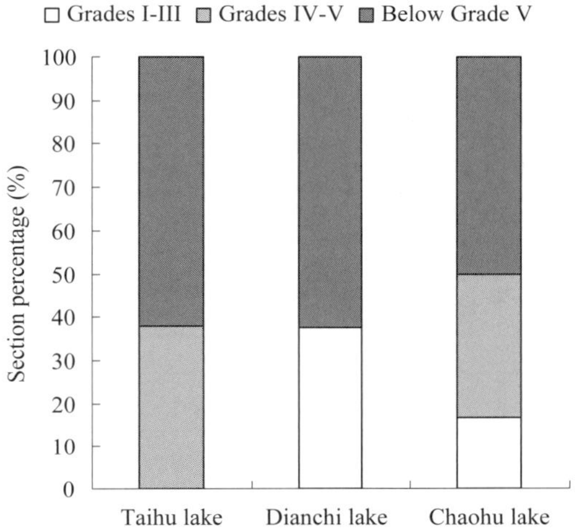

FIGURE 2-18. Water quality classification of three lakes.
Note: Cited from the 2008 Report on the State of the Environment in China.

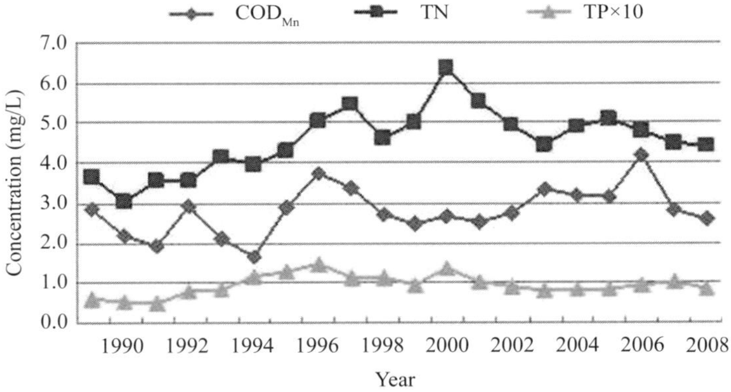

FIGURE 2-19. Water quality of Taihu lake during1990–2008.
Note: Cited from 1990–2008 Report on the State of the Environment in China.

the water quality of monitoring sites worse than Grade V accounted for 62.5% of the total number.

Chaohu lake is located in the Yangtze river delta in southeastern China. It has a mean surface area of 770 km^2, a mean depth of 2.7 m, and a storage capability of 2.1 billion m^3. In recent years, the lake has undergone serious pollution problems. During the warm season each year, the lake is eutrophic with dense cyano bacterial blooms (mainly composed of Microcystis

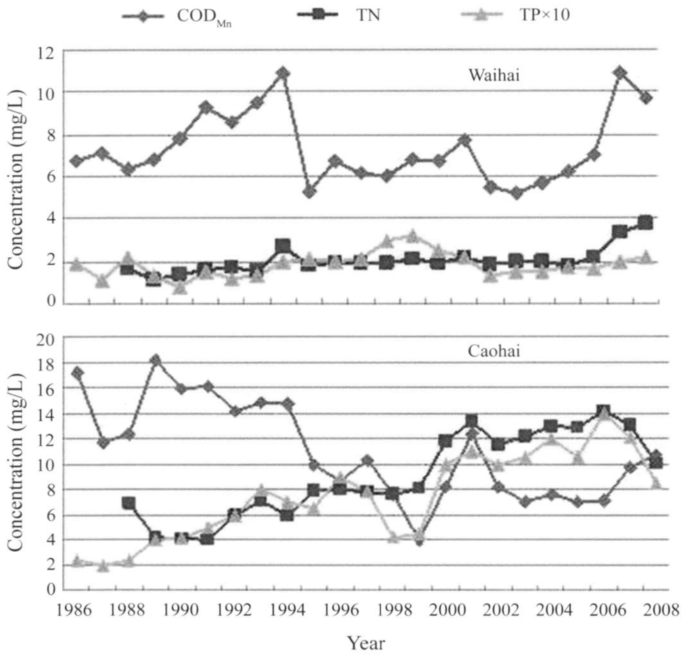

FIGURE 2-20. Water quality of Dianchi lake during1986–2008.
Note: Cited from 1986–2008 Report on the State of the Environment in China.

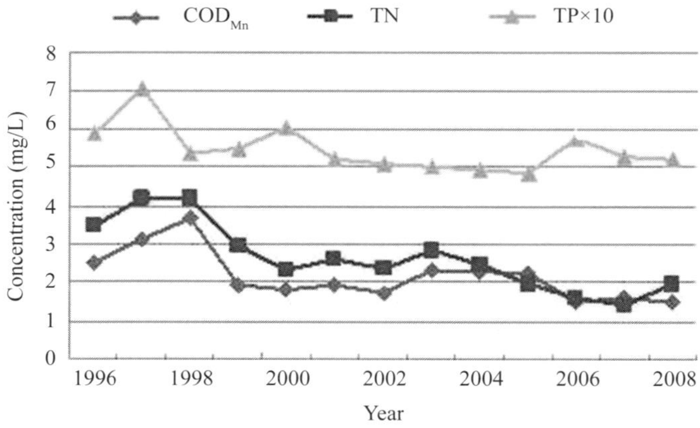

FIGURE 2-21. Water quality of Chaohu lake in 1986–2008.
Note: Cited from 1996–2008 Report on the State of the Environment in China.

and Anabaena) (Xu *et al.*, 2007). In 2008, the number of monitoring sites with water quality worse than Grade V accounted for 50%. The water was mesoeutrophic in the western half and slightly eutrophic in the eastern half. The main pollutants were TP, TN, and oil (**Figure 2-21**).

(4) *Big fresh-water lakes*

Among ten other large fresh water lakes, three, i.e., Baiyangdian, Dalai, and Hongze, have water quality worse than Grade V, and five are slightly eutrophic or mesoeutrophic (**Table 2-17**). The main pollution indicators are TN, TP, and BOD_5. Among the twenty-six lakes monitored, one was hypereutrophic, five were mesoeutrophic, and six were slightly eutrophic, accounting for 3.8%, 19.2%, and 23.0%, respectively.

(5) *Large reservoirs*

Among the ten large reservoirs monitored, Qiandaohu and Danjiangkou had water quality belonging to Grade IV, Yuqiao, Dahuofang, and Songhuahu belonging to Grade V, Laoshan and Menlou belonging to worse than Grade V (**Table 2-18**). The main pollution indicators of these large reservoirs monitored were TN and TP. The eutrophication status of these reservoirs was slight.

TABLE 2-17. Eutrophication status and water quality of ten big fresh-water lakes (Ministry of Environmental Protection, The People's Republic of China, SOE, 2008).

Lakes	Eutrophication index	Eutrophication status	Water quality		Pollutants
			2008	2007	
Dalai	68.7	Mesoeutrophic	>Grade V	>Grade V	pH, COD_{Mn}, TP
Baiyangdian	65.3	Mesoeutrophic	>Grade V	>Grade V	Ammonia, TP, TN
Hongze	55.8	Slightly eutrophic	>Grade V	>Grade V	TN, TP
Nansi	50.8	Slightly eutrophic	Grade IV	Grade V	Oil, TP, TN
Bositeng	50.7	Slightly eutrophic	Grade III	Grade III	–
Poyang	49.4	Mesotrophic	Grade IV	Grade IV	Oil, TP, TN
Dongting	46.6	Mesotrophic	Grade V	Grade IV	TP, TN
Jingbo	40.1	Mesotrophic	Grade IV	Grade IV	COD_{Mn}
Erhai	38.9	Mesotrophic	Grade II	Grade III	–
Xingkai	–	–	Grade II	Grade IV	–

Note: Cited from the 2008 Report on the State of the Environment in China.

TABLE 2-18. Eutrophication status and water quality of ten large reservoirs.

Reservoirs	Eutrophication index	Eutrophication status	Water quality 2008	2007	Pollutants
Laoshan	49.8	Mesotrophic	>Grade V	>Grade V	TN
Yuqiao	46.8	Mesotrophic	Grade V	Grade V	TN
Songhua	45.3	Mesotrophic	Grade V	Grade V	TN, TP
Dongpu	44.2	Mesotrophic	Grade III	Grade III	–
Menlou	40.5	Mesotrophic	>Grade V	>Grade V	TN
Dahuofang	36.7	Mesotrophic	Grade V	Grade V	TN
Qiandaohu	34.1	Mesotrophic	Grade IV	Grade III	TN
Miyun	32.7	Mesotrophic	Grade II	Grade II	–
Danjiangkou	31.9	Mesotrophic	Grade IV	Grade III	TN
Shimen	–	–	Grade II	Grade II	–

Note: Cited from the 2008 Report on the State of the Environment in China.

In 2008, among the 378 key reservoirs with a total water volume of 1.295×10^{11} m^3, 303 with a water volume of 8.23×10^{10} m^3 had Grades I–III water quality, accounting for 80.2% of the total number of reservoirs and 63.6% of the total water volume (**Table 2-19**). Among 347 reservoirs, 106 were eutrophic, accounting for 30.5% of the total number (**Table 2-19**). The main eutrophication indicators of these reservoirs were TN, TP, COD_{Mn}, COD_{Cr}, and ammonia (**Table 2-19**).

(6) *Urban lakes*

Most of the urban lakes studied had poor water quality. With the exception of Kunming lake in Beijing, all lakes were eutrophic and their water quality was worse than Grade V (**Table 2-20**). The main pollution indicators were TN and TP.

(7) *Water environmental situation of groundwater*

The total volume of groundwater resources with less than or equal to 2 g/L mineralization throughout the whole country in 2008 was 812.2 billion m^3, in which the total volume of groundwater resources in plain areas was 173.6 billion m^3 and that in hilly areas was 668.3 billion m^3 (China Water Resources Bulletin, 2008).

(8) *Groundwater pollution*

Groundwater is usually of better quality than surface water, especially from deep confined aquifers. The overall water quality of the groundwater in China is good. According to the Quality Standard for Ground Water (GB/T 14848-93),

TABLE 2-19. Eutrophication status and water quality of the key reservoirs in 2008.

Eutrophication status	Percentage of number (%)	Water quality	Percentage of number (%)	Percentage of water volume (%)
Oligotrophic	0.0	Grade I	6.4	11.9
Mesotrophic	69.5	Grade II	44.2	37.4
Slightly eutrophic	24.7	Grade III	29.6	14.2
Mesoeutrophic	5.2	Grade IV	13.2	33.7
Hypereutrophic	0.6	Grade V	2.4	0.6
		>Grade V	4.2	2.2

Note: Cited from the 2008 Report on the State of the Environment in China.

TABLE 2-20. Eutrophication status and water quality of the five urban lakes.

Urban lakes	Eutrophication index	Eutrophication status	Water quality		Pollutants
			2008	2007	
East lake	63.0	Mesoeutrophic	>Grade V	>Grade V	TP, TN, COD_{Mn},
Xuanwu lake	59.6	Slightly eutrophic	>Grade V	>Grade V	TN, TP
Daming lake	59.5	Slightly eutrophic	>Grade V	>Grade V	TN, BOD_5, TP
West lake	51.8	Slightly eutrophic	>Grade V	>Grade V	TN, Oil
Kunming lake	49.3	Mesotrophic	Grade IV	Grade III	TP, TN

Note: Cited from the 2008 Report on the State of the Environment in China.

67% of China's groundwater belongs to Grades I–III and can be used as drinking water, 17% can be adopted for agricultural or industrial use, and can be suitable for drinking after proper treatment (China Water Resources Bulletin, 2008).

In recent years, the groundwater pollution has become more serious, especially in large and medium cities. Among the 195 cities monitored, 97% have different degrees of groundwater pollution, and 40% have an increasing tendency toward groundwater pollution. There are sixteen provincial capitals in northern China and three in southern China that tend to have higher groundwater pollution. Of the seventy-six cities monitored, the deep groundwater quality has deteriorated in four cities located mainly in northeast and northwest China, has remained stable in sixty-eight cities, and has improved in the remaining four cities (Bulletin of Groundwater Status in Key Cities and Regions of China, 2007).

According to an assessment on the groundwater quality of 641 wells monitored in eight provinces (autonomous regions and municipalities directly under control of the central government), Beijing, Liaoning, Jilin, Shanghai, Jiangsu, Hainan, Qinghai, and Ningxia, wells with Grade I or II water accounted for 2.3%; Grade III, 23.9%; and Grade IV or V, 73.8% (China Water Resources Bulletin, 2008).

Serious groundwater pollution has occurred in heavy industrial areas and developing oil field areas of northeastern China. In northern China, groundwater pollution has generally occurred in both rural and urban areas, especially in several large cities, such as Beijing, Taiyuan, and Hohhot. Groundwater pollution in southern China is slight overall, but serious in cities and industrial and mining areas, particularly in the Pearl and Yangtze river delta areas. In northwestern China, the groundwater is comparatively clean, and pollution was found around the industrial and mining areas and in cities such as Lanzhou and Xi'an.

(9) *Over-exploitation of groundwater resources*

During the past three decades, the annual exploitation of groundwater resources in China has continuously increased at a rate of 2.5 billion m^3 per year, which has caused many serious environmental problems, such as land subsidence, saltwater intrusion, and an insistent decline in the groundwater table owing to arbitrary extraction.

According to the statistics, the amount of groundwater areas experiencing overexploitation exceeded 6.2×10^5 km^2; and the number of cities, approximately 60. In 2007, 212 groundwater descent funnels with a total area of about 1.5×10^5 km^2, including 136 shallow and 65 deep ones, and 11 karst funnels, were formed and are mainly located in northern and northeastern China (**Figure 2-22**) (Bulletin of Groundwater Status in Key Cities and Regions of China, 2007). Among the 136 shallow groundwater descent funnels, the amount of area for 43 funnels expanded, while that for forty-three funnels decreased, and the other 50 funnels remained stable.

(10) *Sea water quality*

(10.1) Quality of national sea areas

In 2009, of all national sea waters, the amount of polluted waters increased compared with 2008. A total of 1,46,980 km^2 of the water areas have not reached a clean water quality standard, of which 70,920 km^2 are comparatively clean, 25,500 km^2 are slightly polluted, 20,840 km^2 have mid-level pollution, and 29,720 km^2 were heavily polluted (Bulletin of Marine Environmental Quality of China, 2009). Generally, the Bohai sea and East China sea are more seriously polluted than the Yellow sea and South China sea. The polluted waters are mainly distributed in Liaodong, Bohai, Laizhou, and Hangzhou bays, the Yangtze and Pearl river estuaries, and the partial coastal waters of a number of large and medium-sized cities (**Figure 2-23**). The main pollutants of the sea waters are inorganic nitrogen, active phosphate, and oil.

(10.2) Sea water quality of four seas

Bohai sea-The area of polluted water in Bohai sea is 12,580 km^2, which is a decrease of 230 km^2 since 2008 (**Table 2-21**). Heavily polluted areas mainly include Liaodong, Bohai, and Laizhou Bays. The main pollutants are in organic nitrogen, active phosphate, and oil.

Yellow sea-The area of polluted water in the Yellow Sea is 15,240 km^2, which is an increase of 7,110 km^2 since 2008 (**Table 2-21**). The heavily polluted areas are mainly distributed in partial coastal waters near Jiangsu province and Dalian bay. The main pollutants are inorganic nitrogen, active phosphate, and oil.

East China sea-The area of polluted water in the East China sea is 37,360 km^2, which is an increase of 4,890 km^2 since 2008 (**Table 2-21**). The heavily polluted areas are mainly distributed in the Yangtze river estuary, Hangzhou bay, and the coastal waters of Xiangshan port and Leqing bay. The main pollutants are inorganic nitrogen and active phosphate.

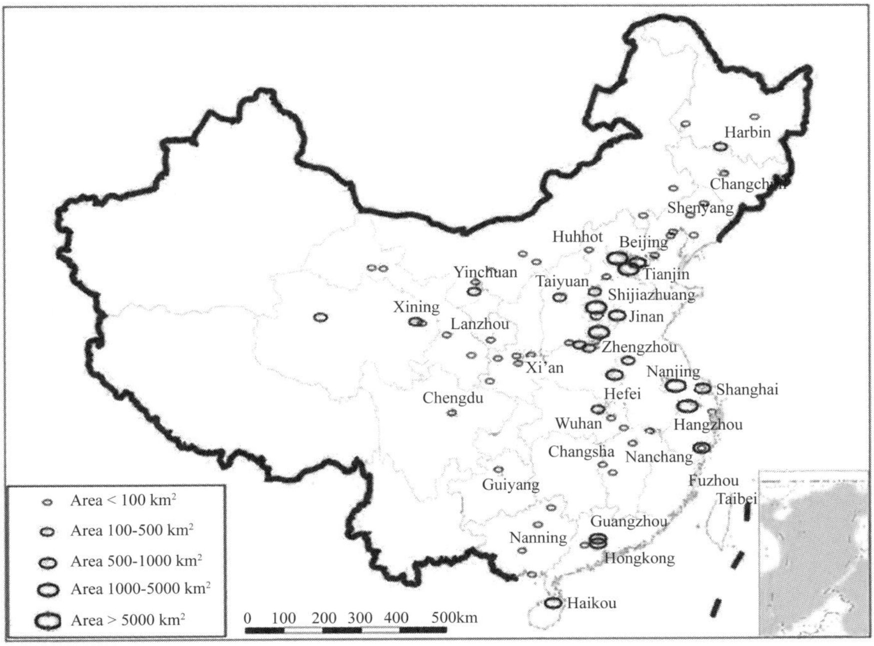

FIGURE 2-22. Sketch map of the distribution of shallow groundwater descent funnels in 2007.
Note: Cited from the 2007 Bulletin of Groundwater Status in Key Cities and Regions of China.

TABLE 2-21. Areas in various waters failing to reach the quality standard of clean waterduring 2004–2009.

Sea	Year	Comparatively clean area (km^2)	Polluted area (km^2)			
			Slightly Polluted	Medium Polluted	Heavily Polluted	Total
Bohai sea	2004	15,900	5,410	3,030	2,310	10,750
	2005	8,990	6,240	2,910	1,750	10,900
	2006	8,190	7,370	1,750	2,770	11,890
	2007	7,260	5,540	5,380	6,120	17,040
	2008	7,560	5,600	5,140	3,070	13,810
	2009	8,970	5,660	4,190	2,730	12,580
Yellow sea	2004	15,600	12,900	11,310	8,080	32,290
	2005	21,880	13,870	4,040	3,150	21,060
	2006	17,300	12,060	4,840	9,230	26,130
	2007	9,150	12,380	3,790	2,970	19,140
	2008	11,630	6,720	2,760	2,550	12,030
	2009	11,250	7,930	5,160	2,150	15,240
East China sea	2004	21,550	13,620	12,110	20,680	46,410
	2005	21,080	10,490	10,730	22,950	44,170
	2006	20,860	23,110	8,380	14,660	46,150
	2007	22,430	25,780	5,500	16,970	48,250
	2008	34,140	9,630	6,930	15,910	32,470
	2009	30,830	9,030	8,710	19,620	37,360
South China sea	2004	12,580	8,570	4,360	990	13,920
	2005	5,850	3,460	470	1,420	5,350
	2006	4,670	9,600	2,470	1,710	13,780
	2007	12,450	3,810	2,090	3,660	9,560
	2008	12,150	6,890	2,590	3,730	13,210
	2009	19,870	2,880	2,780	5,220	10,880
Total	2004	65,630	40,500	30,810	32,060	103,370
	2005	57,800	34,060	18,150	29,270	81,480
	2006	51,020	52,140	17,440	28,370	97,950
	2007	51,290	47,510	16,760	29,720	93,990
	2008	65,480	28,840	17,420	25,260	71,520
	2009	70,920	25,500	20,840	29,720	76,060

Note: Cited from the 2004–2009 Bulletin of the Marine Environmental Quality of China.

South China Sea-The area of polluted water in the South China Sea is 10,880 km^2, which is a decrease of 2,330 km^2 since 2008 (**Table 2-21**). The heavily polluted areas are mainly distributed in the Pearl river estuary. The main pollutants are inorganic nitrogen, active phosphate, and oil.

(11) *Red tides in seawaters*

Inorganic nitrogen, phosphorous, and oils are the main pollutants in China's seawaters, and have caused serious eutrophication. A total of 68 red tides occurred during 2009, which is the same number as in 2008. In addition, the total red tide area was about 14,100 km^2, which is similar to that in 2008. A total of six red tides exceeded 500 km^2, totaling 9,120 km^2. Most red tides occurred in the East China Sea. The number and area of red tide occurrences in the East China Sea accounted for 63% and 46% of the total amounts, respectively (**Table 2-22a**). The occurrence of red tides from 2000 to 2009 is shown in **Figure 2-24**. Although displaying a decreasing trend over the past few years, the number of occurrences and the total area of red tides are still considerable.

In 2009, the predominant organisms caused red tides were mostly consisted of *Noctilucascintillans, Heterosigmaakashiwo, Skeletonemacostatum, Kareniamikimotoi*, and *Prorocentrumtriestinum* (**Table 2-22b**).

TABLE 2-22a. Marine red tide events in various seawaters during 2007–2009.

	Number of red tide events			Cumulative area (km^2)		
Sea areas	**2007**	**2008**	**2009**	**2007**	**2008**	**2009**
Bohai sea	7	1	4	672	30	5,279
Yellow sea	5	12	13	655	1,578	1,878
East China sea	60	47	48	9,787	12,070	6,554
South China sea	10	8	8	496	60	391
Total	82	68	68	11,610	13,738	14,102

Note: Cited from the 2007–2009 Bulletin of the Marine Environmental Quality of China.

TABLE 2-22b. The occurrence of red tides and the predominance of algae in 2009.

Algae	Occurrence number	Areas (km^2)
Noctilucascintillans	13	1,833
Heterosigmaakashiwo	4	4,514
Skeletonemacostatum	8	943
Kareniamikimotoi	7	359
Prorocentrumtriestinum	5	360

Note: Cited from the 2009 Bulletin of the Marine Environmental Quality of China.

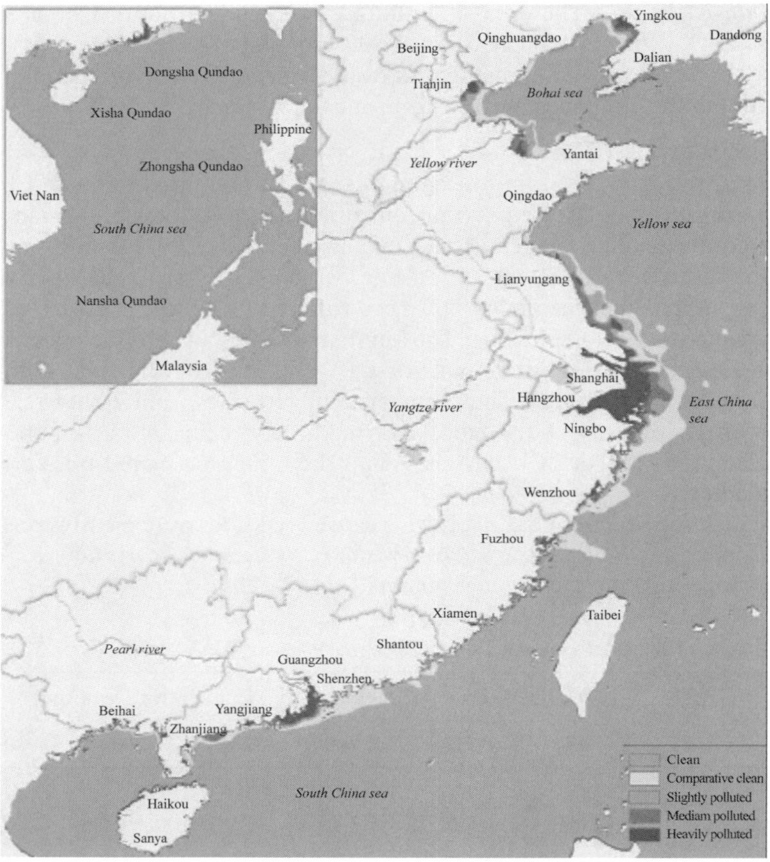

FIGURE 2-23. Sketch map illustrating the distribution of polluted coastal seawaters.
Note: Cited from the 2009 Bulletin of the Marine Environmental Quality of China.

2.3.5 *Water pollution sources and control in China*

2.3.5.1 Wastewater treatment

Wastewater is still the principal pollution source of China's water resources. The annual production of industrial and domestic wastewater in China is shown in **Table 2-23**. In 1995–2008, the amount of domestic wastewater increased gradually along with urbanization. After 1998, the annual production of domestic wastewater exceeded that of industrial wastewater. The discharge amount of COD and ammonia from municipal domestic wastewater was also much larger than that from industrial wastewater, accounting for 65.4% and 76.6% of the total amounts in 2008, respectively (**Table 2-23**).

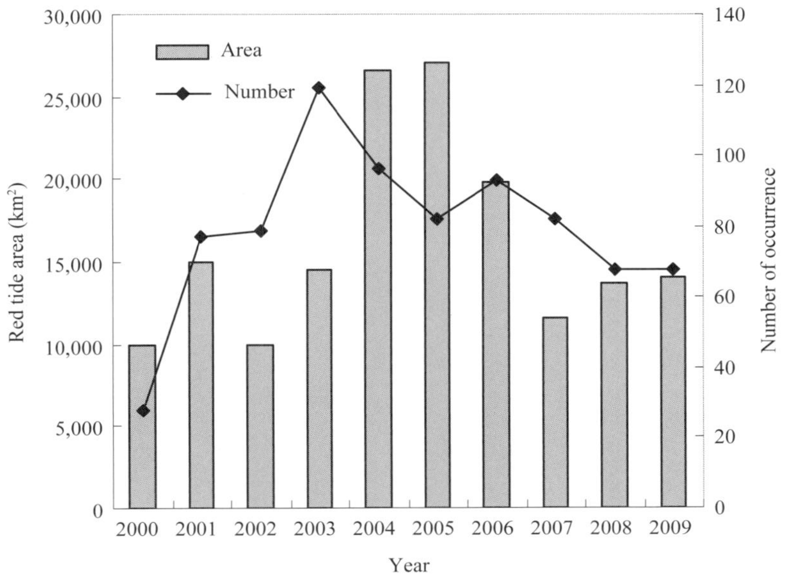

FIGURE 2-24. The occurrence of red tides in Chinese sea waters in 2000–2009.
Note: Cited from the 2000–2009 Bulletin of the Marine Environmental Quality of China.

In recent years, China has made a significant effort toward water pollution control, and its wastewater treatment rate has quickly increased. The discharge amounts of COD and ammonia decreased 4.4% and 4.0% from 2007 to 2008, respectively. The reuse rate of industrial water gradually increased to 83.8% in 2008. The percentage of industrial wastewater up to the discharge standards was higher than 90% from 2004 to 2008. In addition, the treatment rates of municipal domestic wastewater doubled between 2001 and 2008 (**Figure 2-25**).

However, the treatment rates of the total municipal wastewater and municipal domestic wastewater were 70.2% and 57.4% in 2008, respectively (**Figure 2-25**). A large amount of wastewater lower than the discharge standards was discharged into water environments without sufficient treatment, and this is one of the most important reasons for the deteriorating water environment in China.

2.3.5.2 Non-point pollution

Non-point pollution is mainly from agricultural land, animal production, and rural life, and has been considered another important source of water pollution. Pollution arising from non-point sources accounts for a majority of the contaminants in streams, lakes, and other water bodies undergoing

TABLE 2-23. Annual discharge of wastewater, COD, and ammonia in recent years.

Year	Waste-water discharge (10^8 t)			COD discharge (10^4 t)			NH_4-N discharge (10^4 t)		
	Domestic	Industrial	Total	Domestic	Industrial	Total	Domestic	Industrial	Total
1995	133.7	281.6	415.3	610.3	1,622.9	2,233.2	–	–	–
1996	–	205.9	–	–	–	–	–	–	–
1997	189.0	227.0	416.0	684.0	1,073.0	1,757.0	–	–	–
1998	194.8	200.5	395.3	695.0	801.0	1,496.0	–	–	–
1999	203.8	197.3	401.1	697.2	691.7	1,388.9	–	–	–
2000	220.9	194.2	415.2	740.5	704.5	1,445.0	–	–	–
2001	227.7	200.7	428.4	799.0	607.5	1,406.5	–	–	–
2002	232.3	207.2	439.5	782.9	584.0	1,366.9	86.7	42.1	128.8
2003	247.6	212.4	460.0	821.7	511.9	1,333.6	89.3	40.4	129.7
2004	261.3	221.1	482.4	829.5	509.7	1,339.2	90.8	42.2	133.0
2005	281.4	243.1	524.5	859.4	554.7	1,414.2	97.3	52.5	149.8
2006	296.6	240.2	536.8	886.7	541.5	1,428.2	98.8	42.5	141.3
2007	310.2	246.6	556.8	870.7	511.1	1,381.8	98.3	34.1	132.4
2008	330.0	241.7	571.7	863.1	457.6	1,320.7	97.3	29.7	127.0

Note: Cited from the 2000–2008 Report on the State of the Environment in *China*.

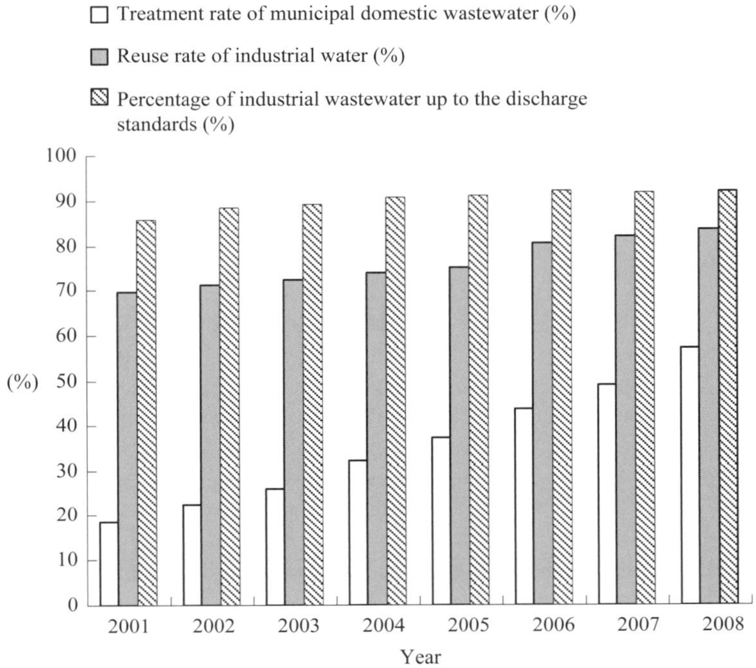

FIGURE 2-25. Treatment, reuse, and discharge of wastewater during 2001–2008.
Note: Cited from the 2001–2008 Report on the State of the Environment in China.

eutrophication. The effect of non-point pollution sources on the water pollution in China has been recognized in recent years.

In agricultural production, pollutants such as fertilizers and pesticides are carried into streams, lakes, and seas by rain in the form of runoff. China is the world's largest consumer of chemical fertilizers (**Table 2-24**), and its application amount per unit area has increased continuously in recent years, and is about three-times the world average. At the same time, the fertilizer use efficiency in China is only about 30% to 35%, which is much lower than that in developed countries. Unused fertilizers may flow into ground and surface waters through leaching and runoff, and results in water eutrophication. It has been estimated that the amount of unused nitrogen flowing out of the fields in China exceeds $1{,}550 \times 10^4$ tons annually.

There has been recent work in several of the large lakes of China where eutrophication is severe. **Table 2-25** shows the large N inputs in rivers and coastal waters, up to 80% of which come from agriculture. N fertilizers are likely responsible for more than half of these agricultural inputs, and together with phosphate inputs from fertilizer, cause serious eutrophication and red tides.

TABLE 2-24. The use of chemical, N, and P fertilizers during1999–2008 in China.

Year	Chemical fertilizer consumption (10^4 t)	N fertilizer consumption (10^4 t)	P fertilizer consumption (10^4 t)
1999	4,124.3	2,180.9	697.8
2000	4,146.4	2,161.5	690.5
2001	4,253.8	2,164.1	705.7
2002	4,339.4	2,157.3	712.2
2003	4,411.6	2,149.9	713.9
2004	4,636.6	2,221.9	736.0
2005	4,766.2	2,229.3	743.8
2006	4,927.7	2,262.5	769.5
2007	5,107.8	2,297.2	773.0
2008	5,239.0	2,302.9	780.1

Note: Cited from the China Statistical Yearbook 1999–2008.

TABLE 2-25. N inputs and exports into rivers and coastal waters, and the percentage of N coming from agriculture.

	N inputs into rivers		N exports into coastal waters	
River	Annual input (kg N/km^2)	From agriculture (%)	Annual export (kg N/km^2)	From agriculture (%)
Yangtze river	11,823	92	2,237	83
Yellow river	5,159	88	214	24

Note: Cited from van Dretch *et al.*, 2001.

The non-point pollution loads in the upper reach of the Yangtze river basin in 2000 were estimated using an export coefficient model and remote sensing techniques (Liu *et al.*, 2009). The total N load was 1.947×10^6 tons, and the total P load was 8.364×10^4 tons. The important source areas for the nutrients were croplands in the Jinsha river and Jialing river watershed, as well as Chongqing municipality. In the Taihu lake watershed, the contribution of non-point pollution sources to the total N and P amounts exceeded the point pollution sources, accounting for 84% and 83%, respectively (**Table 2-26**) (Zhang *et al.*, 2004).

China is also one of the world's largest consumers of pesticides. In 2007, the total consumption of pesticides was 162.28×10^4 tons. The application amount per unit area was up to 13.33 kg/ha. It has been estimated that about 80% to 90% of the pesticides applied to agricultural fields enter water, soil, and air environments, and cause water pollution.

TABLE 2-26. Total N- and P-contributions from different pollution sources into the water system of the Taihu lake watershed.

Pollution sources	**P contribution (%)**	**N contribution (%)**
Non-point source from agricultural land	19	29
Non-point source from animal production	35	23
Non-point source from living sewage in the transition region between rural and urban areas	22	21
Non-point source from rural life	8	10
Point pollution from industry and civil domestic waste	16	17

Note: N- and P-contributions indicates the ratio of N- and P-discharges from certain pollution types into the water systems to the total N- and P-discharge amounts (Zhang *et al.*, 2004).

Livestock and poultry waste is one of the most important pollutants. The COD from animal wastes exceeds the sum of domestic and industrial wastewater. In 2006, the amount of ***livestock*** and poultry feces in China was up to 2.75 billion tons, which is equivalent to 0.1 billion tons of COD. The expected amount of ***livestock*** and poultry feces will reach 4.5 billion tons in 2010. In the Taihu lake watershed, non-point sources from animal production accounted for 35% of the total P and 23% of total N, respectively (**Table 2-26**). Gu *et al.*, (2008) estimated the pollutants excreted annually by livestock and poultry in the Yangtze delta: 0.17 Mt NH_4–N, 0.16 Mt TP, and 0.42 Mt TN, which is equivalent to 1.86 Mt COD and 1.72 Mt BOD_5, respectively. Among livestock and poultry, pollutants produced by pigs make up the greatest amount, followed by poultry, cattle, and sheep.

Non-point sources from rural life and living sewage in transition regions between rural and urban areas contribute significantly to water pollution. In the Taihu lake watershed, these non-point sources accounted for one-third of the total P and N contributions (**Table 2-26**). The total generation of rural refuse was 1,671,000 t/a, and N and P released from rural refuse were 23,397 t/a and 4,679 t/a, respectively. The rainfall runoff was 450,000 $m^3/(km \cdot a)$. Rural refuse was dumped along the rivers and roads, resulting in increments of N and P in the runoff for 3.3 mg/L and 0.7 mg/L, respectively, exceeding the N and P limits of hyper eutrophication (Liu *et al.*, 2008).

China has recognized the effect of non-point pollution sources on water pollution in recent years. Strategies and countermeasures for alleviating non-point pollution have been suggested (Zhang *et al.*, 2004). However, no necessary and effective measures have actually been carried out. In the future, techniques to control non-point pollution should be developed, and relative policies and regulations should be formulated and implemented.

2.3.6 *Future prospects*

In recent years, the Chinese government has focused on pollution control. The nation's investment into water pollution control has increased continuously. Between 2006 and 2010, this investment amount will reach 250 billion RMB. China formulated and put into practice a plan for the prevention and control of water pollution in key regions, including three rivers (Huaihe, Liaohe, and Haihe), three lakes (Taihu, Dianchi, and Chaohu), major state projects (the Three Gorges Project and the South-North Water Diversion Project), two control areas (a sulfur dioxide control area and acid rain control area), the city of Beijing, and Bohai sea. The government has established a number of key pollution-control projects. All enterprises that discharge pollutants are required to reduce their emission amounts to a certain level, contributing to a reduction in the total amount of pollutants. By the end of 2008, of the 1,834 projects for water pollution prevention and control in key drainage areas included in the Eleventh Five-Year Plan, 773 had been completed, accounting for 42.1% of the total number, and the total investment amount reached 43.75 billion RMB (2008 Report on the State of Environment in China).

The current status of China's water resources is generally not optimistic. Although China is making a great effort to improve the nation's water quality, improvements have yet to be significant as water pollution control is a long-term task. In the future, greater effort is needed to protect water environments. If both point and non-point pollution sources can be effectively controlled, the water environment in China will greatly improve.

2.4 EMERGING WATER POLLUTION ISSUES

2.4.1 *Non-regulated chemical pollutants*

Counter measures against non-regulated chemical pollutants are important for protecting environment. The production volumes of major chemical products around the world have not declined, and those in Asia in particular have actually been significantly increasing. In addition, the trade volume and rate of chemical products in Japan and other Asian countries are also increasing. Most industrial products ultimately end up as waste, and chemicals from these products will be released into the environment. Currently, several countermeasures have been taken at each stage of a chemical's life cycle, for example, management during manufacturing, environmental risk assessments, environmental monitoring, and international cooperation. As the legal system in Japan, a law concerning the examination and regulation of manufacturing of chemical substances (the Japanese Chemical Substances Control Law) and the Chemical Substances Control

Law (the PRTR Law) have been set in place, and newly generated chemicals are being controlled.

However, based on small amounts of new chemicals, the number of application requests for the Chemical Substances Control Law is increasing in terms of both production and imports. On the other hand, the number of chemicals registered in the Chemical Abstracts Service (CAS) in 2010 was more than 40,000,000, and the number is increasing every day. The environmental risks of existing chemicals to tens of thousands of species in Japan are poorly understood at present, and understanding the properties of these chemicals in terms of their influence on the environment and the risks to human health and ecosystems requires urgent attention. This should be 5,638 to keep consistency it is thought that 642 new kinds of chemicals were subject to control, and 5,638 new kinds of chemicals were exempt to the Chemical Substances Control Law as of the end of February 2005, whereas 13,763 new kinds of chemicals have been confirmed. Although only 222 of the existing chemicals are subject to control, 19,896 kinds of chemicals have yet to be determined as whether they should be controlled. Therefore, unregulated chemical contamination of water environments continues as a result.

Although the current condition is not positive, various efforts are being made. International inspections are being conducted for many high priority materials. In addition, the PRTR promotes voluntary environmental risk reduction efforts by businesses by leveraging the data communication that has been achieved through its promotion, as well as by environmental risk assessments and continued long-term environmental monitoring.

For unregulated chemicals, it is necessary to gather information on the progress of risk assessments with a consideration of the trends taking place in other countries.

Based on a precautionary approach, particularly for high-risk substances, it is necessary to take measures for a risk reduction through the introduction of regulatory measures.

PRTR information is being developed and should be utilized. Other regulation concepts such as Whole Effluent Toxicity (WET) are also under review.

2.4.2 *Endocrine disrupting compounds*

Estrogenic activity has been attracting increased attention ever since a sexual disruption in wild fish was reported in England (Jobling *et al.*, 1998). The feminization of wild carp was also found in the Tama river in Japan (Tanaka *et al.*, 2003). Several kinds of artificially synthesized chemicals are reported to have endocrine disrupting activities. Among such chemicals, bisphenol-A (BPA) and nonylphenols (NP) are known to have strong activity and are often detected in water bodies. Natural female hormones, such as E2 (17*β*-estradiol), E1 (Estrone), and E3 (Estriol) are also known to create the same activities in wild aquatic life. Artificially synthesized estrogen EE2 (17α-ethynylestradiol) is also produced as a synthesized hormone. The chemical structures of these materials are shown in **Table 2-27**. Field data surveys on these substances have also been accumulated. In Japan, field surveys for estrogenic compounds in water, sediment, and fish were conducted in 2002 by the Ministry of Land, Infrastructure, Transport, and Tourism. The results are summarized in **Table 2-28**. In the survey, tentative concentration criteria of estrogenic compounds (TCCEC) were proposed as a focused survey point.

The distributions of estrogenic compounds, E1, E2, E3, NP (summation of NP1–NP9) and BPA, through the flow direction of the Yodo river system in Japan were investigated by a field survey of the water and sediment at eighteen different points (see **Figure 2-26**), and total load of these problematic compounds into the river water from a municipal wastewater treatment plant located between sampling points ⑫ and ⑬ was determined in October, 2008

TABLE 2-27. Chemical structure of representative estrogen types.

Group	Compound	Molar weight (g/mol)	Structure
Natural estrogen	17 β-estradiol (E2)	272.4	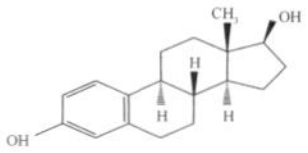
	Estrone (E1)	270.4	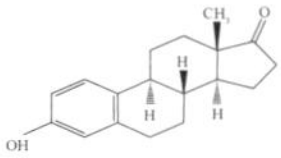
	Estriol (E3)	288.4	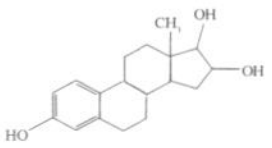
Synthesized estrogen	17 α-ethynylestradiol (EE2)	296.4	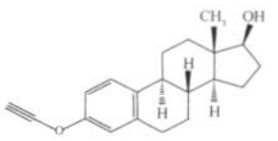
Synthesized chemical	Bisphenol-A (BPA)	228.3	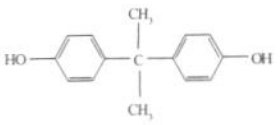
	4-nonylphenol (NP)	220.4	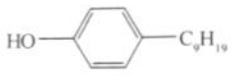

TABLE 2-28. Concentrations of estrogenic compounds in Japan (Ministry of Land, Infrastructure, Transport and Tourism, 2002).

Compound	TCCEC (ng/L)	Points over TCCEC*	Maximum (ng/L)
E1	0.5	31/69	16.7
E2	0.5	4/49	2.2
NP	304	4/65	910
BPA	400	0/46	167

*Tentative concentration criteria of estrogenic compound.

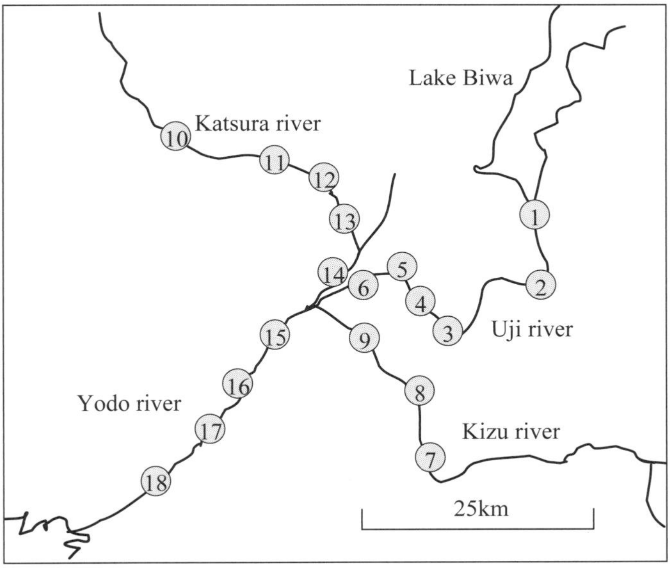

FIGURE 2-26. Sampling points in the Yodo river system, Japan.

(Takabe *et al.*, 2009). The results are shown in **Figure 2-27** and summarized in **Table 2-29**. The concentrations of natural estrogens were in the ranges of 0.43 to 4.1, 0.33 to 3.8, and n.d. to 0.70 for E1, E2, and E3, respectively, which are comparatively high concentrations as compared with the Japanese data overall. The concentrations of NP and BPA were in the range of 130 to 350 and 2.3 to 140 ng/L, respectively. The concentrations of estrogens increased from point ⑫ to point ⑬. The increased amount of each estrogen was calculated and compared with each loading rate from the municipal wastewater treatment plant, as shown in **Table 2-30**. It is shown that the load from the

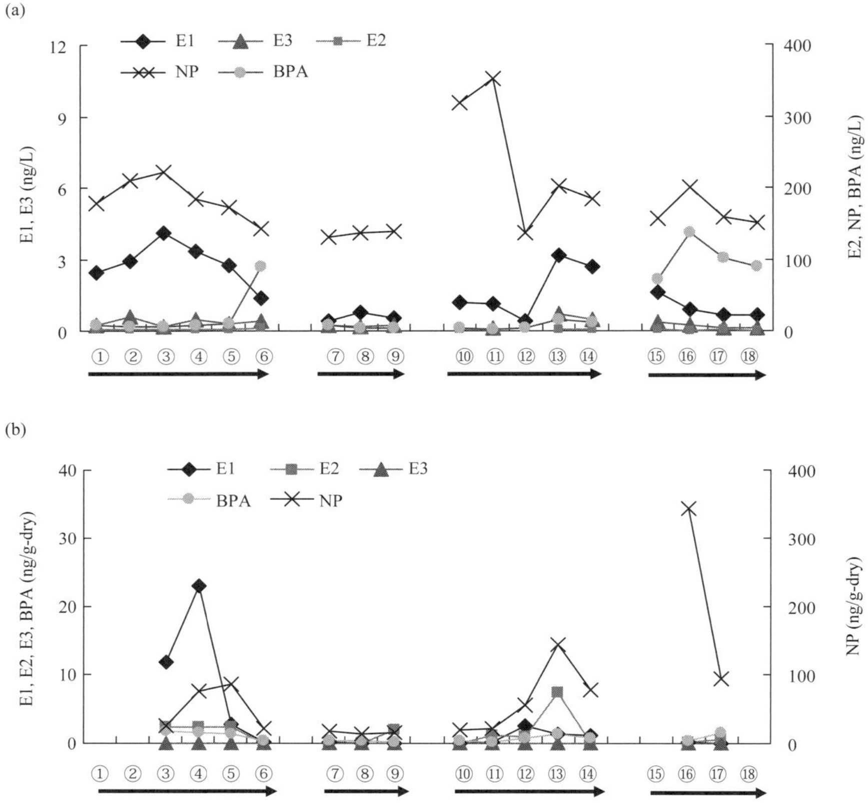

FIGURE 2-27. Concentration of each estrogen through the flow direction.
(a) Water sample (b) Sediment sample

TABLE 2-29. Concentration of each estrogen in the Yodo river system.

	Water (ng/L)				Sediment (ng/g-dry)		
Compound	Mean	Minimum	maximum	Points over TCCEC*	Mean	Minimum	maximum
E1	1.7	0.43	4.1	16/18	3.1	0.064	23
E2	1.7	0.33	3.8	14/18	1.4	n.d.	7.4
E3	0.31	n.d.	0.70	0/18	n.d.	n.d.	n.d.
NP	190	130	350	2/18	72	13	340
BPA	32	2.3	140	0/18	0.85	0.23	1.8

*Tentative concentration criteria of estrogenic compound.

TABLE 2-30. Contribution of estrogen load from a municipal wastewater treatment plant to the increased amount in river waters.

Compound	Loading rate (mg/s)	Increased amount (mg/s)	Contribution ratio
E1	91,000	60,000	150%
E2	38,000	47,000	81%
E3	23,000	13,000	180%
NP	5,200,000	2,900,000	180%
BPA	1,100,000	280,000	390%

treatment plant strongly affected the river concentration, although the loading rate was higher than the increased amount because of the time difference and degradation of the estrogens in the river. However, it is very important to recognize that estrogen can be degraded during the treatment process.

The distributions of estrogenic compounds, E1, E2, 17α-E2 (17a-estradiol), E3, and EE2, through the flow direction in the Shenzhen river in China were investigated by a field survey of six water points in November (dry season) and June (rainy season), 2008, and the results are shown in **Figure 2-28** (Wang, 2011). The concentrations of 17a-E2 were in the range of n.d. to 15.6 ng/L. E1, E2, and E3 were detected within the range of n.d. to 52.9 ng/L. Their concentrations were about ten-times higher as compared with the Yodo river system because of a much higher population in the watershed. EE2 was detected within the range of 25.1 to 444 ng/L in China, while it was not detected in Japan.

2.4.3 *Pharmaceutical and personal care products*

The rapid spread of sewer systems during the 1970s made the aquatic environment a public water space without visible or perceptible pollution in Japan. Recently, however, the occurrence of imperceptible pollutants, e.g., endocrine disrupters (EDs), in an aquatic environment has gradually emerged. In the 1990s, the issue of EDs, academically and socially, became an example of a new toxicity, i.e., the disruption of the endocrine system, even when the concentration is below ng/L. The study revealed that an ingredient in the oral contraceptive ethynylestradiol (EE2) is one of the dominant estrogenic compounds of sewage discharge. Although EE2 consumption in Japan is still small, it was reported that residual EE2 in sewage effluent has contributed to endocrine disruption, especially reproduction and feminization, in fish in European countries (Johnson and Sumpter, 2004). Other pharmaceuticals and personal care products (PPCPs) are also considered as emerging contaminants and bioactive compounds in aquatic organisms because they basically have specific medicinal properties at low concentration. PPCPs are mainly discharged from

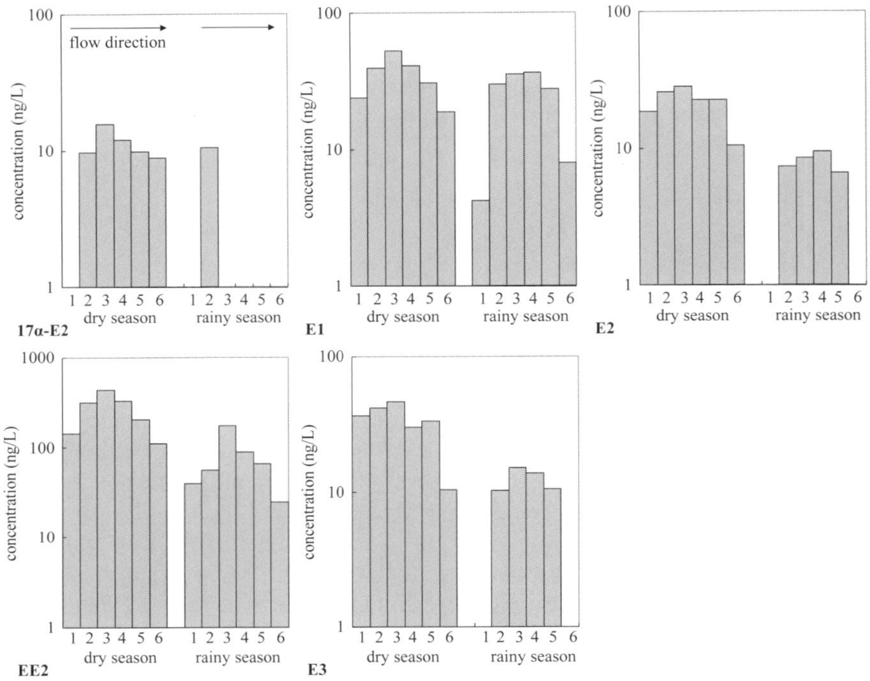

FIGURE 2-28. Concentration of different estrogen types in the Shenzhen river, China (Wang, 2011).

our daily activities and hospitals, and are treated in sewage treatment plants (STPs). Treatment at STPs is designed for solid, hydrophobic, volatile and biodegradable materials. Therefore, conventional STPs using settling, biological, and disinfection processes cannot remove hydrophilic (polar) and/or non-biodegradable compounds, which can pass through them. In fact, many studies have reported the low removal efficiencies of PPCPs through conventional treatment (Tanaka *et al.*, 2008) (see section 4.2.3.3). In European countries, studies on the occurrence and governance of/for PPCPs were conducting for ten years owing to their bioactive nature. On the other hand, these studies have only been carried out on the occurrence and sewage and toxicity treatment of around 150 PPCPs for aquatic organisms (Nakada, 2010). For the occurrence, studies were conducted in rivers, basins, coastal areas, and groundwater (Nakada *et al.*, 2008). However, there are no regulations for PPCPs as environmental pollutants or their disposal in Japan, and therefore large amounts of several PPCP types are discharged into aquatic environments every day.

Therefore, the occurrence and attenuation of PPCPs in river waters was studied (Tanaka *et al.*, 2008). In our study, the target compounds were a maximum of 107 PPCPs including antibiotics, non-steroidal anti-inflammatory drugs (NSAIDs), fragrances, and anti-virus drugs. To determine the occurrence of PPCPs in the Yodo and Tone river systems, which supply drinking water to more than sixteen and twenty-seven million people, and receive wastewater from more than four and two million people, respectively, river surveys were conducted six and three times during four seasons (2005 to 2010). Grab water samples were collected from twenty-nine points including the outlets of seven municipal STPs, the downstream points of fourteen tributaries, and eight points along the main river stretch.

In the surveys, ninety-six out of 107 PPCPs monitored were found from discharge from STPs and/or river waters at a maximum concentration of 1829 ng/L (**Figure 2-29**).

To understand the source distribution of water and PPCPs in the basins surveyed, the total water and PPCP loads, which were detected frequently from upstream, tributaries, and STPs in each basin, were calculated by multiplying the water flow with their concentrations. As the result in **Figure 2-30** show, the water from upstream was dominant, corresponding to 87% to the total input, while STP effluents were in relatively small amounts in the Yodo river basin. On the other hand, the water source distribution into the Tone river basin was estimated as shown in **Figure 2-30**. Water flows from STPs, indicated in yellow, were in small amounts corresponding to 2% of the total flow. For the selected PPCPs, a STP was the dominant source in the Yodo river basin, while both a STP and a tributary were dominant in the Tone river basin. This estimation is inconsistent with the low percentage of sewered population and high dilution factor for STP effluent in the Tone river basin. These results suggest that STPs are a major source of many PPCPs in river waters. However, some PPCPs were detected upstream of the rivers or in tributaries where no STP discharge was received, which indicates the existence of sources other than STPs.

In the above discussion, we discussed the occurrence, distribution, and source of PPCPs under normal conditions. In the case of certain disease outbreaks (e.g., influenza), specific drug categories are used for a short time period. Oseltamivir phosphate (OP; Tamiflu) is a pro-drug of the anti-influenza neuraminidase inhibitor oseltamivir carboxylate (OC) and is widely used in Japan. Therefore, we developed a method for the detection of OC in STP discharge and river waters using solid-phase extraction (SPE) followed by liquid chromatography-tandem mass spectrometry (LC-MS/MS) (Ghosh *et al.*, 2010). In addition, we investigated the occurrence of OC in STP effluent and receiving river waters during the 2008–2009 flu season in Kyoto city, Japan (Ghosh *et al.*, 2010). Furthermore, the concentration of OC in STP discharge on a given day was calculated using the following equation:

$$PC_{\mathrm{STP}}(\mathrm{ng/L}) = \frac{(TIC \times D) \times 0.7 \times 0.8 \times 10^{6}}{P \times 300}, \quad [l] \qquad (2\text{-}5)$$

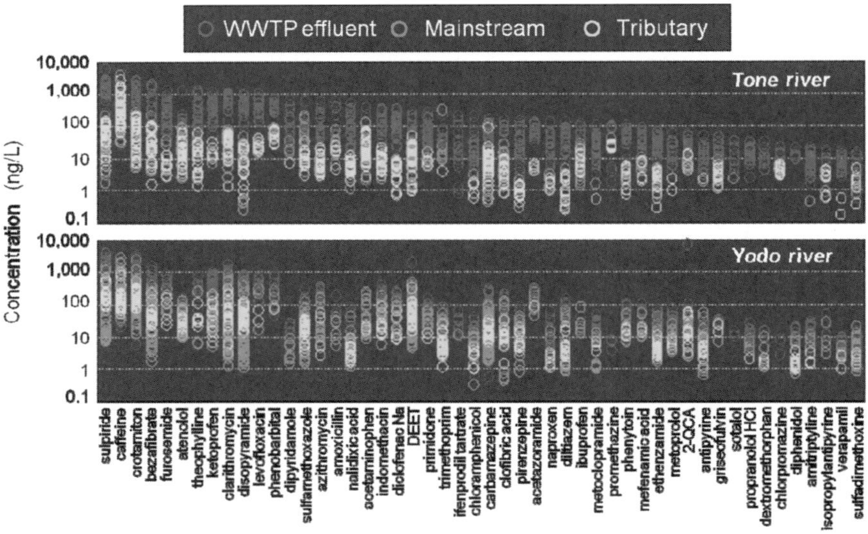

FIGURE 2-29. PPCPs in river waters and STP discharge found from the Tone and Yodo river systems in Japan (Tanaka *et al.*, 2008).

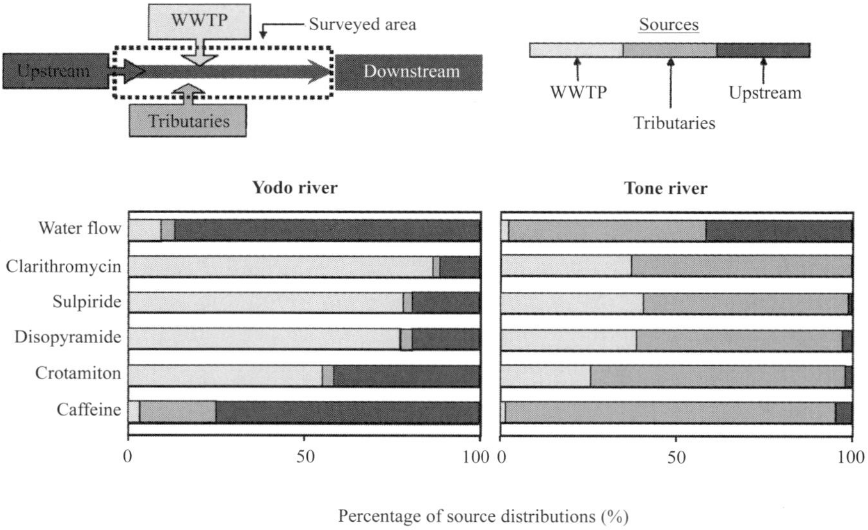

FIGURE 2-30. Percentage of PPCP source distributions in the Tone and Yodo river basins (Tanaka *et al.*, 2008).

where PC_{STP} is the predicted concentration of OC in STP discharge in Kyoto city; *TIC* is the total number of influenza cases per week in Kyoto city during the study period; *D* is the OP dose assuming 85% adults (2 × 75 mg OP/day) and 15% children (2 × 45 mg OP/day), reflecting the population structure of Kyoto city in 2008; 0.7 reflects that, on any day during the week, only five of seven patients will be taking OP because the treatment course takes 5 days; 0.8 represents the 80% of ingested OP that is excreted in an unaltered state; *P* is the total population of Kyoto city (1,389,000); and 300 is the average water use per capita (liters per person per day).

The measured and predicted OC concentrations in the effluent from an STP in Kyoto city are shown along with total influenza cases reported in **Figure 2-31**. The highest concentration of OC was detected and estimated in STP effluent at the peak period of the influenza season with a maximum concentration of up to 300 ng/L. However, OC was present in STP effluent and river waters only during the flu season.

2.4.4 *Water-borne diseases*

A water supply expansion through disinfection using chlorination and sewerage systems has contributed to the prevention of water-borne diseases such as cholera, dysentery, and typhoid. However, not all problems derived from

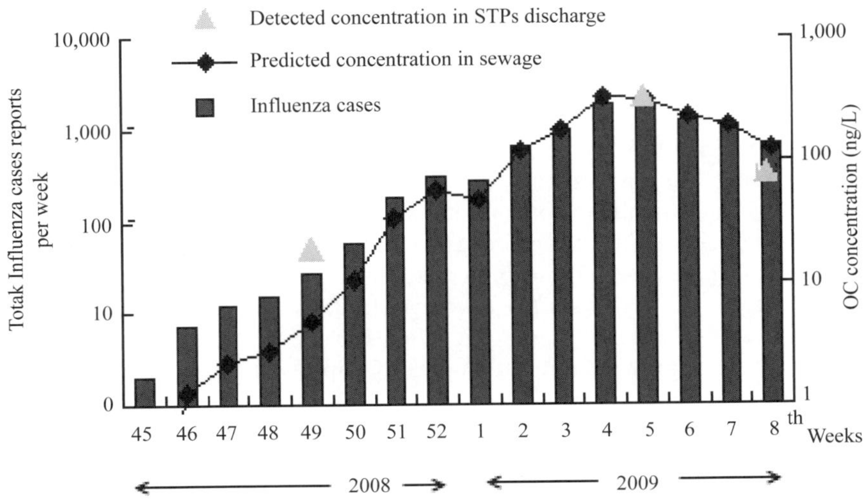

FIGURE 2-31. Total number of influenza cases reported per week along the predicted and maximum detected concentrations of OC in Kyoto city sewage discharges (Ghosh *et al.*, 2010).

microorganisms have been eliminated. These days, in addition to bacteria, viruses and protozoa have attracted attention as new target pathogens to be controlled.

To solve these current issues, it is important to know the occurrence of past incidents. In the following, the reasons why water bone diseases occur and how people have dealt with them are discussed.

In cities, the construction of a water supply system has become essential. If people can drink safe water, public health will be improved, and the number of water-borne diseases will dramatically decrease. A cholera epidemic that took place in 1892 in Hamburg and Altona, which receive their drinking water from the Elbe river, illustrates the effects that the water supply system can have on public health. In Hamburg, 17,000 people out of a population of 640,000 were infected and 8,600 people died. The mortality rate was as high as 134.4 per 10,000 people. On the other hand, although Altona is located downstream of Hamburg, only 500 out of a total population of 150,000 were infected and 300 people died there, and most of the patients were infected in Hamburg. The question remains, why was there such a big difference between these two cities? The answer seems to be in the differences in their water treatment systems. In Altona, slow sand filtration was performed as a water treatment method, while the drinking water in Hamburg was directly supplied from the Elbe river. Then, in 1893, Reincke advised that river water filtration should be conducted as water treatment, which resulted in the prevention of later outbreaks of disease from drinking water. In Lawrence, Massachusetts, USA, drinking water used to be supplied directly from the Mermac river. In 1893, H.F. Mills decided to add filtration to the water treatment process, and the mortality rate attributed not only to typhoid but to other diseases as well decreased. These past cases demonstrate that the purification of the water supply leads to the prevention of water-borne diseases. Sedwick and MacNutt defined this phenomenon as the 'Mills-Reineke phenomenon' as a sign of respect. During that era, drinking water sanitation was not perfectly secured, as water treatment processes mainly consisted of sand filtration and did not include a disinfection process. However, sand filtration can remove most microorganisms, and morbidity did decrease. This phenomenon led to an increase in the water supply coverage. In Japan, a water supply system was adopted at the beginning of the Meiji Era. After the construction of a modern water supply system, public health improved dramatically in Japan (**Table 2-31**).

Infection through contact with polluted water is called a 'water-borne transmission (infection)', and diseases caused by water-borne transmissions are defined as 'water-borne diseases.' Water-borne transmissions have occurred through bathing, showering, swimming, watering plants, and drinking. Many microorganisms exist in the environment, including in drinking water sources.

TABLE 2-31. Changes in the death rates from water-borne and other general diseases after the construction of the modern water supply system in Japan (Kaneko M., 1997).

	Death rate from typhoid, dysentery, and cholera per 10,000 people			General death rate per 10,000 people		
City	Before	After	Increase	Before	After	Increase
Tokyo	7.93	3.83	−4.10	205.5	186.8	−16.9
Osaka	19.44	8.95	−10.49	243.5	241.2	−2.3
Yokohama	93.46	11.74	−81.72	433.7	226.9	−206.8
Kobe	30.69	8.03	−22.66	262.5	245.3	−20.2
Nagasaki	44.40	14.17	−30.23	236.2	176.1	−60.1
Sasebo	11.71	1.99	−9.72	120.8	117.9	−2.9
Akita	3.61	2.42	−1.19	161.6	209.4	+47.8
Okayama	20.94	4.34	−14.60	172.2	158.3	−13.9
Hiroshima	40.53	4.22	−36.31	214.3	188.6	−25.7
Shimonoseki	5.86	5.24	−0.62	194.6	163.6	−31.0

If water treatment fails, the drinking water can become contaminated, posing a risk to public health. These days, with the expansion of international interactions among people and food, more water-borne diseases are caused by pathogens from foreign countries. Microorganisms causing slight cases of diarrhea and opportunistic infections are attracting greater attention. Therefore, we need to control a wider range of microorganisms than ever before. Pathogens are generally classified into three categories: protozoa, bacteria, and viruses. The characteristics of these pathogens will be explained in the following section.

2.4.4.1 Protozoa

In Japan, although the number of disease outbreaks derived from ascarids and hookworm has dramatically decreased, problems related to *Cryptosporidium* have arisen. Some protozoa form oocysts and cysts during their life-cycles and are eluted through patient feces, polluting water environments. The infection route is mainly fecal-oral infection through water and foods polluted by feces of infected patients and animals. Many protozoa, as typified by *Cryptosporidium* and *Giardia*, have resistance to chlorine because their oocysts and cysts block chlorine penetration. Therefore, we should review the conventional water treatment processes and enhance the sanitary safety of drinking water. UV and membrane filtration are effective for the control of protozoa owing to their sensitivity to UV exposure and 1–10 mm size.

(a) *Cryptosporidium*

Cryptosporidium belongs to Sporozoa and have attracted significant attention. After parasitism, *Cryptosporidium* multiplies through asexual reproduction and produces many oocysts through sexual reproduction. The symptoms of *Cryptosporidium* infection, called cryptosporidiosis, are acute stomachaches and watery diarrhea. The cure for cryptosporidiosis has yet to be established and patients must rely on only natural healing.

In Milwaukee, a group infection of *Cryptosporidium* occurred in 1993, when about 400,000 people were infected. In Japan, 8,800 people were infected by drinking polluted water in Ogose, Saitama prefecture in 1996. *Cryptosporidium* has a general resistance to chlorine and it takes very high *CT* values of chlorine for its disinfection (**Table 2-32**). On the other hand, only a 2.5 mJ/cm^2 dose of UV exposure is required for a 90% inactivation of *Cryptosporidium* (U.S. EPA, 2006). Therefore, UV disinfection is considered effective for *Cryptosporidium* control.

(b) *Giardia*

Giardia is a type of mastigophoran, and only *G. lamblia* can parasitize humans. The symptom of *G. lamblia* infestation, called lamblias is or giardias is, is fatty diarrhea. *Giardia* has two stages during its life-cycle: vegetative and cyst forms. As with *Cryptosporidium*, chlorination is not effective for the inactivation of Giardia (**Table 2-32**). On the other hand, it takes only a 2.1 mJ/cm^2 UV dose to inactivate 90% of this parasite (U.S. EPA, 2006).

2.4.4.2 Bacteria

Water-borne diseases derived from bacteria have been known to exist for a long time, and the disinfection process during water and wastewa-

TABLE 2-32. *CT* values (mg•min/L) for an inactivation of protozoa (Moselio Schaechter, 2009).

Protozoa	Free chlorine	Chloramine
Giardia lamblia cysts	47–150[a]	2,200[b]
Cryptosporidium parvum	7,200[c]	7,200[d]

a: 2 log inactivation at 5°C.
b: 3 log inactivation at pH 6–9.
c: 2 log inactivation at pH 7 and 25°C.
d: 1 log inactivation at pH 7 and 25°C.

ter treatment has been mainly for the control of bacteria. Typical pathogenic bacteria are *Campylobacter, Vibrio, Shigella*, and *Salmonella*. The main infection route of these bacteria is through drinking polluted water. *Legionella* infects humans while bathing and through the use of humidifiers. Total coliform, *Escherichia coli*, and fecal coliform are used as indicators of fecal contamination. However, these bacteria are more sensitive to disinfectants and other factors in the environment than protozoa and viruses, and therefore a disinfection process and water environment control focusing on bacteria may not be sufficient for the control of other microorganisms.

2.4.4.3 Viruses

Viruses are the smallest of all microorganisms, ranging from 10 to 100 nm in diameter, as shown in **Table 2-33**. They have not been studied as much as other microorganisms owing to a difficulty of analysis. However, molecular biology methods such as real-time RT-PCR have been improved, and water contamination from viruses has been revealed. Chlorination, particularly chloramine, is ineffective for the inactivation of viruses, and the removal of viruses is difficult owing to their small size. We should therefore consider the control of viruses in water and wastewater treatment. Typical pathogenic viruses include polio, hepatitis A, adenoviruses, and noroviruses. These viruses infect humans through not only polluted water but also clams and oysters, which accumulate them. Moreover, second infections such as droplet infections occur. It should also be noted that influenza viruses excreted from patients are transported by and obtain their drug resistance in waterfowl.

The numbers in the parentheses indicate the number of positive samples to that of the total samples.

TABLE 2-33. Characteristics of typical pathogenic viruses.

Virus	Type	Size (nm)	Symptom
Poliovirus	Single-strained RNA	25–30	poliomyelitis
Coxsackievirus	Single-strained RNA	25–30	hand-foot-and-mouth disease
Hepatitis A virus	Single-strained RNA		Hepatitis A
Norovirus	Single-strained RNA	25–32	vomiting, diarrhea
Adenovirus	Double-strained DNA	70–90	respiratory disease, gastroenteritis
Rotavirus	Double-strained RNA	60–80	vomiting, diarrhea

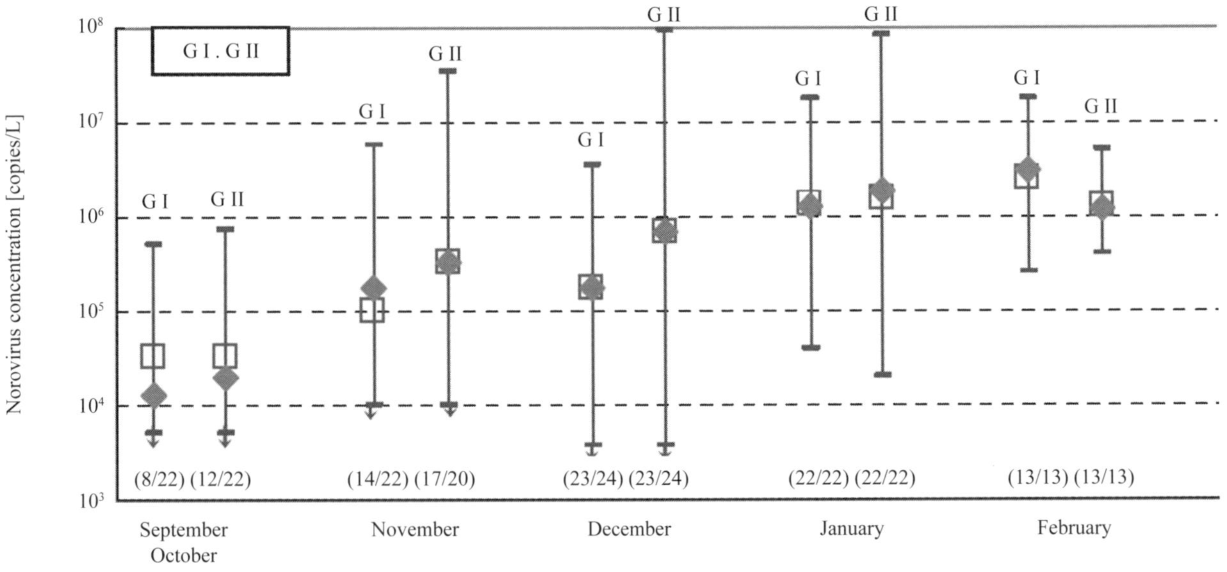

FIGURE 2-32. Occurrence of noroviruses in the influent of sewage treatment plants in Japan (Ministry of Land, Infrastructure, Transport and Tourism, 2010).

NOTE: The numbers in the parentheses indicate the number of positive samples to that of the total samples.

2.4.5 *Natural products*

Hazardous organic substances are also produced naturally through biological activities. Typical examples are toxins produced by algae. For such toxins, the frequent occurrence of eutrophication of water bodies has made the problem more serious. Toxins produced by cyanobacteria (blue-green algae poisoning, i.e., cyanotoxin) and dinoflagellate algae (ciguatera toxin, saxitoxin, and okadaic acid) are known issues.

Anatoxin-a, anatoxin-a(s), and aphantoxin produced by blue-green algae are well known as neurotoxic, while microcystin and Shirindorosupamopuchin are hepatotoxin. Their chemical structures are summarized in **Table 2-34**, which also shows the species of algae produced. Among toxic chemicals, microcystin has become a widespread problem around the world. Microcystins are heat stable and have been detected in wide areas ranging from forests to agricultural and municipal fields. Microcystin is a cyclic peptide with seven amino acids, which constitute the differences among microcystin-RR, microcystin-YR, and microcystin-LR. About seventy homologues are known at present. Among them, microcystin-LR is considered the most toxic, with an LD_{50} level of 50–100 mg/kg.

The maximum microcystin content in algae is about 10 mg/g-dry weight (Kaya, 2002). The integration density of blue-green algae in the same water areas, and the concentration of microcystins, varies greatly by location. It was reported that the microcystin concentration in several lakes around Japan are 104 mg/L (Fujimoto, 2002; Hagiwara, 2004; Park, 2008). Bivalves can accumulate toxins of blue-green algae (Williams *et al.,* 1997), and the possibility of migrating these toxins to fish has been pointed out (Williams *et al.*, 1997). There are two types of blue-green algae: one that can produce toxins, and another that cannot. To quantify toxin-producing algae, real-time PCR techniques have been developed and utilized to determine the concentration of microcystin-producing miocrocystis cells, as shown in **Figures 2-33** and **2-34** (Ha *et al.*, 2009). The World Health Organization (WHO) has created guidelines for microcystin-LR concentration in drinking water of 1.0 mg/L or less (WHO 1998). Because microcystins are usually located in cells, most can be removed, as shown in **Figure 2-35** (Tsuno *et al.*, 2011). However, ozonation or activated carbon adsorption is an effective treatment when microcystins are released into an extracellular environment. The removal efficiency through chlorination depends on the operational pH level.

Though HPLC or LC/MS is useful for the analysis of individual microcystin congeners, there are more than sixty kinds of microcystin congeners, andtherefore the microcystin-LR equivalent value is required to express the whole toxicity of microcystins. The LC method is not the best for an evaluation of amicrocystin-LR equivalent value because the measurement of all microcystin congeners is required. On the other hand, an antigen-antibody reaction method (an enzyme-linked immunosorbent assay, or ELISA), or an enzyme inhibitory activity method (a protein phosphatase assay), is under consideration; however, it was reported that some

TABLE 2-34. Properties of toxic substances associated with blue-green algae.

Type of toxin	Toxins	Chemical structure	LD_{50}	Species of alga
neurotoxin	anatoxin-a,	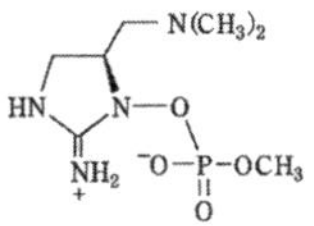	200 mg/kg (intraperitoneal administration: mice)	Anabaena flos-aquae
	anatoxin-a(s)	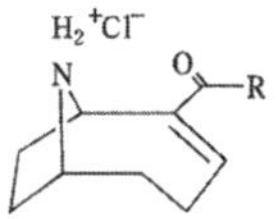	20 mg/kg (intraperitoneal administration: mice)	Anabaena flos-aquae
	Aphantoxin	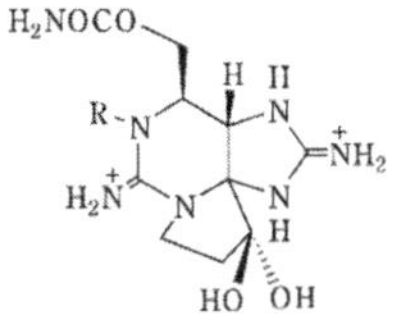 mixture of neosaxitoxin (R=H) and saxitoxin (R=OH)		Aphanizomenonflos-aquae

hepatotoxin	Microcystin	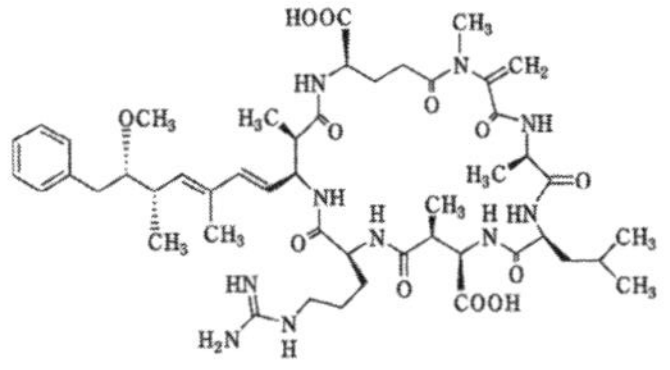 (Microsystin-LR)	LR: 50–100 YR: 100–400 RR: 400–600 (intraperitoneal administration: mice) (Rinehart K, *et al.*, 1994)	Anabaena flos-aquae Anabaena spiroides Microcystisaeruginosa Microcystisviridis Oscillatoriaagardhii Nostic
	Nodularin	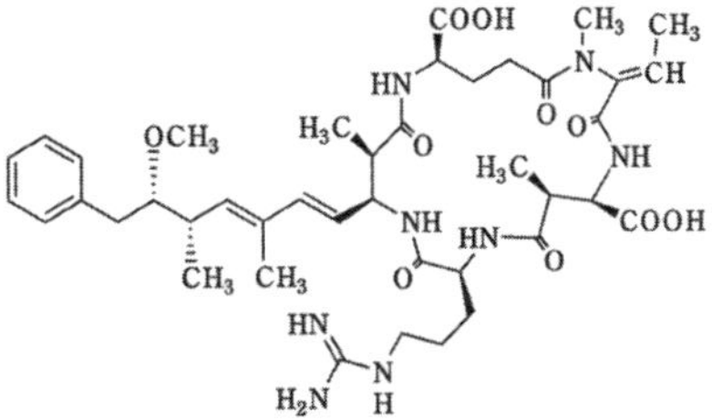		Nodulariaspumigena
	Cylindrosper-mopsin	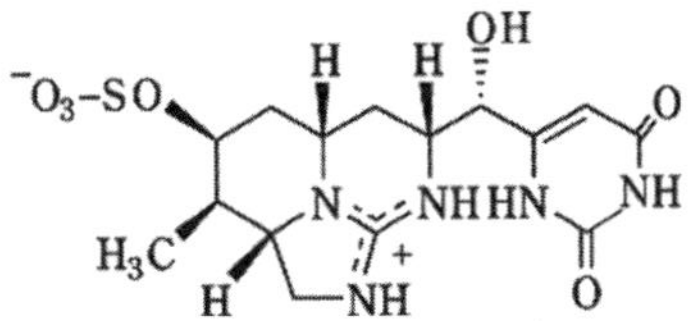	200 mg/kg (intraperitoneal administration: mice)	Cylindrospermopsis-raciborskii

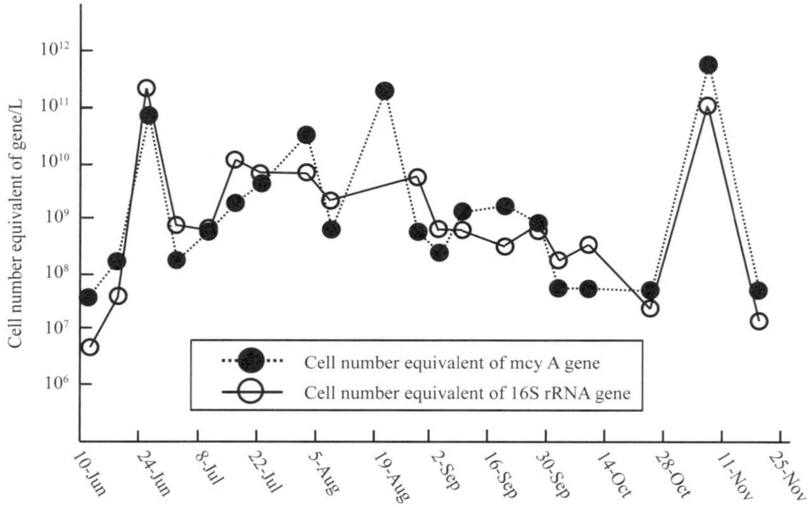

FIGURE 2-33. Variations of Microcystis cell amounts equivalent to mcyA gene and 16S rRNA during the six months (Ha *et al.*, 2009).

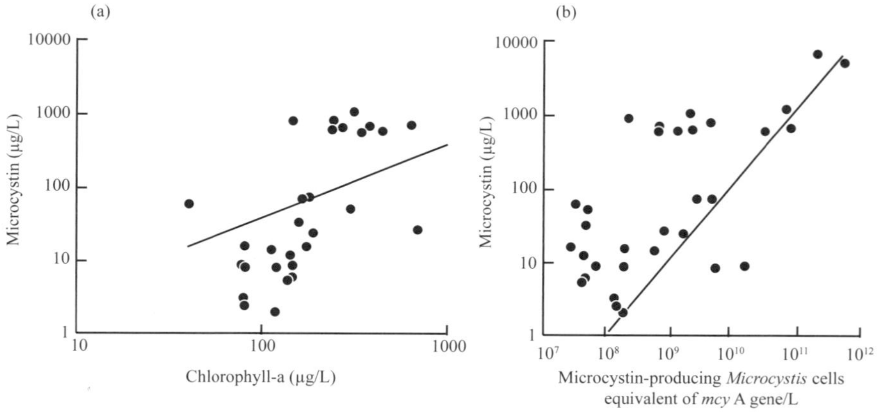

FIGURE 2-34. (a) Correlation between microcystin concentration (mg/L) and chlorophyll-a concentration (mg/L) ($y = 0.36x$, $R^2 = 0.54$, where the equation and R^2 were obtained through a regression analysis). (b) Correlation between microcystin concentration (mg/L) and microcystin-producing Microcystis cells equivalent of mcy A gene/L ($y = 1.05 \times 10^{-8}x$, $R^2 = 0.66$, where the equation, and R^2 were obtained through a regression analysis) (Ha *et al.*, 2009).

problems regarding their accuracy have arisen. Conjugated double bonds have a common part of microcystin congeners. It was reported that the production of 2 methyl -3-methoxy-phenyl acetic acid (MMPB) through a partial oxidation, and its measurement with LC/MS or GC/MS, can be quantified. Based on the measurement principle, the total microcystin amount can be measured in this way; however,

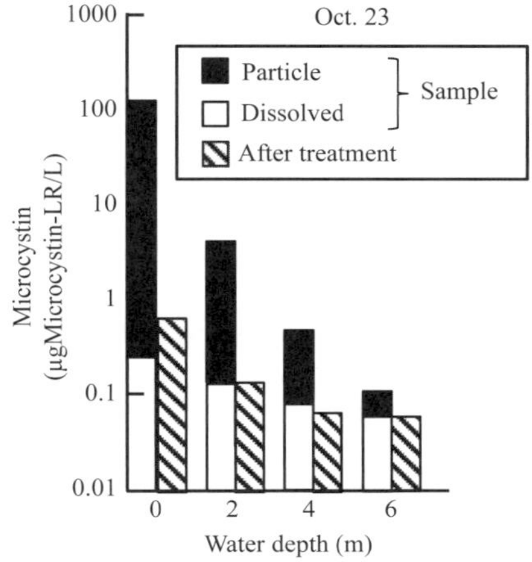

FIGURE 2-35. Removal of microcystin-LR for different dam samples, October 23 (Tsuno *et al.*, 2011).

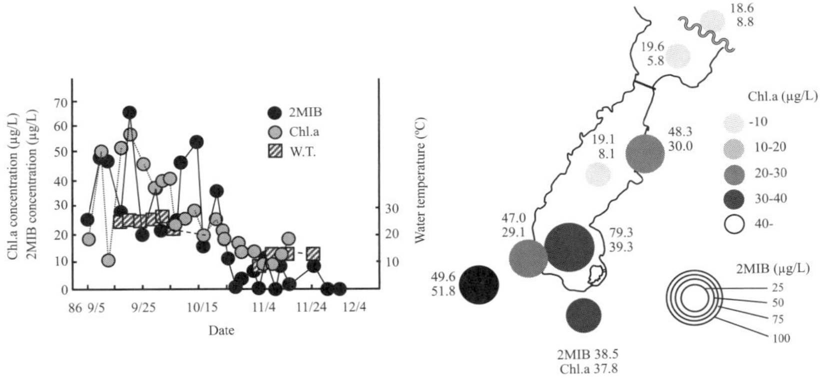

FIGURE 2-36. Example of 2-MIB concentration in Lake Biwa (Somiya *et al.*, 1988).

this method does not show microcystin-LR equivalents, and it was pointed out that the recovery of MMPB at the pre-treatment stage is not very high (Kaya, 2002). The establishment of the evaluation itself is one important issue. Other examples of annoying, but not toxic, compounds are 2-MIB and geosmin, which are produced by Phormidiumtenue and Anabaena macrospore, respectively, and cause moldy odor problems in drinking water when their concentrations are higher than 10 ng/L. Their production with the growth of blue-green algae, and a 2-MIB concentration example from Lake Biwa, is shown in **Figure 2-36** (Somiya *et al.*, 1988).

2.4.6 *Ecosystem preservation*

More than any other species, humans attempt to modify their physical environment to meet their immediate needs, but in doing so, are increasingly disrupting, and even destroying, the biotic components that are necessary for physiological existence.1 Since the industrial revolution, the accelerated development of science and technologies has been enriching our lives, and we have changed the ecosystem more rapidly and extensively than during any comparable period of human history. However, at the same time, many environmental problems have been raised, such as air and water pollution, global warming, species extinction, and a decrease of biodiversity. Since humans are members of the ecosystem, we cannot survive without natural bounty. Therefore, in recent years, ecosystem preservation has been drawing a high degree of attention from all over the world behind the dramatic development of industry activities.

The term 'ecosystem' indicates a material system that consists of a biological community and a physical environment. A biological community consists of a producer, composer, and decomposer, and a physical environment contains an atmosphere, water, soil, light, and so on. Living organisms and their nonliving environment are inseparably interrelated and interact with each other (Odum, 1971). **Figure 2-37** shows a diagram of an ecosystem.

Human-induced eutrophication is an example of the destruction of an ecosystem by humans. Eutrophication is an increase in nutritive salts such as nitrogen and phosphorus in an enclosed water area. An increase of these nutritive salts accelerates anomalously the growth of the producer, i.e., phytoplankton. Phytoplankton use a lot of oxygen and generate an anoxic environment. To make matters worse, some algae produce toxic chemicals. In such an environment, fish and other aquatic organisms cannot survive, and thus eutrophication can cause their sudden death. Although eutrophication occurs both artificially and naturally, human-induced eutrophication through domestic wastewater and agricultural drainage has become a significant problem owing to the concentration of populations and industry.

Not only nutrient salts but also various chemicals are finally discharged into a water environment through domestic wastewater and industrial effluents. Therefore, it is important to know the effects of these released chemicals on aquatic organisms and aquatic ecosystems. As mentioned previously, since an ecosystem is intricately formed, considering an entire ecosystem for an assessment of environmental effects is very difficult. In ecotoxicology, the response of populations and communities to stress is more important than the responses of individuals, and as chronic studies on multi-species systems under natural conditions remain available, a practical approach requires a stepwise procedure starting with single-species tests under laboratory conditions and establishing concentration/effect relationships with indefined time limits (OECD Guideline for testing of Chemicals). Algae, daphnids, and fish are commonly used for ecotoxicological tests globally because they play important roles in an ecosystem. Detailed reasons why these aquatic organisms are often used for such tests are as follows.

First, algae are producers that grow by taking in nutrient salts and performing photosynthesis, and are located at the bottom of the food chain. There-

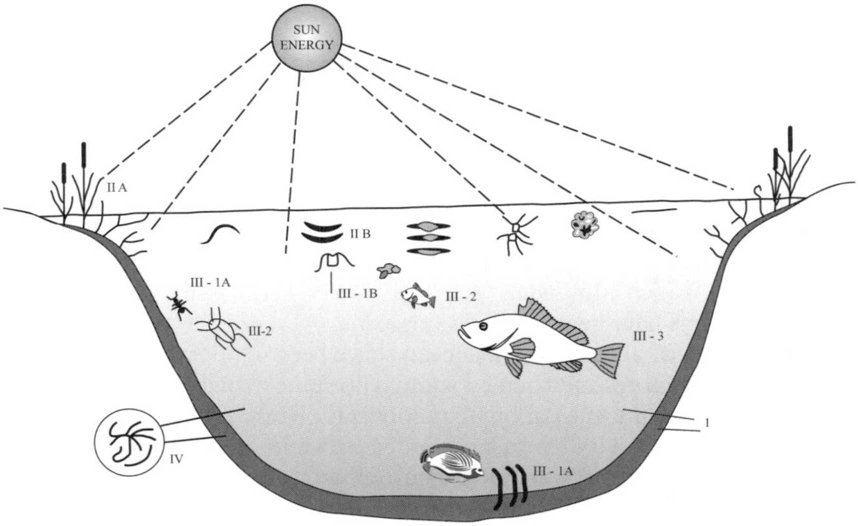

FIGURE 2-37. Diagram of a pond ecosystem. The basic units are as follows: I, abiotic substances, i.e., basic inorganic and organic compounds; IIA, producers, i.e., rooted vegetation; IIB, producers, i.e., phytoplankton; III-1 A, primary consumers (herbivores), i.e., bottom forms; III-1B, primary consumers (herbivores), i.e., zooplankton; III-2, secondary consumers (carnivores); III-3, tertiary consumers (secondary carnivores); and IV, saprotrophs, i.e., bacteria and fungi decay. The metabolism of the system runs on energy from the sun, while the rate of metabolism and relative stability of the pond depend on the rate of inflow of materials from rain and the drainage basin where the pond is located (Odum, 1971).

fore, when algae are inhibited from growing owing to water pollution, the entire food chain will be influenced. It is therefore important to examine the effects on algae. In addition, algae have a short life-cycle, and we can assess the influence of chemicals on them over several generations within a short test period.

Second, daphnids are zooplankton and primary consumers, and are plant-eating link producers and higher-level carnivores in the food chain; furthermore, they are sensitive to a change in water quality. Since daphnids also have a relatively short life-cycle, they can be used in reproduction tests. Daphnids are crustaceans, and can therefore predict the effect on edible crustaceans such as shrip.

Finally, fish are located the top level in an aquatic ecosystem, and are an abundant food source. A high amount of toxicity data on fish has already been reported, and fish are important indicators of environmental pollution.

Since many remarkable environmental pollutants have poor water solubility and tend to be adsorbed by soil when discharged into a water environment, tests using benthos such as chironomidae larvae are needed. In addition, assessing the effects on decomposers is necessary because they also play an important role in an ecosystem.

The occurrence of pharmaceuticals and personal care products (PPCPs) in a water environment has become an increasing concern in recent years. This contamination is mainly due to their incomplete removal from wastewater (Ternes, T.A., 1998; Ying, G.G., *et al.*, 2009). Recently, increased attention has been paid to the potential adverse impacts of PPCPs on aquatic organisms, as well as to the widespread problem of drug resistance bacteria.

The toxic effects of antibacterial agents, levofloxacin (LVFX) and clarithromycin (CAM), which are widely detected as PPCPs in many countries, on aquatic organisms have been studied (Yamashita, N., *et al.*, 2006). Ecotoxicity tests using bacteria, algae, and crustacean shave been conducted. A Microtox test using marine fluorescent bacteria showed no acute toxicity from LVFX and CAM. From the results of a daphnia immobilization test, LVFX and CAM did not show an acute toxicity to crustaceans. Meanwhile, an algal growth inhibition test revealed that LVFX and CAM have a high toxicity to microalgae. The phytotoxicity of CAM was about 100-fold higher than that of LVFX from a comparison of the EC50 (Median Effective Concentration) value, as shown in **Figure 2-38**. From the daphnia reproduction test, LVFX and CAM also showed chronic toxicity to crustaceans. The concentrations of LVFX and CAM in an aquatic environment were compared using the PNEC (Predicted No Effect Concentration) to evaluate the ecological risk (**Figure 2-39**). The ecological

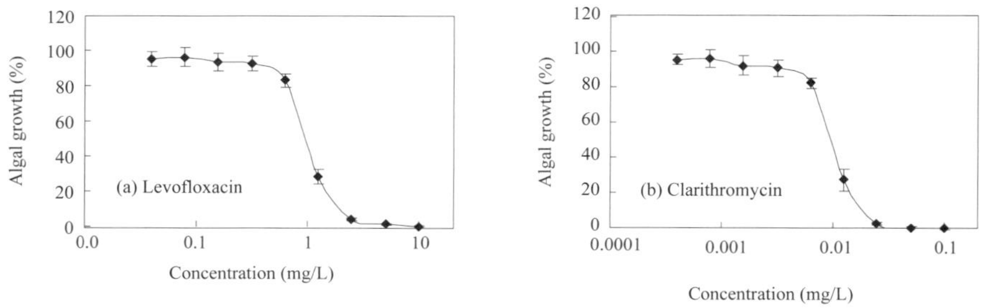

FIGURE 2-38. Effects of levofloxacin and clarithromycin on algae (*Pseudokirchneriella subcapitata*).

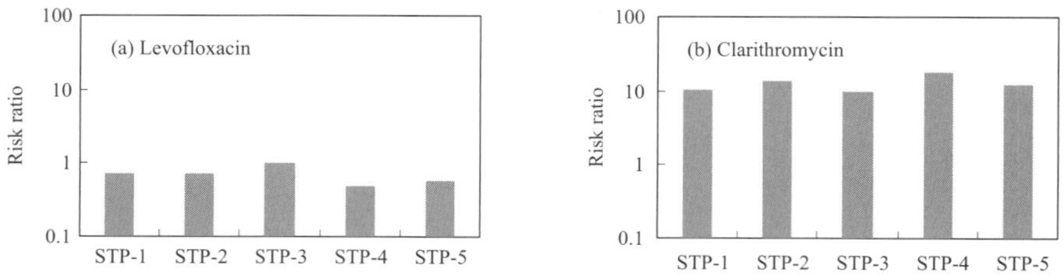

FIGURE 2-39. Risk ratios of levofloxacin and clarithromycin.

risk of LVFX is considered to be low, but that of CAM is higher, suggesting that CAM discharged into an aquatic environment after therapeutic use may affect organisms in that environment.

Since various aquatic organisms have been exposed to different chemicals in water environments, it is preferable to assess the combined impacts on such organisms. Therefore, in America, a Whole Effluent Toxicity (WET) method was introduced, which enable us to conduct a comprehensive assessment of water quality using a bioassay. In Japan, the introduction of the WET method is under review. Meanwhile, in Germany, continuous bio-monitoring has been conducted to detect the presence of harmful chemicals at an early point and limit the extent of damage they may cause. Bio-monitoring is useful for monitoring the water quality in rivers, and protect both aquatic organisms and human health (Wakabayashi A., 2004).

We are now facing various environmental problems, and human activities are affecting ecosystems. We must keep in mind that we are members of an ecosystem, and the development of ecosystem preservation is therefore necessary for the future.

2.5 INTERNATIONAL WATER POLLUTION ISSUES

2.5.1 *Global issues*

Typical global issues are the protection of ocean-scale environments. The dumping of waste, including oil from ships, and the problems of persistent organic pollutants are typical examples. To solve these problems, several international conventions have been established, such as the International Ocean Protection Convention, MARPOL Convention, and POPs Convention. Invader organisms problems through ballast water discharged from ship are also in issue.

Problems in international rivers and lakes, including water resources and water pollution issues, are also important. Therefore, comprehensive and integrated water management, including all watersheds from mountains to the seas, are under discussion. International sea areas must also be conserved by associated countries. In these areas, the floating waste problem is an issue along with water pollution.

2.5.2 *Common issues*

Problems occurring in each area but common to the each area, including the effects, causes, and countermeasures of pollution are also classified as international water pollution issues, and should be overcome commonly and cooperatively. For this reason, almost all water pollution issues are considered to be international water pollution issues. Several international organizations have been established and are currently active. Water pollution in lakes and enclosed coastal seas is a typical example.

REFERENCES

1986–2008 Report on the State of Environment in China. State Environmental Protection Administration of China, 1987–2009, (in Chinese).

1999–2008 China Statistical Yearbook. National Bureau of Statistics of China. (in Chinese).

2000–2009 Bulletin of Marine Environmental Quality of China. State Oceanic Administration, 2001–2010, (in Chinese).

2007 Bulletin of groundwater status in key cities and regions of China. Ministry of Land and Resource of China, 2008. (in Chinese).

2008 annual report of water source quality of China. Ministry of Water Resources of China, 2010. (in Chinese).

2008 China Water Resources Bulletin. Ministry of Water Resources of China, 2009. (in Chinese).

Berger, P.S., *et al.*, (2009), *Encyclopedia of Microbiology (Third Edition)*, Academic Press.

Environmental Quality Standards for Surface Water (GB 3838–2002). State Environmental Protection Administration of China.

Fujimoto, N. (2002), Manual of Measures against Lake Eutrophication, Ministry of Environment, Japan (http://www.env.go.jp/earth/coop/coop/document/mle2_e/mle_e.pdf)

Ghosh, G.C., *et al.*, (2010), Oseltamivir Carboxylate, the Active Metabolite of Oseltamivir Phosphate (Tamiflu), Detected in Sewage Discharge and river Water in Japan, *Environmental Health and Perspectives*, 118(1), pp. 103–107.

Gu, P., *et al.*, (2008), Diffusion Pollution from Livestock and Poultry Rearing in the Yangtze Delta, China, *Environmental Science and Pollution Research*, 15(3), pp. 273–277 (in Chinese).

Ha, J.H., *et al.*, (2009), Quantification of toxic Microcystis and evaluation of its dominance ratio in blooms using real-time PCR, *Environmental Science and Technology,* 43(3), pp. 812–818.

Hagiwara, T., *et al.*, (2004), Concentration of Microsystin in Urban Park Pond Water, *Seikatsu Eisei*, 48(5), pp. 269–275 (in Japanese).

Jobling, S., *et al.,* (1998), Widespread Sexual Disruption in Wild Fish, *Environmental Science and Technology*, 32(17), pp. 2498–2506.

Johnson, A.C., *et al.,* (2004), Removal of endocrine-disrupting chemicals in activated sludge treatment works, *Environmental Science and Technology*, 35(24), pp. 4697–4703.

Joo, H.H., A study on characteristics of microcystin-producing cyanobacterial bloom and microcystin production using real time PCR, *Doctorial Thesis of Kyoto University*.

Kaneko, M. (1997), Water Quality Sanitation, Gihoudo Shuppan.

Kaya, K. (2002), *Analysis of blue-green algae poison*, Bunseki, pp. 436–441. (in Japanese).

Land Infrastucture and Transport, Study Report of committee for control of norovirus in sewage works, March 2009, Land Infrastucture and Transport, 2009. www.mlit.go.jp/common/000116092.pdf

Liu, R.M., *et al.*, (2009), Estimating Nonpoint Source Pollution in the Upper Yangtze river Using the Export Coefficient Model, Remote Sensing, and Geographical Information System, *Journal of Hydraulic Engineering*, 135, pp. 698–704.

Liu, Y.D., *et al.*, (2008), Calculation of contribution value of non- point source pollution of rural refuse in Tai Lake Region, *Journal of Agro-Environmental Science*, 27(4), pp. 1442–1445 (in Chinese).

Matsuoka, S., *et al.*, (2004), Determination of Estrogenic Substances in Coastal Seawater and river Water in Japan Using Four Types of in vitro Assay, *Journal of Japan Society on Water Environment*, 27, pp. 811–816.
Ministry of Environmental Protection, The People's Republic of China, SOE (2008) http://english.mep.cn/standards_reports/soe/soe2008/.
Ministry of Land, Infrastructure, Transport and Tourism: Survey of Estrogenic Compounds All Over Japan in 2002, (http://www.mlit.go.jp/kisha/kisha03/05/051024_.html)
Ministry of Land, Infrastructure, Transport and Tourism (2010), Report on virus measures in sewage works.
Ministry of the Environment Government of Japan (2010) Annual Report on the Environment, the Sound Material-Cycle Society and the Biodiversity in Japan.
Ministry of the Environment, Government of Japan (2011), 2010 Annual Report on the Environment, the Sound Material-Cycle Society and the Biodiversity in Japan, Our Responsibility and Commitment to Preserve the Earth—Challenge 25.
Ministry of the Environment, Government of Japan: http://www.env.go.jp/en/water/wq/pamph/index.html
Moselio Schaechter (Editor): Encyclopedia of Microbiology (Third Edition), Academic Press, 2009.
Nakada, N. (2010), Research progresses in the occurrence, treatment and aquatic toxicity of pharmaceuticals in the aquatic environment in Japan, *Journal of Japan Society on Water Environment*, 55(5), pp. 147–151.
Nakada, N., *et al.,* (2008), Occurrence of PPCPs in the aquatic environments in Japan and Tropic Asian countries, *Journal of Water and Waste,* 50(7), pp. 559–569. (in Japanese).
Nishimura, F., *et al.*, (2009), Distribution of POPs and estrogenic compounds in the Yodo river system, *Proceedings of The 17th Seminar of JSPS-MOE Core University Program on Urban Environment.*
Nojiri, K., *et al.*, (2005), Analysis of estrogen in river water and sediment sample from Kamo river with GCINCI-MS, *Report of center for Environmental Science in Saitama*, 6, pp. 136–141.
Odum, E.P. (1971), *Fundamentals of ecology 3rd ed.*, Philadelphia: Saunder.
OECD Guideline for testing of Chemicals, Summary of Considerations in the OECD Expert Group on Ecotoxicology.
Okayasu, Y., *et al.*, (2008), Conditions of dissolved oxygen concentration for effective removal of natural and estrogens in wastewater treatment process, *Proceedings of IWA World Water Congress and Exhibition*, CD-ROM.
Park, H.D. (2008), Dynamics of the microsystin in aquatic ecosystem, Bulletin of the Plankton Society of Japan, 55(1), (in Japanese).
Rinehart, K.L., *et al.*, (1994), Structural and biosynthesis of toxins from blue-green algae (cyanobacteria), *Journal of Applied Phycology*, 6, pp. 159–176.
River Bureau of Ministry of Land, Infrastructure, Transport and Tourism: Report on Conditions of EDCs in water environment, 2002.
Somiya, I., *et al.*, (1988), Fate of odorous compounds in the growth process of blue-green algae, *Proceedings of 22nd water pollution Research*, pp.185–186.
Takabe, Y., *et al.*, (insert year of publication here), Concentration distribution of POPs and estrogenic compounds in the Yodo river system, *Environmental & Sanitary Engineering Research*, 23(3), pp. 224–227.
Tanaka, H., *et al.*, (2003), Estrogenic activity of river water and feminization of wild carp in Japan, *Proceedings of WEFTEC*, 2003.

Tanaka, H., *et al.*, (2008), Occurrence and physicochemical treatment of PPCPs, *Environmental Technology*, 12, pp. 834–839.
Tanaka, H., *et al.*, (2008), Occurrences of Pharmaceuticals and Personal Care Products and Development of Physicochemical Treatment Technologies for Their Reduction, *Environmental Conservation Engineering*, 37(12), pp. 834–839.
Ternes, T.A. (1998), Occurrence of drugs in German sewage treatment plants and rivers, *Water Research*, 32(11), pp. 3245–3260.
Tsuno, H. (1991), Phosphorus in Water Pollution Issues, *Journal of Water*, No. 5, pp. 22–27.
Tsuno, H., *et al.*, (1979), Simulation of Vertical Distribution of Water Quality in Lake Yunoko, *Proceedings of 13th Symposium on Water Pollution*, pp. 51–52.
Tsuno, H., *et al.*, (2005), Applicability of Mussel as a Bioindicator for Monitoring of PCB in Sea Water, JSCE, No.797/VII-36, pp. 71–79.
Tsuno, H., *et al.*, (2011), Fate of hazardous organic substances produced by *Microcystis* sp. in conventional water treatment process, *Proceedings of 4th IWA-Aspire conference*, pp.1–8 (CD-ROM), 2–6 October, 2011.
U.S. Environmental Protection Agency (EPA) (2006) Ultraviolet Disinfection Guidance Manual for the Final Long-term 2 Enhanced Surface Water Treatment Rule, Office of Water (4601), EPA 815-R-06-007, 1-4-1-7, 2006.
van Drecht, G., *et al.*, (2001), Global pollution of surface waters from point and nonpoint sources of nitrogen, *Science World Journal*, 6 (suppl. 2), pp. 632–641.
Wakabayashi, A. (2004), Currents of science and policy about revision of water quality goal in water environment, *Journal of Japan Society on Water Environment*, 27(1), pp. 2–7.
Wang, L. (2011), *Research on Biotransformation of Typical Estrogens In Shenzhen river and Related Waste Water Treatment Plants*, Dissertation of Tsinghua University.
WHO (1998). *Guideline for drinking water quality, 2nd ed., vol. 2*, Health criteria and Other Supporting information, Geneva.
Williams, D.E., *et al.*, (1997), Andersen R.J.: Evidence for a covalently bound form of microcystin-LR in salmon liver and Dungeness crab larvae, *Chemical Research in Toxicology*, 10(4), pp. 463–469.
Williams, D.E., *et al.*, (1997), Evidence for a covalently bound form of microcystin-LR in salmon liver and Dungeness crab larvae, *Chemical Research in Toxicology*, 10(4), pp. 463–469.
Xu, J., *et al.*, (2007), Stable carbon isotope variations in surface bloom scum and subsurface seston among shallow eutrophic lakes, *Harmful Algae*, 6, pp. 679–685.
Yamashita, N., *et al.*, (2006), Effects of antibacterial agents, levofloxacin and clarithromycin, on aquatic organisms, *Water Science and Technology*, 53(11), pp. 65–72.
Ying, G.G., *et al.*, (2009), Occurrence and removal of pharmaceutically active compounds in sewage treatment plants with different technologies, *Journal of Environmental Monitoring*, 8(11), pp. 1498–1505.
Zhang, W.L., *et al.*, (2004), Estimation of agricultural non-point source pollution in China and the alleviating strategies. I. Estimation of agricultural non-point source pollution in China in early 21 century, *Sci Agr Sin*, 3(7), pp. 1008–1017 (in Chinese).

3

Urban Water Management for Sustainable Development

This chapter addresses issues of water supply in urban areas. First, the framework and regulations for drinking water quality management are presented. The current problems in drinking water treatment are then discussed along with several proposals for optimizing or upgrading treatment systems. Finally, plans for securing water safety and water reuse for urban sectors are discussed.

3.1 MANAGEMENT FRAMEWORK OF URBAN WATER QUALITY

3.1.1 *Water quality standards and health risk management*

3.1.1.1 Summary of quality standards

Needless to say, tap water must be a safe source of drinking water. In addition to safety, tap water needs to be free from unpleasant odors and flavors that may cause anxiety in the public. Various pathways that have the potential for leading to water contamination are shown in **Figure 3-1**. For such a wide variety of pollution sources, new knowledge and technologies need to be continuously introduced that enable water utilities to consistently supply safe tap water that people can trust.

Water quality standards have been established to ensure tap water safety and control required tap water properties (e.g., taste, odor, and color) (Health Science Council, 2003a; 2003b). The quality standards of national drinking water in many countries owe a debt to the World Health Organization (WHO) Guidelines for Drinking Water Quality (WHO guidelines, hereafter) (World Health Organization, 2008a). The guidelines for individual chemicals, together with data from important scientific and epidemiological research and animal testing are the basis for the WHO guideline values. Therefore, the WHO guidelines are considered to be the most reliable source of information on the effects of chemical use on human health. Of course, Japan has contributed greatly

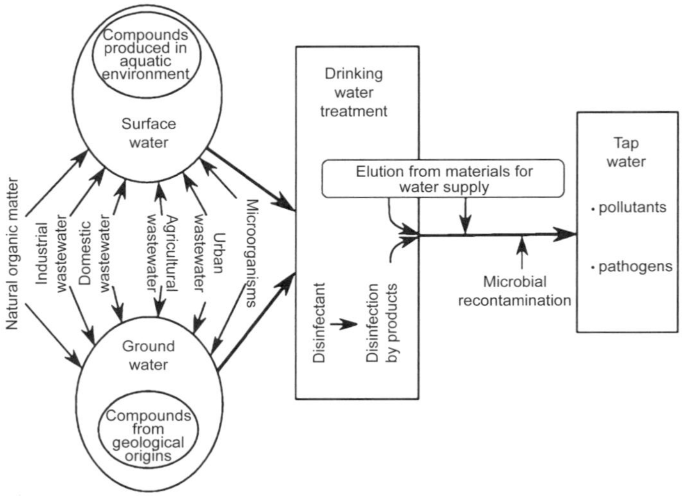

FIGURE 3-1. Types of pollution found in drinking water.

to the establishment of these guidelines. As in Japan, the WHO guidelines are currently used as an important reference when establishing drinking water standards in many countries around the world.

Japan's current drinking water standards are based on those enacted in 2004, although subsequently revised in certain areas. These standards will be successively amended in the future when necessary. For the latest version of water quality standards, one should refer to the website for the waterworks division of the Ministry of Health and Welfare.

The current quality standards for Japanese drinking water as of 2010 are shown in **Table 3-1**. These standards are the criteria set by law, and water suppliers are required to meet them. Items whose concentrations exceed (or are likely to exceed) 10% of their assessment values in finished water (see later sections for more details) are regulated. There are 50 regulated items, as shown in **Table 3-1**.

These regulated items are classified into two categories: items related to human health, and those related to the acceptability of tap water. One of the characteristics of the current quality standards on drinking water is that these two categories are not differentiated. The items related to the acceptability of tap water in **Table 3-1** include zinc (these items may be difficult to distinguish for people who have no expertise in the area of drinking water). The items related to human health are essentially items related to tap water safety, and the items related to acceptability are those related to reliability. The current regulation framework does not differentiate between these two categories as they are equally important. In other words, the current framework is attempting to achieve both tap water safety and user reliability of water supplies. A good example is the addition of geosmin and 2-methylisoborneol (typical musty-odor compounds) to the list.

TABLE 3-1. National quality standards of drinking water in Japan.

	Items	Standard (mg/L)	Category
1	Standard plate count bacteria	100/mL	Indexes for pathogens
2	*E. coli*	Not detected	
3	Cadmium	0.003	Inorganic compounds and heavy metals
4	Mercury	0.0005	
5	Serenium	0.01	
6	Lead	0.01	
7	Arsenic	0.01	
8	Cromium (VI)	0.05	
9	Cyanate and cyanogen chloride	0.01	
10	Nitrate and nitrite Nitrogen	10	
11	Fluoride	0.8	
12	Boron	1	
13	Carbon tetrachloride	0.002	Organic compounds
14	1,4-dioxane	0.05	
15	*cis*-1,2-dichloroethylene and *trans*-1,2-dichloroethylene	0.04	
16	Dichloromethane	0.02	
17	Tetrachloroethylene	0.01	
18	Trichloroethylene	0.03	
19	Benzene	0.01	
20	Chlorate	0.6	Dinfection by products
21	Chloroacetic acid	0.02	
22	Chloroform	0.06	
23	Dichloroacetic acid	0.04	
24	Dibromochloromethane	0.1	
25	Bromate	0.01	
26	Total trihalomethane	0.1	
27	Trichloroacetic acid	0.2	
28	Bromodichloromethane	0.03	
29	Bromoform	0.09	
30	Formaldehyde	0.08	

(*Continued*)

TABLE 3-1. Continued.

	Items	Standard (mg/L)	Category
31	Zinc	1	Color
32	Aluminium	0.2	
33	Iron	0.3	
34	Copper	1	
35	Sodium	200	Taste
36	Manganese	0.05	Color
37	Chloride ion	200	Taste
38	Calsium, magnesium etc. (hardness)	300	
39	TDS	500	
40	Anionic surfactants	0.2	Foam
41	Geosmin	0.00001	Musty smells
42	2-methylisoborneol	0.00001	
43	Non-ionic surfactants	0.02	Foam
44	Phenols	0.005	Smell
45	TOC	3	Taste
46	pH	5.8–8.6	Basic properties
47	Taste	Normal	
48	Odor	Normal	
49	Color	5	
50	Turbidity	2	

In addition to the 50 regulated items, 27 other items are classified as management items (as of 2010). These were not included in the list of items regulated for toxicological or other reasons, but are of potential importance for the quality management of drinking water due to their frequent detection and/or intensive use.

Furthermore, 44 items are listed as monitoring items. These items are not included in the list of regulated items due to a lack of toxicological information and/or information regarding the range of concentration in finished water. Water utilities and related agencies are required to collect information on these 44 items so that more appropriate decisions can be made during the revisioning process of standards in the near future.

TABLE 3-2. Chinese quality standards for drinking water (Ministry of Health of the People's Republic of China and Standardization Administration of China, 2006).

Category	Item	Standard value	Unit
Microorganisms[1]	Total coliform	ND	MPN/100 mL or CFU/100 mL
	Thermotolerant coliform bacteria	ND	MPN/100 mL or CFU/100 mL
	Escherichia coli (E.coli)	ND	MPN/100 mL or CFU/100 mL
	Total colony count	100	CFU/mL
Toxicological Indices	Arsenic	0.01	mg/L
	Cadmium	0.005	mg/L
	Chromium (Hexavalent)	0.05	mg/L
	Lead	0.01	mg/L
	Mercury	0.001	mg/L
	Selenium	0.01	mg/L
	Cyanide	0.05	mg/L
	Fluoride	1	mg/L
	Nitrate	10 20 (Ground water source)	mg/L as N
	Chloroform	0.06	mg/L
	Carbon tetrachloride	0.002	mg/L
	Bromate (when ozone is applied)	0.01	mg/L
	Formaldehyde (when ozone is applied)	0.9	mg/L
	Chlorite (when chlorine dioxide is applied)	0.7	mg/L
	Chlorate (when chlorine dioxide is applied)	0.7	mg/L
	Color	15	Color units
	Turbidity	1.0–3.0	NTU
	Taste	None	–
	Odor	None	–

(Continued)

TABLE 3-2. Continued.

Category	Item	Standard value	Unit
	Visibile substance	None	–
	pH	6.5–8.5	–
	Aluminium	0.2	mg/L
	Iron	0.3	mg/L
	Manganese	0.1	mg/L
Sensory traits and physical indicators	Copper	1	mg/L
	Zinc	1	mg/L
	Chloride	250	mg/L
	Sulfate	250	mg/L
	Total dissolved solids	1000	mg/L
	Total hardness (as $CaCO_3$)	450	mg/L
		3	
	COD	5 (for source water with COD > 6 mg/L)	–
	Volatile phenols	0.002	mg/L
	Detergents	0.3	mg/L
	Alpha particles	0.5	Bq/L
Radionuclides[2]	Beta particles and photon emitters	1	Bq/L

1: When total coliform is positive, either *E. coli* or thermotolerant coliform bacteria needs to be measured.
2: When these values are exceeded, radionuclide analysis must be performed.

The Chinese quality standards on driking water are summrized in **Table 3-2**. In addition to these listed items, there are 64 other non-regular items, which are similar to the management and monitoring items in the Japanese guidelines. Contrary to the Japanese legal framework for the quality management of drinking water, which is essentially based on the Water Supply Act, several laws (e.g., Environmental Protection Act and Food Hygiene Act) are the basis for these standards.

In **Table 3-3**, the regulations on disinfection byproducts (DBPs) are compared among various countries including Japan and China (European Union, 1998; California Department of Public Health, 2006; Ministry of Health of the People's Republic of China and Standardization Administration of China,

TABLE 3-3. Comparison of the standards and guidelines on disinfection byproducts. (European Union, 1998; California Department of Public Health, 2006; Ministry of Health of the People's Republic of China and Standardization Administration of China, 2006; Ontario Ministry of the Environment, 2006; USEPA, 2006; Health Canada, 2007).

	Japan (Standard items, management items, monitoring items)	China (regular items and non-regular items)	USA (federal standards)	Canada (guidelines)	EU water directive	WHO guidelines
Total THMs	0.1	1	0.1	0.1	0.1	e
Cholorform	0.06	0.06	–	–	–	0.3
Bromodichloromethane	0.03	0.06[h]	–	0.016	–	0.06
Dibromochloromethane	0.1	0.1[h]	–	–	–	0.1
Bromoform	0.09	0.1[h]	–	–	–	0.1
Chloroacetic acid	0.02	–	–	–	–	0.02
Dichloroacetic acid	0.04	0.05[h]	–	–	–	P0.05
Trichloroacetic acid	0.2	0.1[h]	–	–	–	0.2
HAA5[a]	–	–	0.06	–	–	–
Bromate	0.01	–	0.01	0.01	0.1	P0.01
Formaldehyde	0.08	0.9	–	–	–	–
Chlorate	0.6	0.7	–	–	–	P0.7
Chloral hydrate	P0.02[b]	0.01[h]	–	–	–	–
Chlorite	0.6[b]	0.7	1	–	–	P0.7
Dichloroacetonitrile	P0.01[b]	–	–	–	–	P0.02

(Continued)

TABLE 3-3. Continued.

	Japan (Standard items, management items, monitoring items)	China (regular items and non-regular items)	USA (federal standards)	Canada (guidelines)	EU water directive	WHO guidelines
N-nitrosodimethyl amine (NDMA)	0.0001[g]	–	0.0001[c]	0.000009	–	0.0001
Cyanogen chloride	0.01[f]	0.07[h]	0.2	0.2	0.05	0.07
Dibromoacetonitrile	0.06[g]	–	–	–	–	0.07
MX	0.001[g]	–	–	–	–	–
2,4,6-trichlorophenol	–	0.2[h]	–	0.005	–	0.2

Unit: mg/L
a: The sum of the concentrations of monochloro-, dichloro-, trichloro-, monobromo-, dibromo-acetic acids.
b: Management items.
c: Notification level (the concentration for which some improvement is required) in the state of California.
d: Provisional standard in the state of Ontario.
e: The sum of the ratios of cacnTHM concentration to individual standards needs to be less than 1.
f: As ronal cyanogen compounds.
g: Monitoring item.
h: Non-regular item.
P: Provisional values.

2006; Ontario Ministry of the Environment, 2006; USEPA, 2006; Health Canada, 2007). Regulated, management, and monitoring items for Japan are listed in this table. The U.S. and some other countries have set regulations on the total concentration of similar types of DBPs (e.g., total trihalomethanes and HAA5). Japan, on the contrary, adheres fundamentally to the style of the WHO guidelines and tends to regulate individual compounds.

3.1.1.2 Acceptable risk

Ideally, tap water should contain no harmful chemicals or microbes. However, it is neither realistic nor practical to pursue such a zero risk policy.

Chemicals and microorganisms are dealt with differently when setting their acceptable risk levels. First, for genotoxic carcinogens, no threshold is assumed, and an incremental lifetime cancer risk of 10^{-5} to 10^{-6} (as annual incremental risk, this value is divided by $70 = 1.4 \times 10^{-7}$ to 10^{-8}) is employed. The values of the drinking water standard in Japan are also equivalent to the 10^{-5} level for this class of compound. The validity of the magnitude of acceptable risk is not based on a certain concept. Rather, it should vary from country to country, and change over time. An incremental cancer risk of 10^{-5} means that a lifetime consumption of water containing the threshold concentration of a chemical leads to one cancer case among 100,000 people.

On the other hand, for non-genotoxic carcinogens and non-carcinogenic substances, the existence of a threshold value is assumed. Also, standard values are determined on the basis of tolerable daily intake (TDI).

The health effects of microorganisms show the following pattern: infection → morbidity → death (worst case). Here, infection is the intrusion and growth of pathogens in the human body, and is not necessarily accompanied by symptoms. Only a small portion of infected people actually proceed to the stage of illness. And a very small subset of infection cases leads to death. In microbial risk assessment, the endpoint is set at the stage of infection. It is also notable that the mortality and severity of microbial diseases vary greatly depending on the type of pathogen. The health outcome of a pathogen also depends on its strain. An annual allowable probability of infection of 10^{-4} (one in 10,000) has been proposed (USEPA, 1989). This risk level has been generally accepted in the developed world. However, similar to the case with chemicals, acceptable risk levels for pathogens should reflect the needs and situations in the target country/region.

3.1.2 *Risk management of chemicals and determination of standard values*

This subsection overviews the process of setting standards for toxic chemicals. Taking contributions from other sources such as foods and air into account, standard values are set based on the levels at which no adverse health effects occur even with life-long intake. As shown in **Table 3-4**, the first step is to classify

TABLE 3-4. Classification of methods for establishing standards.

<table>
<tr><th></th><th colspan="2">Category</th><th>Methods for setting standards</th></tr>
<tr><td rowspan="3">Hazardous chemicals</td><td rowspan="2">Carcinogens</td><td>Genotoxic</td><td>No threshold concentration is assumed, and standards are set based on the lifetime incremental cancer risk level of 10^{-5}</td></tr>
<tr><td>Non-genotoxic</td><td rowspan="2">The existence of threshold concentration is assumed, and standards are set based on tolerable daily intake (TDI)</td></tr>
<tr><td colspan="2">Non-carcinogens</td></tr>
</table>

hazardous chemicals based on their carcinogenicity. Next, those chemicals deemed to be carcinogens are further divided into two categories: genotoxic and non-genotoxic compounds. Classification in this first step is crucial, and determines the following process for calculating the assessment values.

Figure 3-2 shows a simplified concept of the methods use in establishing standard values. First, no threshold is assumed for the case of genotoxic carcinogens. That is, there are no safety levels under this assumption, and it is considered that a certain risk exists for any exposure amount. The standard value for each chemical is set to a level at which the incremental lifetime cancer risk is 10^{-5}. Thus, the standards do not show concentration levels for zero risk. On the contrary, a certain level of risk is accepted. At the same time, however, this level does not cause adverse health effects in any practical sense, and for this reason is called a virtually safe dose (VSD). The results of animal testing are used to determine this level, as shown in **Figure 3-2**. Animal testing is generally conducted at much higher doses than those occurring in everyday life. Thus, the acceptable amount for a chemical corresponding to a 10^{-5} cancer probability is estimated by extrapolating to lower doses using a mathematical model.

For this purpose, a linearized multistage model is usually employed (Yanagawa, 2002; Nakanishi *et al.*, 2003). In this case, the assumption of no threshold level and the use of a linear model to find the VSD corresponding to 10^{-5} result in a safer evaluation. To be on the safe side, the lower end of the 95% confidence interval is used for the evaluation of the VSD. Since the VSD is calculated conservatively, as mentioned above, the safety factor used for the extrapolation from laboratory animals to humans is not further considered. The determined VSD (mg/(kg•day)) is a value equivalent to a 10^{-5} risk level for experimental animals. The same value is used unchanged as the VSD (mg/(kg•day)) for human as well.

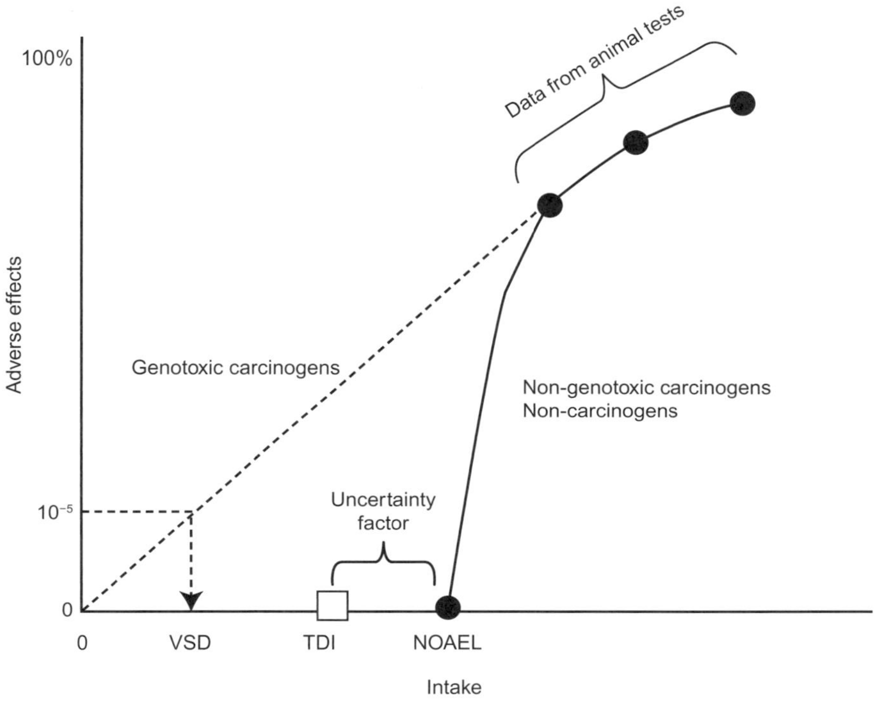

FIGURE 3-2. Concept used for establishing standards.

On the other hand, for non-genotoxic carcinogens and non-carcinogenic substances, a threshold concentration below a level of zero effect is assumed. First, based on the data from animal testing, the maximum dose that shows no significant effects (no observed adverse effect level, or NOAEL) is determined. Depending on the experimental design, NOAEL might not be obtained, and in some cases, only the smallest dose with an observable effect (Lowest Observed Adverse Effect Level, or LOAEL) can be determined.

These values are then divided by an uncertainty factor, which is also called a safety factor. An uncertainty factor of 100 is typically used. This is a product of a factor of 10 for the individual variability of laboratory animals and a factor of 10 for the species differences between human and other animals. However, an uncertainty factor of 1,000 is employed instead when LOAEL is used in lieu of NOAEL, when the quality of the data is insufficient, or when the type of toxicity is severe. If the total uncertainty factor exceeds 1,000, the calculated value is considered provisional due to its extremely high uncertainty.

This value after an adjustment with an uncertainty factor is called the tolerable daily intake (TDI) and is usually expressed in units of mg/(kg•day).

That is, a TDI value is the daily allowable dose for a unit weight of the human body. Finally, TDI values are converted into concentrations for tap water.

The following equation shows how to compute assessment values from TDI values.

$$\text{TDI [mg/(kg•day)]} = \frac{\text{NOAEL or LOAEL [mg/(kg•day)]}}{\text{uncertainty factor (−)}} \tag{3-1}$$

$$\text{Assessment value [mg/L]} = \text{TDI [mg/(kg•day)]} \times \text{body weight [kg]} \times \text{allocation to drinking water (−)} / \text{Water intake per person [L/day]} \tag{3-2}$$

In Japan, the water intake per day per person is assumed to be two liters, and the average weight is set at 50 kg (note, an average of 60 kg is used in the WHO guidelines).

In general, the intake of chemicals through food is dominant over the intake through tap water. For this reason, the quality standards for drinking water should reflect the percentage of exposure through drinking water. A 10% TDI is typically assigned as the allocation for drinking water. For DBPs, 20% is used as DBPs are generated during the disinfection process for drinking water. However, for a continual revision of the quality standards for drinking water, further study and discussion are necessary for many aspects of this issue.

In this subsection, an outline of the process used for setting the quality standards of drinking water are discussed. Toxicological tests are continuously being carried out for suspected toxic chemicals, and for those chemicals having undergone toxicological evaluation, the decision of whether to add them to the standards is made. As a result, the number of regulated items is increasing. Of course, the process of setting these standards is not perfect, and many extensive research efforts are being devoted to this area of study (Matsui *et al.*, 2010).

One major problem of the TDI approach is that the slope of the dose-response curves is not considered. That is, only the value corresponding to NOAEL or LOAEL is used, and other information on the response to higher doses is discarded. The benchmark dose method has been proposed as a solution to compensate for this drawback (Yoshida and Nakanishi, 2006). The concept of this method is summarized in **Figure 3-3**. First, data obtained from animal experiments are fitted to a suitable dose-response curve (e.g., multistage models, but not limited to linearized multistage models). Based on this curve, the dose (e.g., ED_{10}, estimated dose for a 10% response) corresponding to an effect level (in this example, a 10% excess response) is determined. For increased safety, the lower limit at a 95% confidence interval (LED_{10}, lower limit of the estimated dose for a 10% response) is then calculated. This value is the benchmark dose. In addition, this method is also applicable for carcinogens with no threshold. That is, as shown in **Figure 3-3**, after finding ED_{10} and LED_{10}, the curve is linearly extrapolated to the origin with LED_{10}

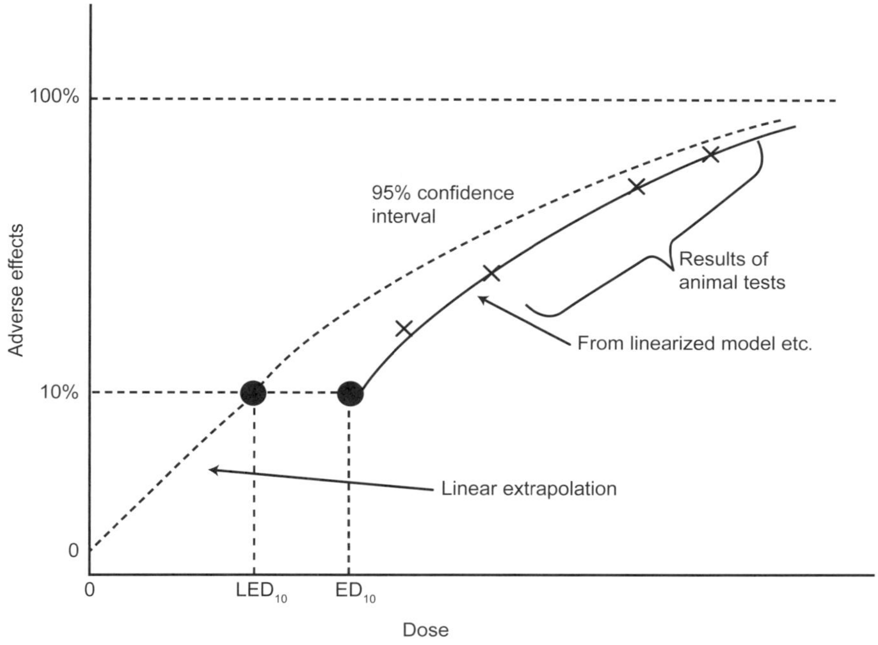

FIGURE 3-3. Concept of the bench mark dose method.

as the starting point, assuming no threshold. The dose corresponding to a 10^{-5} cancer risk can then be calculated. This method is recommended in the USEPA's new guidelines for risk assessment of carcinogens, published in 2005, and is used to derive standard guidelines by the WHO and the Japanese government (USEPA, 2005).

3.1.3 *Risk management of microbial infection*

3.1.3.1 Target microorganisms (Kaneko, 1997; 2006)

First, microorganisms of potential concern for the quality management of drinking water are discussed below.

(A) *Bacteria*

Waterborne pathogenic bacteria are listed in **Table 3-5**. *Vibrio cholerae*, *Shigella*, and *Salmonella typhi* are classical examples of bacteria causing waterborne diseases. Nowadays, waterborne transmissions of pathogenic *Escherichia coli*, *Campylobacter*, *Aeromonas*, and *Salmonella* are more common than *Salmonella typhi*.

In addition to bacteria that cause enteric infections through water intake, sufficient attention should be paid to *Legionella* infections, which are caused

TABLE 3-5. Waterborne pathogenic bacteria.

Pathogenic *Escherichia coli* (i) enterophathogenic *E. coli*: EPEC (ii) enteroinvasive *E. coli*: EIEC (iii) enterotoxigenic *E. coli*: ETEC (iv) enterohemorrhagic *E. coli*: EHEC (v) enteroadherent *E. coli*: EAEC
Campylobacter
dysentery bacillus
Salmonella
Vibrio and Cholera vibrio
Yersinia
Aeromonas
Plesiomonas
Clostridiam perfringens
Legionella

through the inhalation of aerosols. *Legionella* is a gram-negative and aerobic short rod. *Legionella* is important in that it causes severe pneumonia and has a high mortality rate (Aikawa *et al.*, 1997). To prevent the growth of *Legionella* in water supply systems, treatment and distribution systems need to be kept clean, and a heterotrophic bacteria count is a good indicator for this purpose. A heterotrophic bacteria count of 2,000 CFU/mL (note, this is a provisional value) has been set as a management target in Japan.

(B) *Virus*

Viruses that infect humans through tap water exposure are primarily limited to those taken into the body through the mouth, infecting the intestinal system. Major viruses causing waterborne transmission are shown in **Table 3-6**.

The number of reports on the detection and quantification of viruses in water is limited. However, it is best to assume that viruses are present in source waters, as norovirus, enterovirus, and adenovirus have been already identified in the aquatic environment in Japan (Yano, 2006; Sano and Ueki, 2006). In the U.S., many unexplained waterborne infections are attributed to viruses.

(C) *Protozoa*

Table 3-7 lists the protozoa involved in water pollution (Kaneko, 1997). Some types of protozoa form resistant oocysts or cysts during their life cycle, which are discharged into the environment through the feces of the host, polluting aquatic environments. *Cryptosporidium* and *Giardia* are typical examples of

TABLE 3-6. Major viruses related to waterborne infections.

Viruses	Related diseases
Rotavirus	gastroenteritis
Norovirus	gastroenteritis
Sapovirus	gastroenteritis
Astrovirus	gastroenteritis
Enterovirus	aseptic meningitis etc.
Hepatitis A virus	acute hepatitis
Hepatitis E virus	acute hepatitis
Adenovirus	pharyngoconjunctival fever
Poliovirus	flaccid palsy etc.
Coxsackievirus	flaccid palsy etc.

TABLE 3-7. Protozoa related to waterborne infections.

	Protozoa	Route of infection
From feces	*Balontidium coli*	Oral intake
	Cryptosporidium spp.	
	Entamoeba bisfoltica	
	Giardia Lamblia	
	Isospora belli	
	Microspora belli	
	Microsporidia	
	Tosoplasma gondii	
From environment	*Acanthomoeba* spp.	Nasal infection
	Balamuthia moundrillaris	
	Naegleria folweri	

such protozoa, and are commonly found in source waters such as rivers in Japan (Hosaka, 2007).

Protozoa forms cysts (with a spore-like structure), which are highly resistant to chlorine disinfection and chemicals. It is this resistance that makes protozoa so significant. For example, conventional chlorination in drinking water treatment cannot inactivate *Cryptosporidium*. In addition, conventional water treatments such as rapid sand filtration systems may not be able to achieve the complete removal of these oocysts and cysts.

In Japan, it can be said that outbreaks of *Cryptosporidium* triggered the introduction of membrane filtration into the treatment processes of drinking water (JWRC, 2005). Unlike sand filtration, a complete removal of particles is expected with the use of a membrane filter. Also, UV disinfection is now well recognized as an alternative disinfection process for the inactivation of protozoa (Ministry of Health, Labour and Welfare, 2007a).

Thus far, microorganisms that can potentially cause waterborne diseases have been described. Among them, infections of pathogenic *E. coli*, *Campylobacter*, and norovirus are the most common, according to a recent report (Yamada and Akiba, 2007). These organisms occur in various types of water supply systems: municipal, simple, and private water supplies; water tanks; small water supply; wells; and springs. The primary cause of their existence is inadequate disinfection. Moreover, *Giardia* and *Cryptosporidium* infections have been reported and, in some cases, have led to the shutdown of water supplies.

3.1.3.2 Quantitative microbial risk assessment

Similarly to chemical risks, it is necessary to set an acceptable risk level for microbial infections and create sophisticated risk assessment and management processes. The Netherlands is the world's most advanced country in this regard, as outlined below.

It is well known that chlorination is not used for water disinfection in the Netherlands (Smeets *et al.*, 2009; Itoh, 2010a). In 2005, the last chlorination facility was replaced with UV treatment. It should also be noted that quantitative microbial risk assessment (QMRA) in the Netherlands has been included in the management framework of drinking water quality to ensure microbial safety. The current drinking water quality standards in the Netherlands have been in effect since 2001 (Staatsblad van het Koninkrijk der Nederlanden, 2001). **Table 3-8** shows items related to microoganisms. The four items (three of them are regulations for operational management) mentioned above, including *Aeromonas*, are listed. In addition to these items, there is a regulation stating

TABLE 3-8. Microbial requirements for drinking water in the Netherlands.

Items	Standards
Annual microbial infection risk must be managed less than 10^{-4} for Enterovirus, *Cryptosporidium*, and *Giardia* etc. by QMRA.	
E. coil and enterococcus	0 CFU/100 mL
Aeromonas[a]	must be kept at less than 1000 CFU/100 mL
Heterotrophic bacteria[a]	must be kept at less than 100 CFU/mL
Coliform and *Clostridium*[a]	0 CFU/100 m/L

a: management standards

that the annual risks of infection to enteric viruses, *Cryptosporidium* and *Giardia*, as evaluated by QMRA, must be below 10^{-4}. The Netherlands is the only country that has this kind of regulation based on QMRA.

For this regulation, each water company is required by registration to provide a quantitative risk assessment for microbial infections on its tap water (mainly for surface water sources). And each utility is obligated to provide safe water with an annual risk of infection of below 10^{-4}. The QMRA techniques were not developed in the Netherlands, and there are many contributors to its establishment all over the world (Hass *et al.*, 2001). However, the Netherlands is extremely unique, and most advanced in terms of puting QMRA into practice.

Figure 3-4 shows an example of risk evaluation for a water purification plant in the Netherlands. Enteric viruses, *Campylobacter*, *Cryptosporidium*, *Giardia*, were the target microorganisms used in this study. Required log removal is the log removal calculated from the actual concentrations in a source water, and the concentration needed to meet a 10^{-4} risk level. The bars on the right show the cumulative log removal by the water treatment plant in terms of water storage, coagulation, sedimentation, rapid and slow sand filtration, and UV

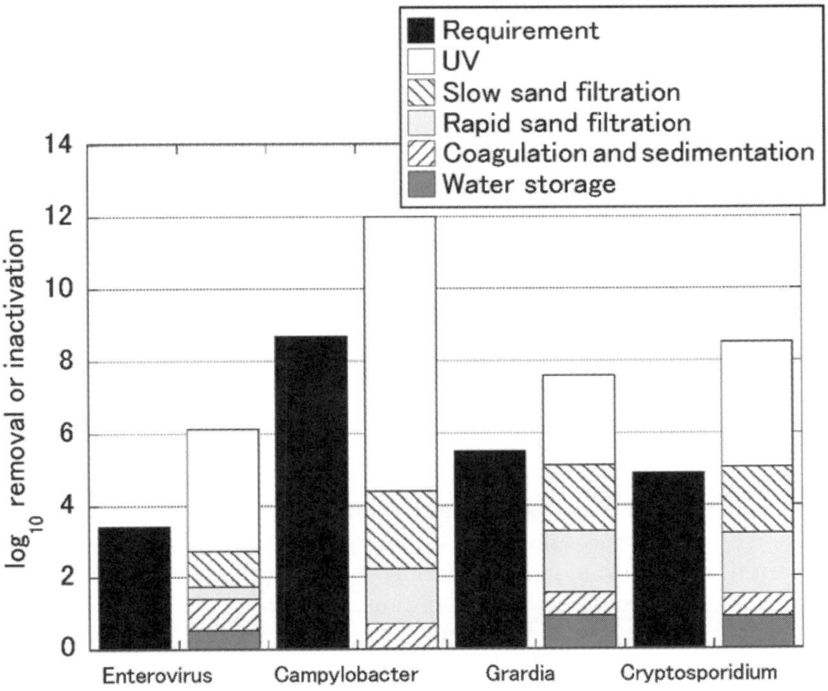

FIGURE 3-4. Example of QMRA (Smeets *et al.*, 2009).

treatment. As an example, since 'Required log removal' corresponds to the removal efficacy for a risk level of 10^{-4}, a log removal of 2-log greater than the required level indicates a risk level of 10^{-6}. Thus, it can be said that the risks associated with these four pathogens are well managed in this water treatment plant even without chlorination.

Itoh performed a QMRA for a water purification plant in the Netherlands. The treatment flow of this plant was as follows: coagulation → water storage (residence time 89 days) → rapid sand filtration → ozonation → softening → granular activated carbon → slow sand filtration. Chlorination was not included in this system, and slow sand filtration was regarded as a 'final disinfection process'. In this assessment, the annual risk of *Campylobacter* infections was evaluated based on *E. coli* data as the indicator organism.

Since chlorination was not performed, tap water was the resulting product after slow sand filtration. Measurements showed zero *E. coli* in most cases, although the average probability of *Campylobacter* infections per year was calculated as 1.68×10^{-3}/(person•year). It is notable that the annual probability of infection was greater than the target value (10^{-4}) even though almost all *E. coli* measurements showed zero presence. Based on these results, the critical control points (ozonation, in this water treatment plant) were also identified through a sensitivity analysis. In addition, combined with an uncertainty analysis, points requiring further information were determined (Itoh, 2010b).

EU countries have systematically developed a QMRA scheme in the MICRORISK project, and the current generation of QMRA techniques is ready for practical use (Medema, *et al*., 2006).

3.1.4 *DALYs and risk management in the future*

This section describes the health risks of exposure to microorganisms and chemicals. To discuss the future management of these two different types of risks, let us once again discuss the topic of disinfection.

It is well understood that chlorination reduces the risk of microorganisms in tap water, but also leads to certain cancer risks caused by chemical substances such as trihalomethanes. This dilemma is an essential element of disinfection technologies. This concept is summarized in **Figure 3-5**.

It would be good if a quantitative evaluation of disinfection efficacy, determination of the optimal disinfectant dosage, and selection of an alternative disinfectant were possible based on the concept shown in **Figure 3-5**. The balance between microbial and chemical risks caused by disinfectants has long been discussed. The dashed line in the figure shows the ideal properties of an alternative disinfectant. That is, it would be ideal if low doses of such a disinfectant could reduce microbial risks without the production of significant byproducts.

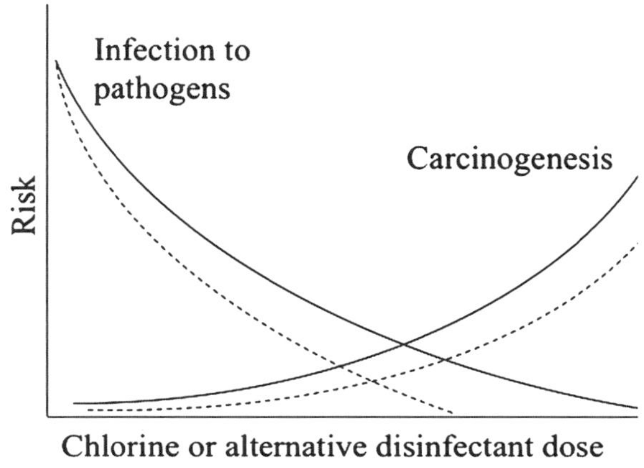

FIGURE 3-5. Risk trade-off between pathogens and DBPs.

However, because the contents and qualities of different risks to drinking water are completely different, it is not easy to evaluate and compare them.

Let us start with microbial risks. The health effects of microorganisms generally follow the order of infection → morbidity → death. Infection is the invasion and growth of microorganisms in the human body, but is not necessarily accompanied by symptoms. In fact, only a small portion of those infected actually show symptoms. In addition, only a very limited number of cases lead to human death.

For a conversion of infection events to the risk of death, it has been common practice to assign 1% or 0.1% as the mortality of infected individuals (sometimes individuals with symptoms) with reference to the morbidity and mortality data after infection (Kaneko, 1996; Fewtrell *et al.*, 2003). However, this practice limits the scope of risk assessment to death, and leaves various health effects unevaluated. In many microbial infection cases, and the following onset cases, patients recover from diarrhea and fever after a week or so, and death is very rare. Thus, if only the risk of mortality is assessed, only a small portion of health effects of microbial infections are captured, and an underestimation of the health risk is quite likely. As mentioned earlier, the mortality and severity of microbial diseases vary greatly depending on the type of pathogen. The health outcome after exposure to a pathogen also depends on the strain.

On the other hand, for chemical toxicity, an evaluation is conducted based on the occurrence of lesions, although chemicals also induce various types of health effects. The severity of the disease is not usually included in the evaluation. One exceptional evaluation is when setting the uncertainty factor for the calculation of TDI. In this process, an extra factor of 10 for serious diseases (e.g., cancer) is added (note, for non-cancer lesions, a factor other than 10 is used in certain cases) (Health Science Council, 2003b).

When converting health effects into mortality risk, cancer is considered to be lethal in principle. Thus, cancer risk tends to be regarded as mortality risk (a different approach could be taken). For non-carcinogenic effects, they are not counted as (mortality) risk.

Disability-adjusted life years (DALYs) are indexes used to overcome the above problems. The concept of DALYs integrates the evaluations of different health effects caused by a specific disease. Also, DALYs can be used as a common index to compare the effects of different diseases (Fewtrell, 2003; Murry and Lopez, 1996). In particular, this concept has been used to quantify the global burden of disease (GBD), which is also one of the key indicators for understanding global environmental issues (World Health Organization, 2008b).

DALYs are expressed as a sum of the years of life lost (YLL) due to early death, and years lived with a disability (YLD). YLD is weighed based on the severity of the disability, and DALYs are recorded in years.

$$\text{DALYs} = \text{YLL} + \text{YLD} \tag{3-3}$$

YLL indicates the total years of life lost in a population, as represented by the following equation.

$$\text{YLL} = \sum_i e^*(a_i) \sum_j d_{ij}, \tag{3-4}$$

where i is the index for different age groups, j is the index for different patient groups, d_{ij} is the number of fatal cases for each age group, and $e^*(a_i)$ is the average life expectancy of each age group.

YLD is expressed, as in the following equation, using the average duration of an illness and the weighting factor used to reflect the severity of the disease in a population (i.e., from 0 (completely healthy) to 1 (death)).

$$\text{YLD} = \sum N_j L_j W_j, \tag{3-5}$$

where j is the index for different patient groups, N in the number of patients, L is the disease duration, and W is the severity of the disease. The severity of the disability is classified into seven grades, as shown in **Table 3-9**.

Therefore, in an evaluation using the concept of DALYs, in addition to deaths caused by disease and injury, the quantification of health burdens less severe than death is possible.

It is foreseeable that the standard and guideline values for chemicals and microorganisms in drinking water will be integrated in the future under the concept of DALYs.

Actually, the concept of DALYs has already been successfully applied to the issue of disinfection (**Figure 3-5**). In this example, the inactivation of

TABLE 3-9. Grade of disability and disease examples (Murry and Lopez, 1996).

Grade	Weight	Example
1	0.00–0.02	Facial pallor, low weight
2	0.02–0.12	Diarrhea, adenoiditis
3	0.12–0.24	Adenoiditis, anginaw
4	0.24–0.36	Amputation, loss of hearing ability
5	0.36–0.50	Down syndrome
6	0.50–0.70	Depression, blindness
7	0.70–1.00	Mental illness, dementia, paralyzed extremities

chlorine-resistant *Cryptosporidium* through ozonation is addressed. Note that bromate ions, a carcinogen, are generated during ozonation. For this trade-off, Havelaar *et al.*, attempted to estimate these different types of risks in the Netherlands using the concept of DALYs (Havelaar *et al.*, 2003). The outcome of this evaluation has also been useful in forming a strategy against *Cryptosporidium* in the process of setting drinking water quality standards in Japan (Health Science Council, 2003) (Health Science Council, 2003a; Itoh and Echigo, 2008).

First, the health effects of *Cryptosporidium* infections per person were determined to be 1.03×10^{-3} DALYs. This is the sum of YLL (0.0937×10^{-3}) and YLD (0.937×10^{-3}), Here, YLD is one order of magnitude larger than YLL. That is, the health effects due to diarrhea are much more significant than the mortality from *Cryptosporidium* infection. This is a good example of the underestimation of microbial risk that occurs when only mortality risk is considered.

Havelaar *et al.*, (2003a) then evaluated the DALYs due to bromate ions from ozonation. Bromate ions are a genotoxic carcinogen, and the standard value (and the WHO guideline value) for this compound is set based on the incremental cancer risk for a lifelong ingestion of 10^{-5} (1/100,000 persons). In this evaluation, they estimated the DALYs of renal cell cancer caused by bromate ions in drinking water at the same level as the standard. As a result, it was calculated as 1.4×10^{-6} DALYs. In this case, YLD is much smaller than YLL, and YLD was negligible. In other words, an acceptable cancer probability of 10^{-5} corresponds to 1.4×10^{-6} DALYs (note that this is an annual value).

Now, suppose that the concentration of *Cryptosporidium* in raw water is 1 oocyst/10 L. If the water is not treated, the annual probability of infection is 1.5×10^{-1}/year (the dose-response relationship from Hass was used) (Itoh and Echigo, 2008). As noted above, the health effects per infected individual is 1.03×10^{-3} DALYs, and thus the annual health impact per capita is calculated to be 1.5×10^{-4} DALYs. This is the risk of *Cryptosporidium* for a concentration in drinking water of 1 oocyst/10 L.

This value of 1.5×10^{-4} DALYs is much larger than that of renal cell cancer from bromate (1.4×10^{-6} DALYs). In other words, the use of ozonation to inactivate *Cryptosporidium* is beneficial even when the concentration level of bromate ions is at the standard value.

The following discussion is also possible. For source water with a *Cryptosporidium* concentration of 1 oocyst/10 L, if 2-log removal (99% removal) by ozonation or other measures is achieved, the risk level will be reduced to the level equivalent to the DALYs corresponding to a 10^{-5} cancer risk (note that in this example, DALYs is for renal cell cancer). That is, if the same risk management level for carcinogens is required for *Cryptosporidium*, the treatment system (e.g., filtration) has to achieve 2-log removal of *Cryptosporidium* for the source water in this example.

The above example showed that the comparison between different risk factors (i.e., chemicals and microorganisms) is possible with DALYs. It was also possible to set the target level for treatment of pathogens in drinking water. It can be concluded that we now have a systematic way to set acceptable risk levels (i.e., standard values).

In QMRA, while the primary focus is an estimation of the probability of infection, a unifying approach is also taken with DALYs in some cases (Fewtrell *et al.*, 2003; Havelaar and Melse, 2003). For example, Itoh estimated the DALYs within a distribution area based on the probability of *Campylobacter* infections of 1.68×10^{-3}/(person•year) (Itoh, 2013). Since the population of the distribution area was approximately 360,000, the annual incidence of infections was estimated to be the annual probability of infection (1.68×10^{-3}) × 360,000 persons = 605 infections. From this value, the DALYs for this area was determined to be 2.78. Also, the cost associated with the occurrence of gastroenteritis caused by *Campylobacter* in this area was estimated to be €220,000, with the cost per case as €370 and the number of incidents.

Also, in the WHO guidelines on water quality for the water reuse of sewage water, DALYs are used as an index as well as an acceptable probability of infection (World Health Organization, 2011).

It is a general trend that the expression of risks is shifting from an evaluation of infection probability to a direct expression of health effects. One should note that the index for microbial safety is evolving in the following order: microbial concentration → infection probability → DALYs as shown in Figure 3-6 (Douglas, 2009).

It is generally accepted that the annual target level of 10^{-6} DLAYs per person is appropriate (World Health Organization, 2011). This level is considered to be approximately equivalent to the incremental cancer risk of 10^{-5} due to

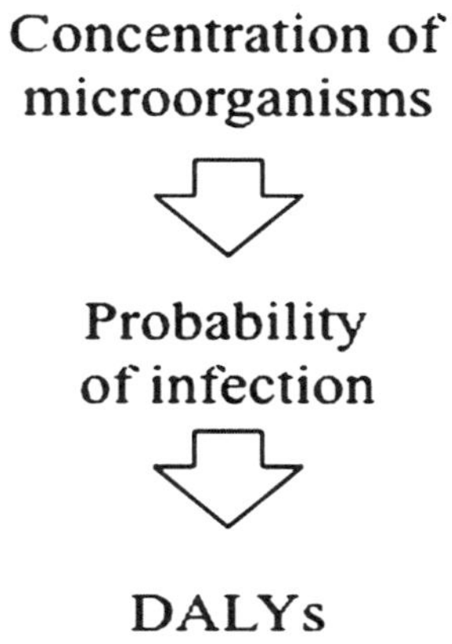

FIGURE 3-6. Transition of microbial index for drinking water.

genotoxic carcinogens. One example is renal cell cancer as described above. For pathogenic microorganisms, if their principal health effect is diarrhea and the fatality rate is low, the above target level corresponds to a disease probability of 10^{-3}. A comparison between the infection risk of *Cryptosporidium* and the magnitude of DALYs confirms this relationship.

In this context, the probability of infection per year of 10^{-4} is one order of magnitude more strict than the incremental cancer risk level of 10^{-5}. Note that the health effects from cancer are chronic, and those from microbial infection are acute. In general, in terms of risk perception, people tend to find acute and catastrophic risks more unacceptable and serious than chronic risks if the actual levels of these two risk types are the same (Okamoto, 1992). The acceptable risk levels for microbial infection and cancer have been separately estimated, but we may have inexplicitly taken the characteristics of our risk perception into account.

The use of DALYs is more popular in the field of food safety assessment than in the water sector. WHO and FAO (UN Food and Agriculture Organization) are conducting a joint project, in which they are attempting to quantify GBD using DALYs for a selected set of microorganisms and chemicals (World Health Organization, 2008). In the field of safety assessment of drinking water, for example, WHO has been working on the quantification of DALYs for several pathogens and chemicals (Haverlaar and Melse, 2003).

The quantification of DALYs itself is challenging. Setting systematically the drinking water standards/guidelines on pathogens and chemicals under the concept of DALYs is not an easy task. However, we are certainly moving toward this direction and should therefore pay sufficient attention to the progress of this field of study.

3.2 TECHNOLOGIES FOR WATER SUPPLY IN URBAN AREAS

3.2.1 *Current status and problems of drinking water treatment technology*

Japan's water supply system covers of 97.5% of the population, and safe drinking water is available for almost all people. On a per volume basis, 77% of Japan's water is treated by rapid sand filtration, and 3.6%, 18.5%, and 0.7% is processed by slow sand filtration, disinfection only, and membrane filtration, respectively (note, treatment technologies are classified by the type of filtration) (Japan Water Works Association, 2007). Among these treatment technologies, membrane filtration is being rapidly introduced these days, although the current volume undergoing processing is still quite small. Membrane technologies are expected to be applied to various aspects of drinking water treatment.

Currently, the most common water treatment system in Japan is a rapid filtration system. In this system, small particles and colloids collide to form larger particles with the addition of a coagulant such as aluminum sulfate. Coagulants form multivalent ions in water, neutralize the electrical repulsion between particles, and bridge between particles as insoluble compounds. Grown particles (flocks) are removed through gravitational settling in sedimentation tanks. This process removes most flocks, and the rest are almost completely removed through rapid sand filtration. As the final step, chlorine is added to inactivate pathogens passing through the sand filtration, and to prevent microbial regrowth in the distribution system. This system is basically able to produce clear and microbiologically safe water. However, one should note that some small colloids and ionic species (i.e., compounds fully dissolved in water) are not removed. Also sufficient attention should be paid to the fact that sand filtration does not physically 'block' particles and flocks using smaller pores, and is a stochastic process in terms of particle removal.

A rapid sand filtration system is sufficient for the production of drinking water if the raw water quality is high enough. However, this system cannot remove micropollutants discharged from human activities and musty odor compounds. It is also not capable of controlling dissolved organic matter, which is a precursor of disinfection byproducts. Thus, in many cases in Japan, additional processes to overcome these problems are necessary. Collectively, systems that utilize these additional unit operations are called advanced drinking water treatment systems. Ozonation and activated carbon treatment are the main unit operations in this system. Biological processes can be used as components of advanced water treatment systems, but a combination of ozonation and activated carbon treatment, or activated carbon treatment, alone is typically used. Approximately 26% of the total water treated in Japan is processed through an advanced drinking water treatment system. In addition, 16% of the total volume is processed using ozonation (Japan Water Works Association, 2007; Ministry of Health, Labour and Welfare, 2009).

Activated carbon treatment is basically an adsorption process when used without ozonation. In this process, organic compounds are adsorbed onto the surface of porous activated carbon. This method is inexpensive and efficient. However, since raw water is a complex mixture of various compounds within a wide range of concentration, the competition between target contaminants and other organic matters is unavoidable. In particular, if a large city is located along the upper or middle stream area of a river basin, the downstream area tends to face this situation. In this case, activated carbon alone is insufficient as an advanced drinking water treatment system, and thus a more efficient system is necessary. The most typical configuration for this purpose is the combination of ozonation and activated carbon treatment.

The application of ozone to water treatment was first attempted in France and Germany beginning in the late 19th century, and ozonation was first introduced to a drinking water treatment plant in Nice, France in 1906 (Society on Ozonation, 1997). Since then, for over 100 years, ozonation has been used in many water treatment plants for various purposes. It took more than a half century for Japan to introduce ozonation in drinking water treatment. The first ozonation facility was installed at the Kanzaki water purification plant in Amagasaki in 1973 for controlling musty odor compounds in Lake Biwa and the Yodo river basin (Society on Ozonation, 1997). Since then, many utilities (e.g., the Osaka prefectural government, the City of Osaka, and the Tokyo metropolitan government) have adopted ozonation for better quality drinking water.

Ozonation is used in the field of drinking water treatment for various purposes: decomposition of odorous compounds such as 2-MIB and geosmin, oxidation of the precursors of disinfection byproducts such as trihalomethanes, control of color (i.e., oxidation of manganese, iron, and natural organic matters), oxidation of micropollutants such as pesticides, inactivation of chlorine-resistant pathogens, and an enhancement of coagulation.

Figure 3-7 shows the treatment flows of typical advanced water treatment systems. One or more unit processes is generally added after the coagulation or sand filtration processes are completed. Since ozone is a gas within the temperature range acceptable for water treatment, a device to effectively dissolve ozone gas into water is crucial. Also, the design guidelines of water supply systems in Japan require that ozonation be followed by granular activated carbon treatment (GAC) to quench or remove oxidation products (e.g., hydrogen peroxide and aldehydes). The use of GAC after ozonation is a common practice worldwide, but is not necessarily mandatory. Each flow in **Figure 3-7** has certain advantages and disadvantages: (a) the risk of microorganisms leaking from activated carbon is higher, (b) the treatmnet loads for ozonation and GAC are greater.

Now, let us discuss the chemical reactions in advanced drinking water treatment systems, particularly ozonation. Ozone (O_3) is a strong oxidizing agent, and reacts with various substances in water. There are two types of oxidation

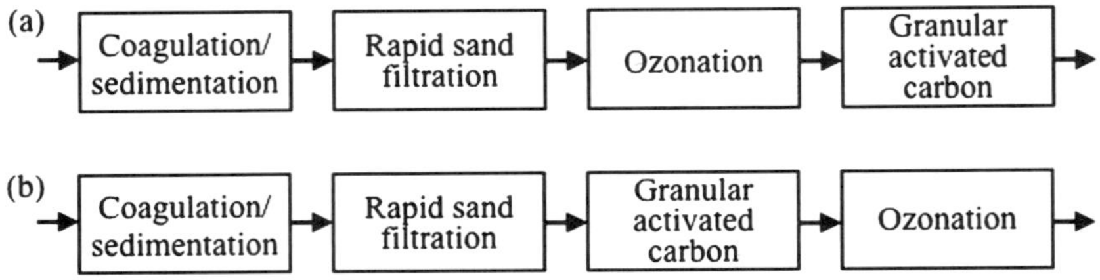

FIGURE 3-7. Examples of treatment flows with ozonation and GAC.

reactions during ozonation: a direct reaction is oxidation with ozone molecules (O_3 itself), and an indirect oxidation is caused by hydroxyl radicals formed through the self-decomposition of ozone. The direct reaction is a selective reaction, and unsaturated compounds (i.e., certain compounds with aromatic rings and double bonds) are oxidized. Note that aromatic compounds with low electron density in the benzene ring do not undergo this direct oxidation reaction. For color removal, disinfection, and the oxidation reaction of dissolved organic matter (effective in reducing chlorination byproducts such as trihalomethanes from the subsequent chlorination), a direct reaction plays a major role. On the other hand, the indirect reaction, an oxidation reaction by hydroxyl radicals, is a non-selective reaction due to the oxidation capability of hydroxyl radicals being stronger than molecular ozone. This indirect reaction is often initiated from the hydrogen abstraction reaction from C–H bonds. Thus, almost all organic compounds are oxidized through this reaction pathway. In Japan, one of the primary objectives of ozonation is the oxidation of odorous substances such as geosmin and 2-MIB. Since these are saturated compounds, they are actually oxidized by hydroxyl radicals. The oxidation of micropollutants is also considered as the indirect reaction by hydroxyl radicals.

Thus, the relative importance of direct and indirect reactions highly depends on the purpose of the ozonation. Also, the percentage of ozone undergoing an indirect reaction changes based on the reaction conditions. For example, the self-decomposition of ozone is promoted at a higher pH, and the proportion of indirect reactions will increase.

Ozonation is a unique unit operation in that it changes the water quality without removing impurities from the water. This feature is different from those of coagulation, a sand filtration process, and activated carbon treatment. In other words, ozonation and chemical oxidation, in a broader aspect, do not generate waste. Also, since dissolved organic matter is decomposed to smaller, more hydrophilic, and more biodegradable compounds ozonation, ozonation promotes the biodegradation of dissolved organic matter by microorganisms living in the activated carbon. Thus, ozonation plays a complementary role with activated carbon treatment. In short, an advanced water treatment system with ozonation and GAC is an effective device to overcome the shortcomings of a rapid filtration system.

However, not all problems are solved using ozonation and GAC. There are still difficulties in ensuring microbiological and chemical safety, as well as in finding a good balance between microbiological and chemical risks and the risks caused by disinfections. This trade-off problem and its solution will be discussed in later sections.

In addition to safety, the aesthetic aspects of water are also important. According to a survey by the Cabinet Office, 60% of the population in Japan does not drink tap water directly (Cabinet Office, 2008). The primary reason for this high percentage is considered due to its odor. The sense of smell is an important factor in determining human behavior. Even though water safety is ensured scientifically, some people do not feel tap water quality is sufficiently reliable and therefore do not drink it directly. In order to maintain sustainable water supply systems, customer satisfaction is also an important issue.

According to the Ministry of Health and Welfare, among the complaints regarding the odor and taste of tap water, 63% are related to earthy and/or musty odors (Ministry of Health, Labour, and Welfare, 2007b). In addition, plant, fish, and putrid odors, along with a wide range of chemical smells, have been reported. An ozone-GAC system is an appropriate way to improve water quality for musty odors and odors originating from the organic matter in raw water. Therefore, ozonation is an important technology in ensuring not only the safety but also the taste and odor of water.

However, in addition to odors caused by raw water pollution, serious attention has to be paid to the chlorinous odor of tap water due to water chlorination. Chlorinous odor has thus far been recognized as evidence of the biological safety of tap water. But, in fact, more than a few people are concerned about chlorinous odor, and thus the safety of finished water, even though they are told tap water is safe to drink (Itoh *et al.*, 2007).

Compounds causing chlorinous odor have not been fully identified, but trichloramine (NCl_3), which is produced in the reactions between hypochlorous acid and amino acids and/or ammonium ions, is considered to be a major contributor along with chlorine itself. Ozonation is effective to some extent in reducing chlorinous odor that occurs after chlorination. This is because a fraction of the precursor (i.e., dissolved organic matter) is oxidized, and because chlorine demand is reduced after ozonation. However, complete oxidation of certain precursors is difficult (e.g., ammonium ion), this can be a major challenge for future advanced treatment systems for drinking water.

In a sense, the problem of chlorinous odor is a result of our high dependence on water chlorination in Japan. Of course, a water supply without chlorination is not easy to achieve and serious consideration is necessary before changing our system. Nonetheless, seeking alternative water treatment and distribution systems that can ensure biological safety with lower residual chlorine is an extremely important issue. The key aspect in such a process is the sophistication of an advanced risk management system and water treatment technology to support this new generation of risk management. Current advanced water treatment systems have overcome the drawbacks of rapid sand filtration systems, but incomplete treatment of organic matter with low molecular weight through ozonation (i.e., the amount of not fully mineralized assimilable organic carbon (AOC) increases) leads to a potential microbial regrowth in water distribution systems. So, what can solve this problem and the chlorinous odor problems do? In the following section, a next-generation technology for the treatment of drinking water that solves these problems is discussed.

3.2.2 *Next-generation drinking water treatment technology*

3.2.2.1 Unit operations

An important technology when considering a next-generation advanced treatment system for drinking water is membrane treatment. Microfiltration (MF) and ultrafiltration (UF) can be introduced as a replacement for rapid sand filtration. In addition to the removal of turbidity, membrane treatment is expected to allow better control of pathogens such as *Cryptosporidium*. Furthermore, nanofiltration (NF) is likely to be introduced gradually for the removal of dissolved organic matter and micropollutants. In addition, reverse osmosis (RO) is the central technology for seawater desalination and the reuse of treated water. The future use of RO in Japan depends on an improvement in treatment efficiency (i.e., cost) and the status of water resources. Given the possible instability of water resources brought about by climate change, water utilities should pay attention to this technology. In addition, forward osmosis (FO, a membrane technology using positive osmotic pressure) as well as functional membranes that mimic cell membranes are being developed (McCutcheon *et al.*, 2006).

Another technology of interest is UV disinfection. UV disinfection was generally uncommon in Japan due to a lack of residual effect and because the reduction of the survival ratio of *Cryptosporidium* is low. However, UV treatment has been found effective in reducing the infectivity of *Cryptosporidium*, and is therefore becoming popular (Drescher *et al.*, 2001). UV treatment essentially does not require any chemicals and is easy to maintain and operate. For this reason, UV treatment is particularly suitable for small-scale facilities. Also, although DBP formation characteristics are not fully understood, UV treatment is expected to produce less DBPs than other disinfection techniques. The Japanese Ministry of Health and Welfare has listed UV treatment as an effective treatment technology for controlling *Cryptosporidium* (Ministry of Health, Labour and Welfare, 2007c).

A more sophisticated control of the flow pattern or mixing condition during disinfection is another issue requiring further attention. The kinetics of chlorine disinfection is often described using the Chick-Watson equation, and it is a very common practice to evaluate and manage disinfection processes using a CT value. However, the concentration distribution (C) and the residence time (T) in actual contactors are hard to determine, and smaller values (i.e., a lower C and shorter T) are generally used to be on the safe side. However, if we consider odor and DBPs, the 'safe side' concept becomes a one-sided assessment. The introduction of a new unit operation is one aspect, but the sophistication of the management used for current systems is also an important and urgent issue for drinking water treatment technology.

3.2.2.2 Upgrading the current advanced drinking water treatment system

Ozone and GAC processes have been a great success for water utilities in downstream areas for the control of musty odor compounds and trihalomethanes. However, looking at the profiles of dissolved organic matter removal, advanced drinking water treatment processes are not always efficient. In particular, hydrophilic acids are not removed by the current process (**Figure 3-8**). While most hydrophilic compounds themselves are not toxic, there is a problem in that they are major precursors of trihalomethanes and haloacetic acids (Imai *et al.*, 2003; Jo *et al.*, 2008). Therefore, considering the dominance of hydrophilic compounds in many raw waters in Japan, the current advanced drinking water treatment process focusing on the removal of humic substances (hydrophobic fraction) is, in a sense, scientifically irrelevant. In addition, unremoved hydrophilic compounds consume residual chlorine and affect the biological stability of water as an AOC component in distribution systems (Hammes *et al.*, 2006). Thus, not only due to the formation of DBPs, but also based on a microbiological and aesthetic perspective, it is

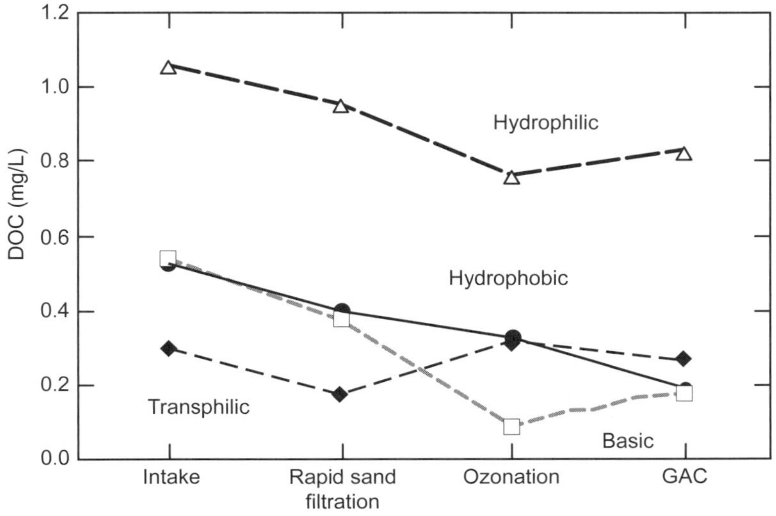

FIGURE 3-8. Example of a DOC profile in drinking water treatment with ozonation and GAC.

difficult to say that our current water treatment system is complete and scientifically ideal.

Given the discussion above, a so-called advanced water treatment process (ozone + GAC system) is not the goal, but rather a transit point to the next generation of advanced driking water treatment systems. One direction we should pursue is the combination of sufficient oxidation under the constraint of bromate ion formation and the removal of ionic species after oxidaiton through ion exchange. In this new treatment system, an advanced oxidation process (note that in the presence of excess hydrogen, peroxide bromate ion fomation is suppressed) converts dissolved organic matter to ionic compounds, and these ions are then effectively removed through an ion-exchange process. Thus far, an ion-exchange process has not been extensively used in the treatment of drinking water due to its cost, but more efficient sytems have been recently developed (Johnson and Singer, 2004). Given the fact that very small oranic compounds among DOM and ammonium ions are assciated with AOC, chlorine consumption, and chlorinous odor, the use of ion-exchange technology may be more effient than membrane treatment.

3.2.3 *Control of disinfection byproducts*

3.2.3.1 Disinfection byproducts (DBPs)

Since the first discovery of trihalomethanes in tap water in the early 1970s, many disinfection byproducts (DBPs) have been discovered in actual tap water and/or laboratory-scale experiments. Until today, approximately 700 related compounds have been identified (Richardson, 1998; Woo *et al.*, 2002; Krasner *et al.*, 2006). Since it is beyond the scope of this paper to cover all of these DBPs here, I will first address several important aspects of the overall DBP issue. I will then focus on common ozonation DBPs, bromate ion (BrO_3^-) in particular, and describe the technologies used for controlling bromate ion in Japan and the related ongoing research projects for better controlling ozonation DBPs. Finally, we will review the research projects on ozonation of DBPs and DBP studies related to ozonation in Japan. This paper mainly discusses ozonation in drinking water treatment, but the authors believe that most issues addressed here also apply to waste water treatment.

The disinfection byproducts regulated in Japan were listed previously in **Table 3-2**. Japan has one of the most stringent regulation systems for drinking water quality in the world. In addition to these compounds, many other compounds (e.g., other haloacetic acids containing bromine, haloacetonitriles, and *N*-nitrosoamine (NDMA)) are registered as monitoring items.

While extensive research efforts have been devoted to the characterization and control of DBPs, our knowledge on the chemical identities of DBPs remain very limited. Even for chlorinated DBPs, the most studied class of DBPs, major individual DBPs account for only 50% of the total organic

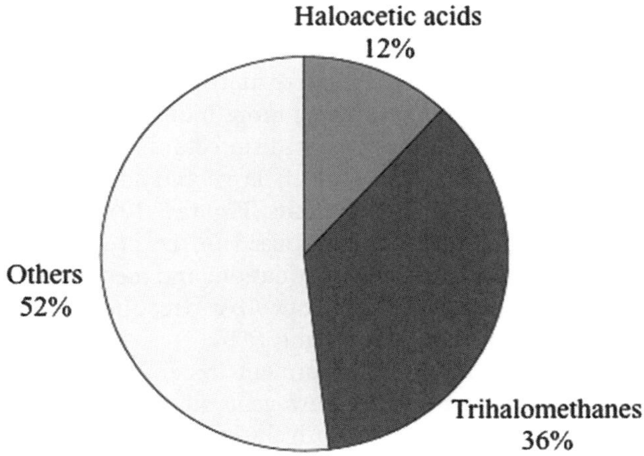

FIGURE 3-9. Contents of DBPs based on TOX (Zhang *et al.*, 2000).

halogen (TOX), as shown **Figure 3-9**. This percentage is smaller in DBPs from other disinfectants including ozone, and if evaluated on a carbon basis, would be even much smaller (Zhang *et al.*, 2000).

Due to its extensive use and the historical background of the DBP issue, halogenated, especially chlorinated, compounds have been a focus of related research and administrative efforts (this may be true for the problems of micropollutants in general). However, non-halogenated DBPs are also produced. Actually, oxidation reactions without the inclusion of halogen atoms are more common, and the presence of halogenation is rather exceptional even during chlorination (Larson and Weber, 1994).

When the four major disinfectants, chlorine, chloramine, ozone, and chlorine dioxide, are compared, the amount of DBPs is usually considered in the following order: chlorine > chloramines > chlorine dioxide > ozone. This is because the amount of DBPs is typically measured in terms of trihalomethane formation potential or TOX. However, this is not true for all types of DBPs. For example, more aldehyde would be produced from ozonation. Thus, the use of alternative disinfectants does not always reduce DBP formation. Rather, it changes the types of DBPs. In other words, the reaction between disinfectants and dissolved organic matter (DOM) is unavoidable no matter the disinfectant, as pathogenic organisms are also made of organic compounds (Itoh and Echigo, 2008).

3.2.3.2 Control of chlorination byproducts

The basic strategies for controlling disinfection byproducts are classified as follows: (1) optimization of the disinfectant dose, (2) removal of precursors, (3) use of alternative disinfectants, and (4) the removal of DBPs after formation. However, the removal of DBPs after formation is very difficult in general, and other options are usually taken. This is because, after the addition of an oxidizing agent as a disinfectant, an oxidative decomposition reaction of dissolved organic matter occurs, forming more hildrophilic compounds with lower molecular weight. The reactions of disinfectants and dissolved organic matter are a transformation from relatively large and adsorbable compounds to smaller and more hydrophlic compounds (**Figure 3-10**).

Table 3-10 shows the unit operations used in the precursory removal of DBPs. Coagulation, activated carbon, ozonation, and membrane filtration are common techniques. It is known that ozone-GAC treatment can reduce trihalomethanes and haloacetic acids by roughly 90%.

Thus, through ozone and GAC treatment systems, classical DBPs (e.g., trihalomethanes and haloacetic acids) are generally well controlled. The raw waters in Japan generally contain relatively low concentrations of organic compounds (measured as TOC or DOC) compared with other countries/regions. This is a favorable situation in terms of DBP control. However, for disinfection byproducts, which we know little about as mentioned earlier, further research efforts are needed to better understand the formation characteristics and mechanisms of DBPs, and to better manage the related risks.

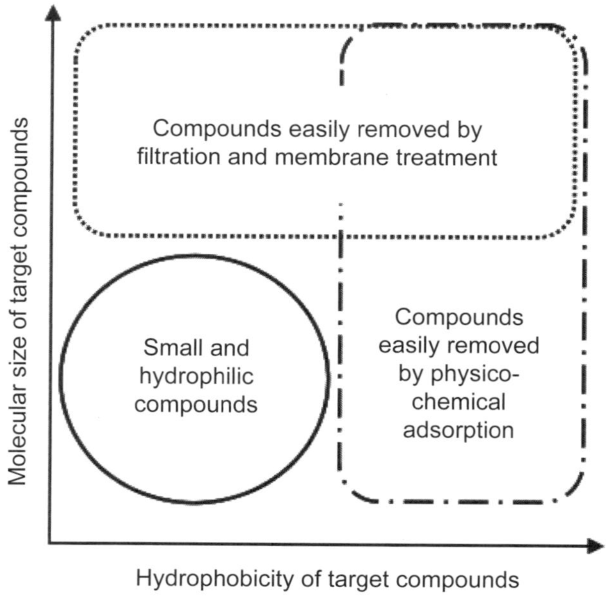

FIGURE 3-10. Removal of pollutants and their properties.

TABLE 3-10. DBP control through the treatment of drinking water (Itoh *et al.*, 2008).

Process	Effect	Ability	
		for THMs	for HAAs
Enhanced coagulation	Precursor removal	40–60%	40–60%
Powder activated carbon	Precursor removal	up to 60%	up to 60%
Granular activated carbon	Precursor removal	up to 80%	80%
	DBP removal		
Ozone-BAC	Precursor removal	80% or above	80% or above
Nanofiltration	Precursor removal	80% or above	80% or above
	DBP removal	up to 60%	
Biological treatment	Precursor removal	10–20%	20–30%
Intermediate chlorination	Minimization of DBPs	20–40%	20–40%
Chlorine dioxide	Minimization of DBPs	10–70%	10–50%
Chloramination	Minimization of DBPs	40–60%	40–60%
Lower chlorine dose	Minimization of DBPs	10–20%	20–30%

Recently in Japan, chlorate ion (ClO_3^-) was added to the list of standard items. Chlorate ion enters tap water primiarly as an ipurity of sodium hypochlorite, and is different from other DBPs. The use of highly pure sodium hypochlorite, a short storage time, and temperature control are known to be an effective methods for chlorate control.

3.2.3.3 Ozonation and disinfection byproducts

As mentioned above, ozonation has become an important technique to control chlorination DBPs, but ozoantion itself produces DBPs as it is also a strong oxidizing reagent. In the following section, the formation mechanism and control methods for ozone treatment of disinfection byproducts are reviewed. While we focus on ozonation byproducts, particularly bromate ion, keep in mind that a very similar discussion (i.e., the trade-off between DBP formation and loss of oxidation capability and/or importance of the total toxicity of water) can be applied to other types of DBPs.

(A) *Bromate ion*

Bromate ion (BrO_3^-) is a major ozonation byproduct, and controlling bromate ion is a major challenge in optimizing ozonation systems. The chemistry and kinetics of bromate ion formation in the absence of DOM are relatively well understood (e.g., von Gunten, 2003). In this sequential reaction, bromide ion (Br^-), the precursor in source water, is oxidized several times and converted into bromate ion. Both molecular ozone and hydroxyl radicals are involved in this reaction sequence. While various reaction pathways are possible, the following order is kinetically preferred under the reaction conditions commonly found in drinking water treatment practice: oxidation by molecular ozone, oxidation by hydroxyl radicals, a disproportionation reaction, and oxidation by molecular ozone (von Gunten, 2003).

Bromate ion formation tends to increase with an increase in pH, ozone dose, and bromide ion concentration. The presence of ammonium ion and DOC generally reduces bromate ion formation for fixed ozone doses. For example, bromate ion formation is roughly doubled when the pH is elevated from 7 to 8. DOC generally reduces bromate ion formation by consuming molecular ozone and hydroxyl radicals, but in some cases, certain DOC fractions act as radical promoters. For this reason, it is not surprising to observe significant differences among DOM isolates with respect to bromate ion formation for the same ozone dose and pH.

It is somewhat agreed that if bromide ion concentration in source water is above 100 μg/L, controlling bromate ion below 10 μg/L (the WHO guideline value and Japanese Standard) becomes difficult (von Guten, 2003). For a lower range, appropriate control of the ozone dose can be used to effectively control bromate ion. Common techniques for bromate control are summarized in **Table 3-11** (Acero, 2001; Pinkernell and von Gunten, 2001; Shiba *et al.*, 2001; Buffle *et al.*, 2004; Takada, 2004; Somiya *et al.*, 2006). Among these strategies, the control of ozone dose and aqueous phase ozone concentration are common in Japan. These options are relatively simple to install and have been proven successful. However, one should note that these options essentially limit the ozone dose. When the quality of the source water is high enough, these options are sufficient to meet all standards. However, these strategies may not be able to handle abrupt changes of odor compounds in source water (e.g., 2-MIB and geosmin).

The same thing can be said for other treatment options. That is, most of these strategies sacrifice (some of) the functions of ozonation. For example, if pH control is used, the oxidation of unsaturated micropollutants is suppressed along with bromate formation because fewer hydroxyl radicals are formed at a lower pH. For the same reason, hydrogen peroxide is not a good choice if one intends to use ozone as a disinfectant because a sufficient concentration of molecular ozone for disinfection is hard to achieve in the presence of hydrogen peroxide. Thus, an appropriate option for bromate ion control depends on the purpose of ozonation.

For optimizing ozonation processes, an appropriate evaluation of the CT values of ozone and hydroxyl radicals is also crucial. Traditionally, ozone CT values are determined in a conservative way. That is, an excessive amount of ozone tends to be introduced, resulting in higher bromate formation. Developing a simple but more reasonable way to estimate CT values under practical situations is a major challenge for the ozone industry.

TABLE 3-11. Summary of bromate control strategies.

Strategy	Mechanism	Remark	References
Dose control	Limits ozone available for bromate ion formation	Practically applied to drinking water treatment in Japan	Takada, 2004
Control of aqueous ozone concentration	Monitors an control ozone concentration and minimize ozone dose for achieving the purpose of ozonation	Practically applied to drinking water treatment in Japan	Shiba *et al.*, 2004
pH control	Suppresses radical chain reaction	• First system is being installed in Japan • 60% reduction is possible (pH 8 to 6)	Pinkernell and von Gunten, 2001
Ammonium addition	Seavenges hypobromous acid (HOBr), an intermediate	• Not common in Japan • 80% reduction possible • May results in trichloramine formation (odor compounds after chlorination)	Pinkernell and von Gunten, 2001
Chlorination/ chloramination	Converts bromide ion to organic bromines	• 85% reduction possible • Major drawback is potential formation of toxic organic bromines	Buffle and von Gunten, 2004
Hydrogen peroxide	Reduces HOBr back to bromide ion while promoting radical chain reaction	• 80% reduction possible while maintaining the oxidation capability • Not suitable when molecular ozone is required (e.g., disinfection)	Acero *et al.*, 2001, Somiya *et al.*, 2006

Bromate formation can be reduced through chlorination or chloramination prior to ozonation, as bromide ion is easily converted into HOBr, and HOBr is highly reactive to DOM. However, one should be careful when using this option, as organic bromine and organic chlorine (including trihalomethanes and haloacetic acids) are formed. In particular, organic bromine is known to be toxic in general, and we should pay serious attention to the formation of this class of DBP (Echigo *et al.*, 2004a). We suggest putting this option on hold until more information on the safety of this pretreatment is collected. In short, the use of chlorination/chloramination for controlling bromate ion should currently be avoided.

While the options listed in **Table 3-11** are mainly used to control bromate formation during ozonation (Amy and Siddiqui, 1999; Kimbrough and Suffet, 2002; Johnson and Singer, 2003; Sanchez-Polo *et al.*, 2006; Echigo *et al.*, 2006), bromide ion (i.e., the precursor of bromate ion) can be removed in various ways. Reverse osmosis, activated carbon treatment, oxidation by electrodes, adsorption to aerogels, and ion-exchange treatments have been applied toward bromide removal. Among these alternatives, ion-exchange is a generally well-established technology, and a promising one for bromate control. In addition, while it might also be possible to control the amount of bromide emitted from human activities, the origins of bromide ion is so diverse (e.g., foods) that regulating a limited number of industiries would not be an effective solution (Miyagawa, 2006).

(B) *Effects of ozonation on the total toxicity of chlorinated water*

As mentioned earlier, controlling bromate ion is one of the key issues when applying ozonation to the treatment of drinking water. However, it is notable that ozonation greatly affects the formation characteristics of other DBPs. This aspect is particularly important in Japan where the use of chlorine as the final disinfectant is mandatory. That is, in addition to bromate ion, it is essential to evaluate the role of ozonation with respect to the total toxicity of finished water. **Figure 3-11** is a good example to illustrate this point (Echigo *et al.*, 2004b). In this experiment, we ozonated, and then chlorinated, a high concentration of humic acid solution in the presence of bromide ion. For this experiment, a high concentration of humic acid solution was used in order to run a chromosomal aberration test without a sample concentration process for an evaluation of the toxicity of the reaction byproduct mixture.

Figure 3-11 clearly shows that even with bromide ion (i.e., a condition under which bromate ion formation is possible), a pretreatment through ozonation reduced the total toxicity after chlorination. That is, under our experimental condition, the positive effect of ozonation (reduction of toxic chlorination byproducts) was much greater than the drawback (formation of bromate ion). Of course, this result is just one example of a bioassay, but suggests a very important point of view. Actually, USEPA is currently running a more comprehensive research project on the toxicity of DBP mixtures.

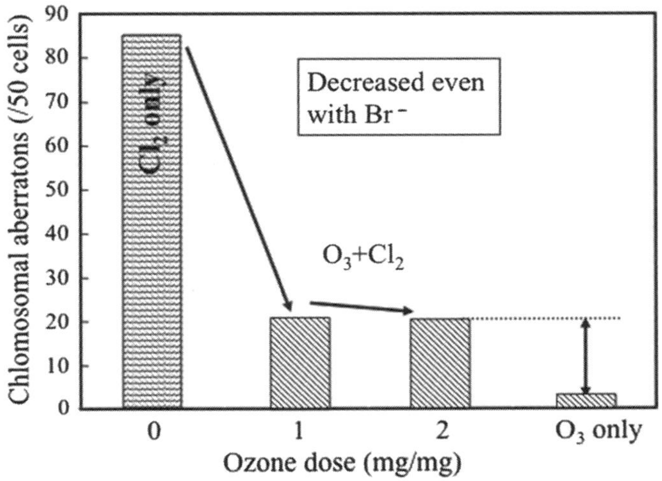

FIGURE 3-11. Effects of ozonation on the total toxicity after chlorination (Conditions: Humic acid, 750 mg C/L; Br^-, 37.5 mg/L; time, 1 day; temp., 20°C; chlorine dose, 1,500 mg Cl/L; pH, 7.0) (Echigo *et al.*, 2004).

In this project, not only carcinogenicity, but also other types of toxicity (e.g., immunotoxicity), are being evaluated (Simmons *et al.*, 2002; 2004).

Itoh, *et al.*, (2011) reviewed the results of studies on the overall toxicity of disinfected water and demonstrated the presence of toxicity in disinfected water that cannot be attributed to the currently regulated by-products. This confirms the importance of estimating the overall toxicity of drinking water. Standard values are never sufficient as golden rules as far as DBPs are concerned. International organizations and authorities charged with reviewing and revising national drinking water standards should collect information on the overall toxicity of disinfected water. Obtaining useful information for actual regulation depends on the progress and success of *in vivo* bioassays that can be used to derive health-based values. Therefore, *in vivo* assays with experimental animals should be given a higher international priority (Itoh *et al.*, 2011).

3.2.4 *Recent topics on disinfection byproducts*

Recent topics on emerging DBPs are listed in **Table 3-12**. Carboxylic acids are not toxic in general, but account for a considerable fraction of AOC (assimilable organic carbon) (Hammes *et al.*, 2006). AOC is a key index to determine the biological stability in distribution systems. In Japan, the possibility of a water supply with less residual chlorine is being investigated. Carboxylic acids are important ozonation byproducts in this context.

TABLE 3-12. Emerging DBPs.

Class	Example	Remark	Reference
Non-halogenated DBPs	Carboxylic acids	Biological stability of finished water	Hammers *et al*., 2006
Iodinated haloacetic acids	Iodoacetic acid	More toxic than bromoacetic acid	Plewa and Wagner, 2004
	Dichloroiodomethane	Frequent occurrence	Cancho *et al*., 2000
DBPs containing nitrogen	NDMA	Highly toxic	Oya *et al*., 2008
	Haloaectonitriles	From basic fraction of	Lee *et al*., 2007
Taste and odor compounds	Trichloramine	Chlorinous odor	Kajino *et al*., 1999
Inpurity in disinfectant	Chlorate ion	Regulated from 2008	Itoh and Echigo, 2008

Some iodinated haloacetic acids are known to be toxic (Plewa and Wagner, 2004), but virtually no information is currently available on their occurrence in Japan. While the total iodine concentration in the source water in Japan is not very high (a few μg/L at most), a more detailed survey is necessary given their toxicity.

N-nitrosodimethylamine (NDMA) has been added to the list of monitoring items in Japan. NDMA is commonly known to form in chloramination and chlorination. However, a recent study revealed that NDMA could be formed during the ozonation of source water in Japan (Oya *et al*., 2008).

As mentioned previously, the aesthetic aspect (i.e., taste and odor) of drinking water is another point of concern. Chlorination is known to produce odorous compounds. In a broader sense, these compound are classified as DBPs. The precursors of chlorinous odor are not fully identified. Trichloramine has been proposed as a possible compound causing chlorinous odor (Kajino *et al*., 1999), but there are multiple reaction pathways for trichloramine formation. Our current knowledge is not sufficient for deciding which pathway is dominant under conditions commonly found in the treatment of drinking water. Also, the effect of ozonation on chlorinous odor is not clear. Currently, several research projects on this issue are underway, both in our research group and in collaborating institutions.

3.3 SECURING A SUSTAINABLE URBAN WATER SUPPLY

3.3.1 *Water safety plan*

Water quality standards are the fundamental framework for ensuring the safety of drinking water. However, it has been recognized that setting the WHO guidelines for drinking water quality and national drinking water standards is not sufficient to ensure the safety of drinking water. For example, in Bangladesh, a wide range of arsenic contamination from geological origins has been found, and cases of adverse health effects related to this contamination have been reported. Also, raw water contamination due to pathogens such as *Cryptosporidium* has been sporadically reported. For these cases, simply requiring compliance with established drinking water standards does not provide a proper solution.

Under these circumstances, the third edition of the WHO Drinking Water Quality Guidelines, published in 2004, has proposed a 'Water Safety Plan' (WSP) for securing a supply of safe water (Ministry of Health, Labour, and Welfare, 2008). In this plan, the concept of HACCP (Hazard Analysis and Critical Control Point), an idea already established in the food industry, was introduced, and hazard analysis and management are performed at all stages of water supply, from the source to the tap.

In response to this revision, the Japanese Ministry of Health and Welfare issued its 'Guidelines for Water Safety Plan' in 2008. **Figure 3-12** illustrates the flow of the planning and its operation.

This water safety plan consists of three elements: (1) the evaluation of water supply systems, (2) the configuration of management measures, and (3) the implementation of the plan. The first step is the organization of a team to establish and promote the WSP. This team should (1) survey the current status of the target water supply system, extract hazards in the system, set the proper risk levels, and perform a hazard analysis. Next, (2) for an effective management configuration, the team should establish management measures, monitoring protocols, and management standards for each hazard. This configuration will be the plan used for normal operation. Finally, (3) the team should determine the rules and protocols for executing the WSP following the PDCA cycle (Plan = planning, Do = implementation, Check = regular inspections, Act = improvement) as well as the document management rules. Also, for the procedures used in emergency situations in which the management rules and standards are violated, it is necessary to evaluate their effectiveness on a regular basis. The WSP itself should be

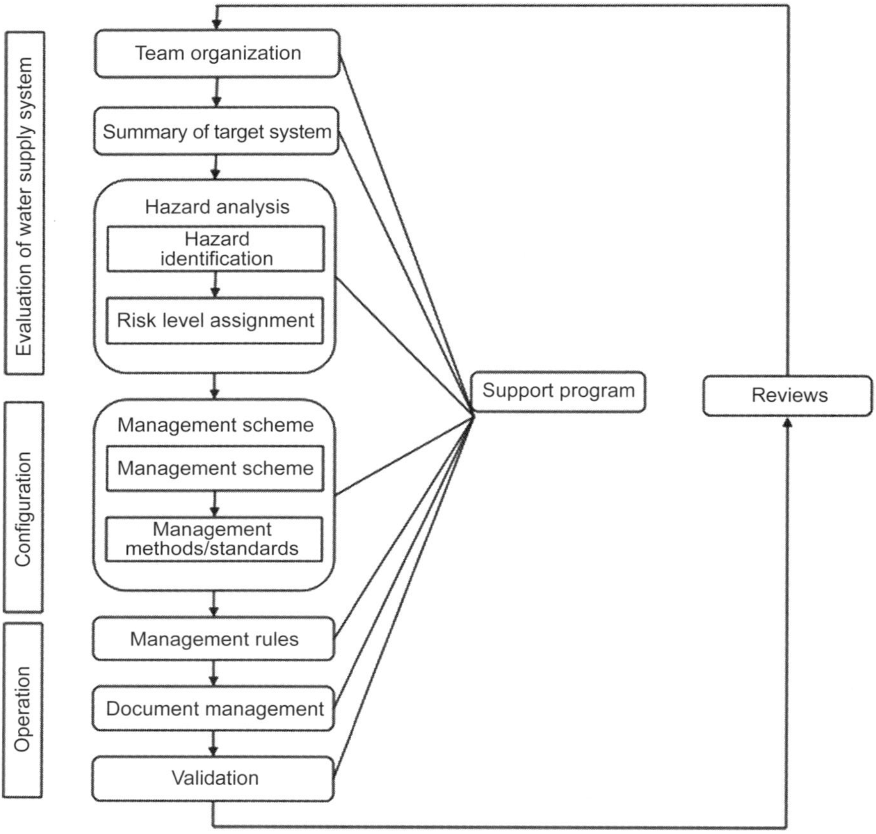

FIGURE 3-12. Concept of a water safety plan.

regularly reviewed as well. Note that the WSP has to include a description of the technical relevance of each element of the plan.

In a hazard analysis, which is the main task in the evaluation process of water supply systems, the frequencies of the events extracted as hazards are determined. For this purpose, an analysis of events of relatively high concentrations based on reference standards and the experiences of operators and other workers will be useful. The impact of each event is then set. If the event is a violation of the water quality standards or a similar situation, several different levels of impact should be assigned depending on the observed concentration.

TABLE 3-13. Example of a risk level matrix (Ministry of Health, Labour, and Welfare, 2008).

		Impact of hazards				
		Negligible	Minor	Major	Serious	Extremely serious
Frequency of hazardous events						
Frequent	Every month	1	4	4	5	5
Common	Once in several months	1	3	4	5	5
Occasionally	Once in several years	1	1	3	4	5
Unlikely	Once in 3 to 10 years	1	1	2	3	5
Rare	Once in more than 10 years	1	1	1	1	5

Based on these considerations, as shown in **Table 3-13**, the risk levels are set in the form of a risk level matrix. If a hazardous event has a great impact and very low frequency, it should be considered a serious event for drinking water. In this matrix, events with low impact are classified as Level 1 regardless of the frequency. On the other hand, events with tremendous impact but low frequency are categorized as Level 5.

Another example of risk management is the use of software called MaRisk A (van Lieverloo *et al.*, 2006; Medema and Smeets, 2009). This software was developed and is used in the Netherlands. The concept of this software is similar to those in **Table 3-13** and the WHO guidelines, in which scores are calculated for the extracted hazardous events from the water source to the tap. The scores are given in the form of frequency and severity of the product. The sum of these scores is the overall risk of a water supply system. If certain measures are taken, the overall score is reduced. This allows us to calculate the cost and benefit of these measures. This evaluation scheme will make it possible to discuss the priorities and cost-effectiveness of such measures.

Water utilities in Japan are setting their water safety plans based on their own particular situations. These WSPs are expected to be implemented in an effective way, and to contribute toward the stable supply of safe drinking water.

REFERENCES

Acero, J.L., *et al.*, (2001), MTBE oxidation by conventional ozonation and the combination ozone/hydrogen peroxide: efficiency of the processes and bromate formation, *Environmental Science and Technology,* 35, pp. 4252–4259.

Amy, G.L., *et al.*, (1999), *Strategies to Control Bromate and Bromide*, American Water Works Association Research Foundation Final Project Report, Denver.

Aikawa, M., *et al.*, (1997), *Modern Infectious Diseases*, Iwanami, p. 222.

Buffle, M.-O., *et al.*, (2004), Enhance bromate control during ozonation: the chlorine-ammonia process, *Environmental Science and Technology,* 38, pp. 5187–5195.

Cabinet Office (2008), *National Survey on Water.*

California Department of Public Health (2006), *A Brief History of NDMA Findings in Drinking Water.*

Cancho, B., *et al.*, (2000), Determination, synthesis and survey of iodinated trihalomethenes in water treatment processes, *Water Research*, 34(13), pp. 3380–3390.

Douglas, I. (2009), Using Quantitative microbial risk assessment (QMRA) to optimize drinking water treatment, *American Water Works Association Water Quality Technology Conference*, Washington State Convention & Trade Center, Seattle, Washington, USA.

Drescher, A.C., *et al.*, (2001), Cryptosporidium inactivation by low-pressure UV in a water disinfection device, *Journal of Environmental Health*, 64(3), pp. 31–35.

Echigo, S., *et al.*, (2004a), Toxicity of chlorinated water in the presence of bromide ion: contribution of brominated disinfection by-products to the toxicity of chlorinated water, *Environmental Engineering Research*, 41, pp. 279–289.

Echigo, S., *et al.*, (2004b), Contribution of brominated organic disinfection by-products to the mutagenicity of drinking water, *Water Science and Technology*, 50(5), pp. 321–328.

Echigo, S., *et al.*, (2006), Bromide removal by hydrotalcite-like compounds in a continuous system, *Water Science and Technology*, 56(11), pp. 117–122.

European Union (1998), Council Directive 98/83/EC on the quality of water intended for human consumption, *Official Journal of the European Communities*, 330, pp. 32–54.

Fewtrell, L., *et al.*, (2003), *Assessment and Management of Risks of Waterborne Infections,* Gihoudou, p. 434.

Haas, C.N., *et al.*, (2001), *Quantitative Microbial Risk Assessment* (translated by Mitsumi Kaneko into Japanese), Gihoudou, p. 452.

Hammes, F., *et al.*, (2006), Mechanistic and kinetic evaluation of organic disinfection by-product and assimilable organic carbon (AOC) formation during the ozonation of drinking water, *Water Research*, 40(12), pp. 2275–2286.

Havelaar, A.H., *et al.*, (2003), Balancing the risk and benefits of drinking water disinfection: Disability adjusted life-years on the scale. *Environmental Health Perspectives,* 108(4), pp. 315–321.

Havelaar, A.H., *et al.*, (2003), *Quantifying Health Risks in the WHO Guidelines for Drinking Water Quality*. A burden of disease approach. Report 734301022, RIVM, Bilthoven, The Netherlands.

Health Canada (2007), *Guidelines for Canadian Drinking Water Quality.*

Health Science Council (2003a), *Revision of Drinking Water Quality Standards* (in Japanese).

Health Science Council (2003b), *Summary on the Revision of Drinking Water Quality Standards* (in Japanese).

Hosaka, M. (2007), 'Contamination of drinking water and water environments by *Cryptosporidium* and *Giardia*' (in Japanese), *Journal of the National Institute of Public Health*, 56(1), pp. 24–31.
Imai, A., *et al.*, (2003), Trihalomethane formation potential of dissolved organic matter in a shallow eutrophic lake, *Water Research*, 37(17), pp. 4284–4294.
Itoh, S. (2013), Effect of the ratio of illness to infection of *Campylobacter* on the uncertainty of DALYs in drinking water, *Journal of Water and Environment Technology*, 11(3), pp. 209–224.
Itoh, S., *et al.*, (2011), Regulations and perspectives on disinfection by-products: Importance of estimating overall toxicity, *Journal of Water Supply: Research and Technology-Aqua*, 60(5), pp. 261–274.
Itoh, S. (2010a), Management of water supply system without using chlorine in the Netherlands, *Journal of Japan Water Works Association*, 79(10), pp. 12–22, (in Japanese).
Itoh, S. (2010b), Effects of water consumption data on sensitivity analysis in quantitative microbial risk assessment, *Journal of Water and Waste*, 52(8), pp. 55–65.
Itoh, S., *et al.*, (2008), *Disinfection Byproducts* (in Japanese), Gihoudou Shuppan, p. 325.
Itoh, S., *et al.*, (2007), Psychosocial considerations on strategies for improving customers' satisfaction with tap water based on causal modeling, *Journal of Japan Water Works Association*, 76(4), pp. 25–37, (in Japanese).
Japan Water Works Association (2007), *Statistics on Water Supply,* 89.
Jo, I., *et al.*, (2008), Importance of hydrophilic dissolved organic matter on haloacetic acid formation, *The 59th National JWWA Conference*, pp. 534–535.
Johnson, C.J., *et al.* (2004), Impact of a magnetic ion exchange resin on ozone demand and bromate formation during drinking water treatment, *Water Research*, 38(17), pp. 3738–3750.
JWRC (2005), *e-Water Guidelines* (in Japanese), pp. 1–109.
Kajino, M., *et al.*, (1999), Odors arising from ammonia and amino acids with chlorine during water treatment, *Water Science and Technology*, 40(6), pp. 107–114.
Kaneko, M. (1996), *Water Sanitary* (in Japanese), Gihoudou, p. 579.
Kaneko, M. (1997), *Water Disinfection*, Japan Environmental Education Center, p. 401.
Kaneko, M. (2006), *Controlling Water Infection through Water Supply*, Maruzen, p. 255.
Krasner, S., *et al.*, (2006), Occurrence of a new generation of disinfection byproducts, *Environmental Science and Technology*, 40(23), pp. 7175–7185.
Larson, R.A., *et al.*, (1994), *Organic Reactions in Environmental Chemistry,* CRC Press, Boca Raton, FL, p. 450.
Lee, W., *et al.*, (2007), Dissolved organic nitrogen as a precursor for chloroform, dichloroacetonitrile, *N*-nitrosodimethylamine, and trichloronitromethane, *Environmental Science and Technology*, 41(15), pp. 5485–5490.
Matsui, Y., *et al.*, (2010), *Report on integrated risk management of drinking water.*
McCutcheon, R., *et al.*, (2006), Desalination by a novel ammonia-carbon dioxide forward osmosis process: Influence of draw and feed solution concentrations on process performance, *Journal of Membrane Science*, 278, pp. 114–123.
Medema, G., Loret, J.F., Stenstrom, T.-A., and Ashbolt, N. (eds.) (2006), *MICRORISK, Quantitative Microbial Risk Assessment in the Water Safety Plan.*
Medema, G., *et al.*, (2009), Quantitative risk assessment in the Water Safety Plan: case studies from drinking water practice, *Water Science and Technology: Water Supply-WSTWS*, 9(2), pp. 127–132.

Ministry of Health, Labour and Welfare (2007a), *Guidelines for Cryptosporidium Control in Water Supply* (in Japanese).
Ministry of Health, Labour, and Welfare (2007b) http://www.mhlw.go.jp/topics/bukyoku/kenkou/suido/jouhou/suisitu/pdf/o5.pdf.
Ministry of Health, Labour and Welfare (2007c), *Control of Cryptospridium in Drinking Water* (Document 0330005).
Ministry of Health, Labour, and Welfare (2008), *Guidelines on Water Safety Plan*, p. 70.
Ministry of Health, Labour and Welfare (2009), http://www.mhlw.go.jp/topics/bukyoku/kenkou/suido/jousui/01.html
Ministry of Health of the People's Republic of China and Standardization Administration of China (2006), *Standards for Drinking Water Quality* (GB 5749-2006).
Murray, C.J.L., *et al.*, (1996), *The Global Burden of Disease and Injury Series.* vols. 1, 2, Harvard School of Public Health on behalf of the World Health Organization and The World Bank, Cambridge, MA.
Nakanishi, J., *et al.*, (2003), *Exercise in Environmental Risk Analysis* (in Japanese), Iwanami Shoten, p. 230.
Okamoto, Kouichi (1992), *Introduction to Risk Psychology* (in Japanse), Saiensu-Sha, p. 161.
Ontario Ministry of the Environment (2006), *Technical Support Document for Ontario Drinking Water Standards, Objectives and Guidelines* (PIBS4449e01).
Oya, M., *et al.*, (2008), NDMA formation potentials from waters processed by advanced drinking water treatment, river water, and waste water treatment, *The 59th National JWWA Conference*, pp. 542–543.
Pinkernell, U., *et al.*, (2001), Bromate minimization during ozonation: mechanistic considerations, *Environmental Science and Technology*, 35, pp. 2525–2531.
Plewa, M.J., *et al.*, (2004), Chemical and biological characterization of newly discovered iodoacid drinking water disinfection byproducts, *Environmental Science and Technology*, 38(18), pp. 4713–4722.
Richardson, S. (1998), Drinking water disinfection by-products, *Encyclopedia of Environmental Analysis and Remediation* (in) Meyers, R.(ed.), 3, Wiley-Interscience, pp. 1399–1422.
Sanchez-Polo, M., *et al.*, (2006), Bromide and iodide removal from waters under dynamic conditions by Ag-doped aerogels, *Journal of Colloid Interface Science*, 306, pp. 183–186.
Sano, D., *et al.*, (2006), Contamination of river and sea water with pathogenic viruses: case of norovirus investigations, *Journal of Japan Society on Water Environment*, 29(3), pp. 130–134 (in Japanese).
Shiba, M., *et al.*, (2001), Management of ozonation by dissolved ozone monitor, *The 55th National JWWA Conference*, pp. 256–257.
Simmons, J.E., *et al.*, (2002), Development of a research strategy for integrated technology-based toxicological and chemical evaluation of complex mixtures of drinking water disinfection byproducts, *Environmental Health Perspectives*, 110(6), pp. 1013–1024.
Simmons, J.E., *et al.*, (2004), Component-based and whole-mixture techniques for addressing the toxicity of drinking-water disinfection by-product mixtures, *Journal of Toxicology and Environmental Health*, Part A, 67, pp. 741–754.
Smeets, P., *et al.*, (2009), The Dutch secret: how to provide safe drinking water without chlorine in the Netherlands, *Drinking Water Engineering and Science*, 2, pp. 1–14.

Society on Ozonation (1997), *Water Reclamation by Ozonation*, Sanyu (in Japanese).
Somiya, I., *et al.*, (2006), Removal of odor compounds by advanced oxidation processes, *The 57th National JWWA Conference,* pp. 254–255.
Staatsblad van het Koninkrijk der Nederlanden (2001), p. 53.
Takada, Y. (2004), Bromate ion formation in Murano water purification plant, *The 55th National JWWA Conference,* pp. 614–615.
USEPA (1989), National Drinking water regulations, filtration disinfection, turbidity, *Giardia lamblia*, viruses, *Legionella*, and heterotrophic bacteria, Final rule, 40 CFR parts 141 and 142, *Federal Register*, 54, p. 27486.
USEPA (2005), *Guidelines for Carcinogen Risk Assessment*, EPA/630/P-03/001F.
USEPA (2006), National Primary Drinking Water Regulations: Stage 2 Disinfectants and Disinfection Byproducts Rule; Final Rule, *Federal Register*, 71(2), pp. 388–493.
van Lieverloo, J.H.M., *et al.*, (2006), Risk assessment and risk management of faecal contamination in drinking water distributed without a disinfectant residual, *Journal of Water Supply: Research Technology -AQUA*, 55(1), pp. 25–31.
von Gunten, U. (2003), Ozonation of drinking water: Part II. Disinfection and by-product formation in presence of bromide, iodide or chlorine, *Water Research*, 37(7), pp. 1469–1487.
Woo, Y.T., *et al.*, (2002), Use of mechanism-based structure-activity relationships analysis in carcinogenic potential ranking for drinking water disinfection by-products, *Environmental Health Perspectives*, 110, pp. 75–87.
World Health Organization (2006), *WHO Guidelines for the Safe Use of Wastewater, Excreta and Greywater*, Volume I-IV, World Health Organization.
World Health Organization (2011), *Guidelines for Drinking-Water Quality, Fourth Edition*, World Health Organization, Geneva, Switzerland.
World Health Organization (2008a), *The Global Burden of Disease, 2004 Update*, World Health Organization, p. 146.
World Health Organization (2008b), *WHO Initiative to Estimate the Global Burden of Foodborne Diseases, A summary document.*
Yamada, T., *et al.*, (2007), Waterborne health hazard cases in the last 10 years, *Journal of the National Institute of Public Health*, 56 (1), pp. 16–23, (in Japanese).
Yanagawa, T. (2002), *Environment & Health Data*, Kyoritsu Shuppan, p. 201 (in Japanese).
Yano, K. (2006), Viral pollution of metropolitan water and shellfish, *Journal of Japan Society on Water Environment*, 29(3), pp. 124–129, (in Japanese).
Yoshida, K., *et al.*, (2006), *Introduction to Environmental Risk Analysis* (Chemical Compounds), Tokyo Tosho, p. 243, (in Japanese).
Zhang, X., *et al.*, (2000), Characterization and Comparison of Disinfection By-Products of Four Major Disinfectants, *Natural Organic Matter and Disinfection By-Products: Characterization and Control in Drinking Water,* in Barrett, S.E., Krasner, S.W. and Amy, G.L. (eds.), American Chemical Society, Washington, DC, pp. 299–314.

4

Control and Management of Municipal Wastewater for Sustainable Development

4.1 FRAMEWORK FOR THE CONTROL AND MANAGEMENT OF MUNICIPAL WASTEWATER

4.1.1 *Source and characteristics of municipal wastewater*

Wastewater generated in a city has diverse sources and can be basically categorized into domestic or industrial wastewater. These wastewaters are treated at a centralized treatment plant such as a municipal wastewater treatment plant or treated on-site using a septic tank. Treated wastewater is then discharged into the aquatic environment. In addition, water from roofs, roadways, agricultural land, urban green spaces, and overflows from rain or snowstorms combined with sewage are additional sources of the wastewater.

Sources of domestic wastewater include not only drinking water but also household effluent from cooking, washing, bathing, cleaning, toilets, and aspersion; commercial wastewater from restaurants, department stores, hotels, pools, and offices; and municipal wastewater from public facilities such as public restrooms. Wastewaters contain various kinds of materials whose concentrations and particle sizes vary depending on the discharge time and materials eluted from pipes or drains. The materials contained in domestic wastewater can be classified into dissolved and solid substances. The latter materials can be divided into suspended and settleable solids, both of which include organic and inorganic materials. In addition, organic as well as volatile materials can be further divided into persistent and biodegradable organic materials. The target materials in conventional sewage treatment are settleable solids for primary treatment, and suspended solids and biodegradable organic materials for secondary treatment. Inorganic materials contained in dissolved substances are usually removed at lesser amounts from conventional treatment. The main components of organic materials are sugar, protein, fat, and surfactants, whereas the main components of inorganic materials included issoluble salts (Na^+, K^+, Cl^-, etc.), ammonia, and phosphates. In a heavily populated area, the contribution of pollutants, especially nutrient salts, is rather high in domestic wastewater even after conventional treatment at a wastewater treatment plant. For example, in Tokyo Bay, about 65% of nitrogen and 62% of phosphorus as annual loading volumes are discharged from wastewater treatment plants (**Figure 4-1**). Therefore, the removal of these nutrients through advanced processes is now ongoing.

The properties of industrial wastewater vary depending on the type of industry, but the quality is regulated based on the effluent standard discussed below. In Japan, strong acids, contaminants including cyanides, fats and oils, harmful materials, and heavy metals are strongly regulated in the water discharged from wastewater treatment plants to generate levels sufficient for satisfying the Water Pollution Control Act. This regulation was intended to prevent the erosion of concrete buildings, outbreaks of poisonous gases, the blocking of pipes, and a decrease in the treatment ability.

In urban areas, drainage from roads and roofs include deposition materials from the air, floating substances produced from factories or vehicle emissions, pavement materials, or tires. These substances also greatly affect the properties of rainwater drainage in an urban area in rainy weather. Although storm water is not a target for treatment in a separate sewer system, the contaminants included in the water sometimes need special attention for the protection of receiving waters such as lakes and enclosed bays.

Agricultural wastewater is generated particularly from artificial ditches used for rice paddy fields and their effluent. In paddy fields, the nitrogen is partly consumed; however, in many cases, the erosion of nutrients, pesticides, and solid materials including soil occur along with seasonal changes. However,

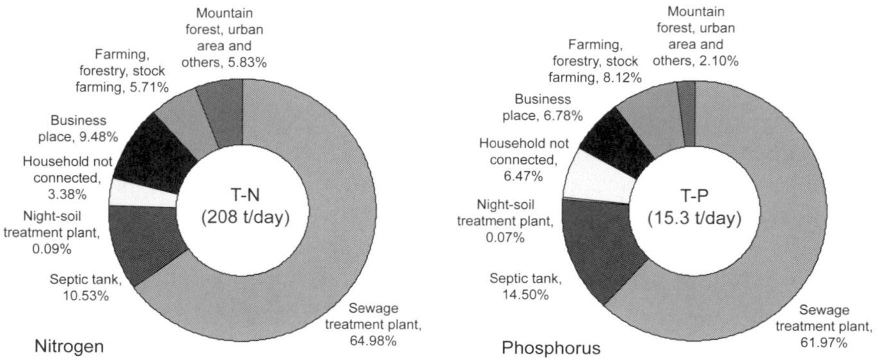

FIGURE 4-1. Loading amounts of nitrogen and phosphorus, in Tokyo Bay (JSWA, 2009b).

agricultural wastewater is not usually collected by public sewers as target wastewater for sewage treatment. In the case of urban farmlands, agricultural wastewater is generated from underground seepage and soil runoff after severe rains. These wastewaters include wasted plants and compost materials and agricultural chemicals, nutrients, and organic materials produced after farming. A combined sewer system may sometimes receive urban storm water that has an agricultural origin.

4.1.2 *Municipal wastewater management system and watershed management*

The Water Pollution Control Act established in 1970 started regulating the concentration of discharge for targeting the public water area. Discharge standards for harmful materials were set to about ten-times higher than the standard levels for the environment to maintain human health. These standards were determined by taking the dilution of public water from rivers into consideration. Uniform standards for living environments are used to regulate the average daily levels of biochemical oxygen demand (BOD), chemical oxygen demand (COD), and suspended solid (SS). The discharge standards for nitrogen and phosphorous in public water effluent are set by the Minister of Environment for only special lakes, enclosed bay areas, and water discharge for levels high enough to cause remarkable phytoplankton growth. Standard regulations for harmful materials were adapted for factories and businesses with discharge amounts larger than 50 m^3/day. However, regulations for BOD, COD, SS, T-N, and T-P levels of less than 50 m^3/day have are not been adopted. The 50 m^3/day restriction can be lowered by each local government. In addition, when an uniform regulation is not adequate to prevent pollution, a lower standard for harmful materials permits special cases in which stricter levels need to be set by each prefecture.

On the other hand, for discharge from small factories and service and residential areas, where a uniform regulation is not adequate, the development of sewage and septic tank systems has been promoted by the national and local governments. The maintenance of water flow and the remediation of sediment in rivers and channels have also been conducted. As a result of these countermeasures, river water quality has been improved, and the rate of achievement from environmental quality standards for BOD has reached as high as 90%. However, the rate of achievement for COD is still as low as 50% and 70% for lakes and seas, respectively. This tendency has been significantly observed in Tokyo Bay, Ise Bay, and Seto Inland Sea where the water areas are enclosed and the population density and industry are high (**Figure 4-2** (Water Pollution Control and Policy Management in Japan, 2009)). To improve the water quality in enclosed water areas, the regulation of simple concentrations

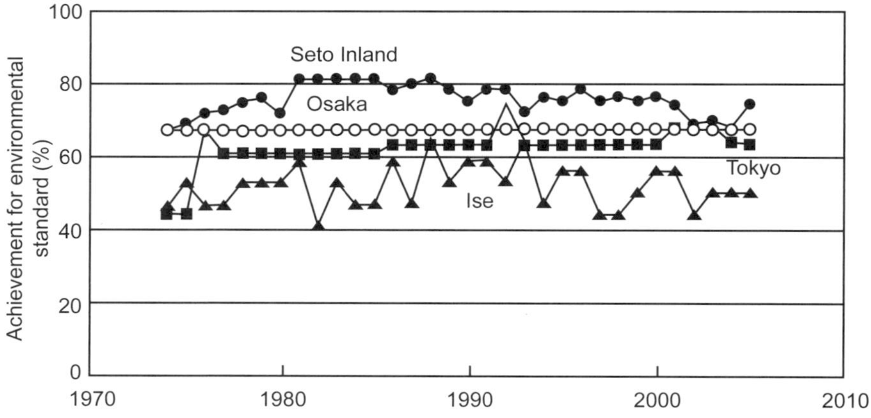

FIGURE 4-2. Transition of achievement for environmental standards for COD in bay areas (JSWE, 2009).

is not sufficient to prevent pollution, and the establishment of a systematic water quality control over the governance line is necessary to reduce the total amount of pollutants, including the internal load, point source, and human sewage. For this reason, the total amount regulation of pollutants was set for Tokyo Bay, Ise Bay and Seto Inland Sea (in Seto Inland Sea, the Law Concerning Special Measures for Conservation of the Environment of the Seto Inland Sea was set for industrial wastewater in 1973). The desired reduction in the total load includes a consideration of the future increase and decrease in the population, and the development of sewage technologies and sewage systems, which is the sixth regulation on the total load, was implemented until 2009. In some water areas, the water quality was improved. However, since the water quality was not completely improved, a seventh regulation on the total load aimed at 2014 is now under consideration. Although regulations for the total amount were carried out by the Laws Concerning Special Measures for Conservation of the Environment of the Seto Inland Sea and by the Water Pollution Control Act, in some lakes and marshes where the water quality is still higher than the standard levels for a water environment, additional regulations were carried out for COD, nitrogen and phosphorus from the specified business from where the total wastewater are discharged larger than 50 m^3/day, and ten lakes including Kasumigaura and Biwa have been selected as representative locations (Law Concerning Special Measures for Conservation of lake Water Quality).

As described above, for the important preservation of water environments, various actions to regulate effluent from various sources have been enacted by central, prefectural, and municipal governments. In addition, to reduce the amount of pollutant loading from cities, businesses, and the public with different levels of diversity, various countermeasures are being carried out by individuals, businesses, industries, agricultural sectors, and river and water management administrators. From this viewpoint, the sewerage system as regulated by the Sewerage Law, amended in 1970, aims at the preservation of water quality, and a comprehensive project, the Comprehensive Basin-wide Planning of Sewerage Systems is going to be developed for each water body. This plan aims at predicting the current and future loading sources in all water areas, estimating the amount of decrease in loading to achieve the standard for environmental preservation, and finally allocating the amount of loading reduction for each sector. The amount of loading that should be reduced by the sewerage system is determined based on the most rational improvement project available in sewage management. This comprehensive plan intends to develop a sewerage system for each river basin by evaluating the effectiveness of the operation by local governments. Further, the time span, area, and method of operation of the plan are being discussed to develop a future sewerage system for local governments by taking into consideration natural conditions such as the rainfall amount, volume of river flow, an understanding and prediction of the amount of sewage and water quality, the costs required to conduct a maintenance of sewerage system, and the effectiveness of the reduction in the amount of pollution.

Therefore, the Comprehensive Basin-wide Planning of Sewerage Systems is important for determining the appropriate sewerage systems in terms of size and treatment level, including advanced treatment of river basins where environmental water quality standards need to be immediately achieved. If the Comprehensive Basin-wide Planning of Sewerage Systems is completed, each public or regional sewerage system designated in the Sewerage Law will need to follow the planning shown in **Figure 4-3**.

4.1.3 *Legal system for municipal wastewater management*

The basis of the environmental policy of Japan consists of the Basic Environment Act. This law was derived from the Basic Law for Environmental Pollution Control (established in 1967) which was established as a countermeasure for seven typical kinds of pollution in Japan (air, water and soil pollution, ground subsidence, noise, vibrations, and offensive odors). There are some important differences between the Basic Law for Environmental Pollution Control and the Basic Environment Act (1993): the incorporation of an ecosystem protection concept and attention to the construction of a low-impact society and global environmental conservation in light of a future global environment design.

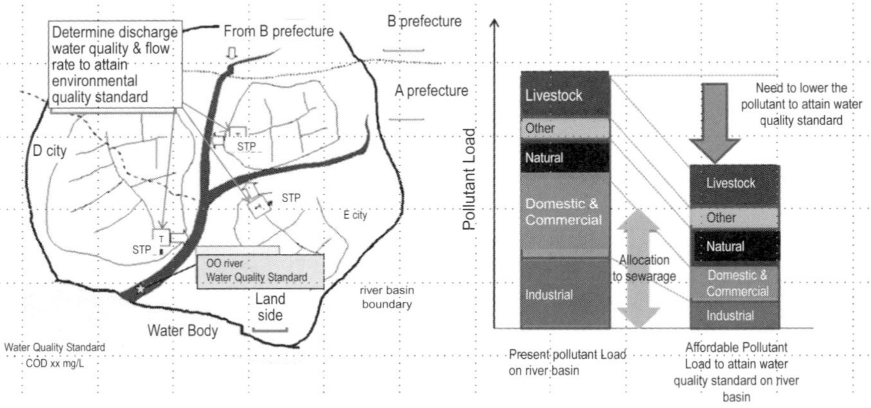

FIGURE 4-3. Concept of comprehensive basin-wide planning of sewerage systems (MLIT, HP).

In the Basic Environment Act, environmental standards related to water pollution are set, which are goals expected to be maintained for human health protection and conservation of our living environment. To attain these goals, effluent standards are set by the Water Quality Pollution Control Act for factories and businesses. For countermeasures to domestic wastewater pollution, sewerage systems have been introduced by the Sewerage Law. Furthermore, the Law Concerning Special Measures for Conservation of lake Water Quality, and the Water Supply Act, provide the conservation of lake environments and public environmental health, respectively (**Figure 4-4** (Water Pollution Control and Policy Management in Japan, 2009)).

Environmental standards for water quality have been instituted to protect water quality for water supplies, fisheries, agriculture, and industry. For water quality items related to the living environment, such as BOD and SS, standards are set depending on the type of water area classified for water use. In addition, the water quality in the source of water supply is regulated taking capacity of conventional water purification treatment into consideration. In 2003, water quality items related to the conservation of aquatic organisms were added, and zinc was included in the environmental standards. The value of zinc is lower in the environmental standards than in the water quality standards for drinking water, which suggests that there is a difference between the viewpoints of conservation for aquatic organisms and human health protection.

In items of environmental standards related to human health (health items), standards regarding heavy metals are uniform throughout the country regardless of the type of water area. Many health items have uniform regulations throughout the country, and because water quality standards are related directly to human health, and ranking raw water quality to a safe level based

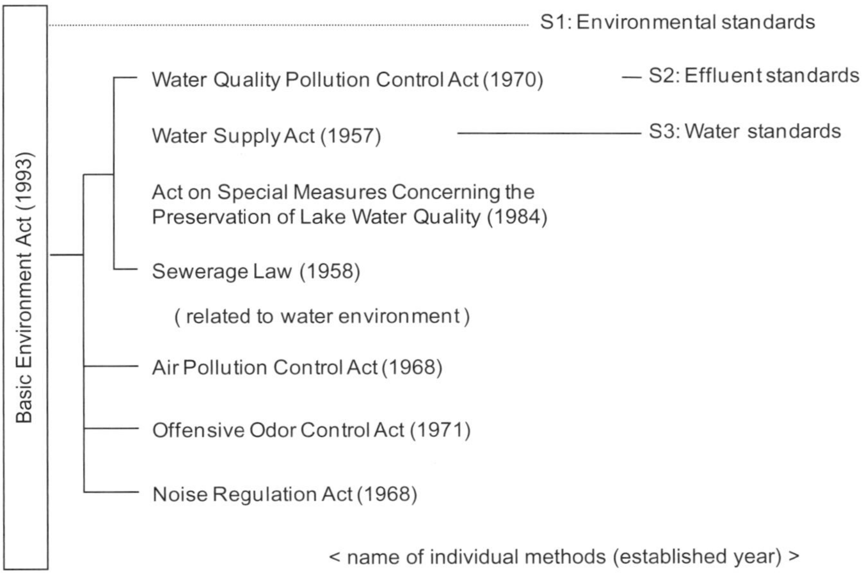

FIGURE 4-4. Laws and standards related to water environments (JSWE, 2009).

on these standards is needed, the value of these health items are similar to the water quality standards for drinking water.

The Water Quality Pollution Control Act regulates the effluent from businesses such as factories. For items of this act related to the living environment, while occasionally differing from one type of business to another, a uniform value is generally applied to each toxic substance standard. On the other hand, since it is difficult to apply effluent standards to domestic wastewater, the increasing use of sewerage systems plays a substitute rule. Sewage treatment plants are positioned to remove items related to the living environment such as BOD, SS, TN, and TP for water quality conservation. Therefore, in sewage treatment plants, more severe standards for the effluent quality are adopted than in common business institutes by the Sewerage Law. These contain regulatory standards for nitrogen and phosphorus that are discharged into enclosed water areas after advanced treatment. Since not only domestic wastewater but also industrial wastewater flows into sewage treatment plants, standards for toxic substances applied to businesses by the Water Quality Pollution Control Act are applied to specified plants as a limit of acceptance into public sewers and sewage treatment plants. In addition, regulations on contamination removal facilities prevent toxic substances from negatively affecting wastewater treatment plants (**Figure 4-5**).

4.1.4 *Advantages of centered systems*

In the Sewerage Law mentioned above, sewerage systems are defined as a total system including pumping and other facilities, such as drain pipes, drain ditches, and other drainage facilities established to exclude raw sewage. As facilities to treat sewage similar to the sewerage systems defined in the Sewerage Law, there are also community plants, rural sewerage systems, individual on-site sewage treatment tanks, and so on. Public and river-basin sewerage systems contained in the law, as well as rural sewerage systems, are categorized as centered sewage management systems, while individual sewage treatment tanks are dispersed sewage management systems. It is important to promote and manage these systems by taking both the characteristics of each system shown below and each area itself into consideration. The choice between centered systems and dispersed systems depends on the total amount of construction costs and the maintenance and operation costs per individual. The features of each system are as follows.

The centered systems are applied in urban areas and agricultural settlements because these systems are more economically advantageous in populated areas.

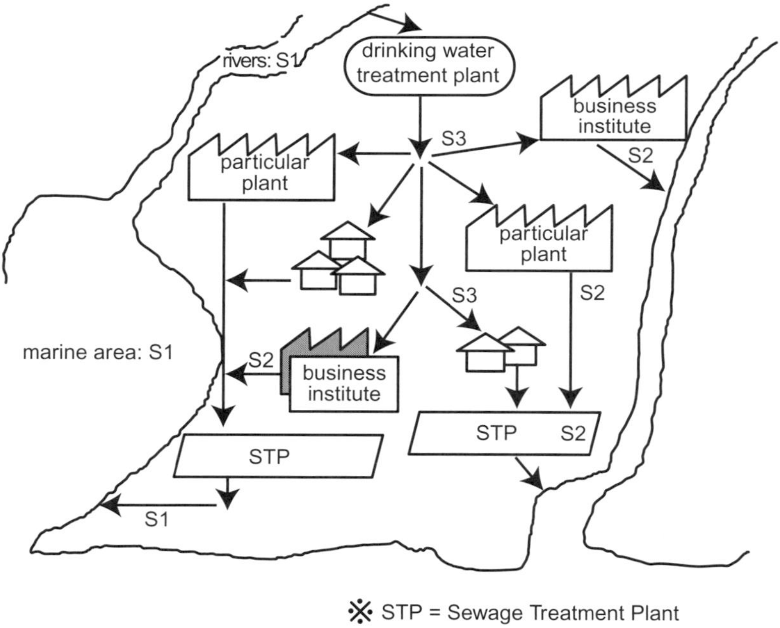

FIGURE 4-5. Adoption of each standard (JSWE, 2009).
Note: S1, S2, S3 stand for the standards referred in **Figure 4-4**.

The target districts in a public sewerage system and the specified environmental preservation of a public sewerage system are urban districts and districts other than urbanized areas, respectively. The target sewage is mainly black and gray waters, industrial wastewater, and commercial wastewater, in addition to rainfall drainage. The objective of these facilities is the collection and transport of sewage, sewage treatment, sludge treatment, and rainwater drainage. The target facility is managed by each local government, and based on the Sewerage Law is partially funded by the central government. At the end of 2007, the sewage usage rate nationwide was 72%, and 91 million people used public sewerage systems.

The purposes of sewerage systems located in rural communities are the water-quality preservation of agricultural drainage, the functional maintenance of agricultural waste facilities, improvements in the rural environment, and the water-quality preservation of public water bodies. The target districts are agricultural settlements inside agricultural promoted areas. The areas are set and maintained by the local government with a grant made possible by the water resource recycling of the sewage from agricultural communities, and through other means. The sewerage systems in rural communities service areas with a population of more than twenty households and less than 1,000 people on average, and treat black and gray water sewage. At the end of 2007, the distribution rate of such systems, which contains sewerage systems for fisheries and forestry, was 3%, affecting 3.7 million people in Japan.

The sewer network in a centered system can drain sewage and rainwater quickly. Such network has recently been expected to be a tool for the recovery of energy and resources from sewage, although the methods for the recovery have not been established.

4.1.5 *Advantages of a dispersed system*

A dispersed system, e.g., a septic tank, is used to treat black and gray or yellow water to preserve aquatic environments and improve public health. The target districts are outside areas where the sewage is or will be treated at a public sewerage system. The system is set and controlled by individuals or local governments (municipalities). The user of the system can receive budget from the municipalities for setting the tanks and promotion work for the maintenance of sewage treatment tank for the municipality. Although there is no limit regarding an appropriate population, the target population is more than twenty households for systems supported by the government. A tank treatment system requires a regular sludge drainage for treatment facility. At the end of 2007, the diffusion rate was 9%, and eleven-million people were using septic tank systems in Japan.

A dispersed system is a type of on-site treatment and is easy to be settled. However, there are certain disadvantages: such systems are small in scale, and because of the short distance from the source, the water quantity and quality varies enormously, the amount of energy required for treatment is relatively high, the management architecture for maintenance such as sludge removal is not efficient, and the actual conditions of the treatment are unclear. Therefore, an improvement of this system in terms of efficiency, more reliable membrane technologies, remote monitoring, and future cooperation with other combined systems is required.

The sludge produced during on-site treatment is usually collected regularly and sent to a night-soil or centralized treatment plant. For sustainability, the recovery of materials and energy from sludge is extremely important and urgent. The integration of centralized and on-site systems is important not only for maintenance but also for sustainability.

4.2 TREATMENT TECHNOLOGIES FOR MUNICIPAL WASTEWATER

4.2.1 *Municipal wastewater treatment*

Typical sewage treatment consists mainly of primary sedimentation, biological treatment, and disinfection. Schematic flow diagrams of typical treatment processes incorporating biological processes are shown in **Figure 4-6**.

The objective of treatment by sedimentation is to remove readily settable suspended solids (SS), and both primary and secondary sedimentations are used. The typical design parameters for conventional activated sludge processes are summarized in **Table 4-1**. The general removal rate for primary sedimentation is 30–50%, whereas a 40–60% removal is achieved for BOD and SS (JSWA, 2009a).

The overall objective of biological treatment is to transform dissolved and particulate biodegradable constituents into acceptable end products. The removal of dissolved and particular BOD, as well as the stabilization of organic matter, is accomplished biologically using a variety of microorganisms, principally bacteria. Microorganisms are used to oxidize dissolved and particulate organic matter into simple end products and additional biomasses, as represented by the following equation for the aerobic biological oxidation of organic matter.

$$3C_6H_{12}O_6 + 8O_2 + 2NH_3 \\ = 2C_5H_7NO_2 \text{ (new cells)} + 8CO_2 + 14H_2O \tag{4-1}$$

A term new, cells, is used to represent the biomass produced as a result of the oxidation of organic matter. The most widely used empirical formula for the organic fraction of cells is $C_5H_7NO_2$. A formulation with phosphorus or

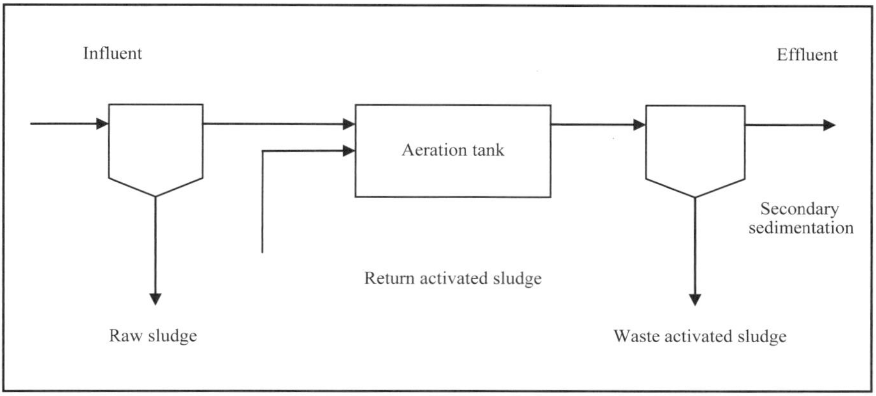

FIGURE 4-6. Typical flow diagram for biological processes.

TABLE 4-1. Typical design parameters for primary and secondary sedimentation tanks (JSWA, 2009a).

Parameter	Primary	Secondary
Overflow rate, $m^3/(m^2 \cdot d)$	35–70 (separate) 25–50 (combined)	 20–30
Depth, m	2.5–4.0	2.5–4.0
Detention time, h	1.5 (separate)	3–4
Weir loading, $m^3/(m \cdot d)$	around 250	around 150

other elements can also be used. The ratio of the amount of biomass produced to the amount of substrate consumed (biomass/substrate in grams) is defined as the biomass yield. For aerobic heterotrophic reactions with organic substrate, the ratio is around 0.6.

The design and operational parameters for typical activated sludge processes are summarized in **Table 4-2**. Those for oxidation ditch processes, which are common for small facilities, are also summarized. These parameters are calculated using the following equations:

$$\text{BOD-MLSS loading rate (kg BOD/(kg MLSS}\cdot\text{d))} = [\text{Influent BOD}] \times [\text{Inflow rate}]/([\text{MLSS}] \times [\text{Volume of aeration tank}]) \quad (4\text{-}2)$$

$$\text{BOD-volume loading rate (kg BOD/(m}^3\cdot\text{d))} = [\text{Influent BOD}] \times [\text{Inflow rate}]/[\text{Volume of aeration tank}] \quad (4\text{-}3)$$

$$\text{Hydraulic retention time (HRT) (d)} = [\text{Volume of aeration tank}]/[\text{Inflow rate}] \quad (4\text{-}4)$$

TABLE 4-2. Typical design parameters for a conventional activated sludge process (JSWA, 2009a).

	Conventional activated sludge process	Oxidation ditch process
BOD-MLSS loading (kg BOD/(kg MLSS•d))	0.2–0.4	0.03–0.05
BOD-volume loading (kg BOD/(m^3•d))	0.3–0.8	0.1–0.2
MLSS (mg/L)	1,500–2,000	3,000–4,000
Sludge age (d)	2–4	15–30
Aeration rate (aeration/flow rate)	3–7	5–8
HRT for aeration tank (h)	6–8	24–36
Sludge return ratio (%)	20–40	50–150

(Red; 1984 edition)

$$\text{Substantial HRT (d)} = [\text{Volume of aeration tank}]/([\text{Inflow rate}] + [\text{Flow rate of sludge return}]) \quad (4\text{-}5)$$

$$\text{Sludge age (d)} = [\text{Volume of aeration tank}] \times [\text{MLSS}]/([\text{Inflow rate}] \times [\text{Influent SS}] \quad (4\text{-}6)$$

$$\text{Solids retention time (SRT) (d)} = [\text{Volume of aeration tank}] \times [\text{MLSS}]/([\text{Flow rate of sludge wasting}] \times [\text{Concentration of waste activated sludge}] + [\text{Outflow rate}] \times [\text{Effluent SS}]) \quad (4\text{-}7)$$

These equations can be generated easily if their meanings and units are clearly understood. Care needs to be taken with the unit conversions, however. The concentrations of some water quality items such as organics, nitrogen, and phosphorus are expressed in 'mg/L,' and the flow rates of real wastewater treatment plants are expressed using m^3/d rather than L/d. Considering that mg/L = g/m^3 and g/L = kg/m^3, the unit conversion is easy. For example, the concentration (kg/m^3) × flow rate (m^3/d) indicates the total mass flow per day (kg/d). The BOD-MLSS loading rate is occasionally called an F/M ratio, but it originally indicated the food-to-microorganism ratio (mg/mg), which has a different unit in Japan. The solids retention time (SRT) is sometimes misunderstood as the sludge retention time. In typical conventional activated sludge processes, the amount of SS mass in the influent is almost the same as that of the sludge generation, and it depends on the addition of coagulants and SRT. A commonly used measure to quantify the settling characteristics of activated sludge is the sludge volume index (SVI). The SVI is the volume of 1 g of sludge after 30 min of settling (mL/g). The SVI is determined by placing a mixed-liquor sample in a 1- or 2-L cylinder, and measuring the settled volume after 30 min and the MLSS concentration of the corresponding sample.

$$SVI = [SV_{30}\,(\%)]/[MLSS\,(mg/L)] \times 10^4 \tag{4-8}$$

A value of 100 mL/g is considered good for settling sludge, and SVI values above 200 are typically associated with bulking sludge with poor settling characteristics. A principal type of sludge bulking is caused by the growth of filamentous organisms. Sludge bulking can be caused by a variety of factors, and introducing a selector under anaerobic conditions can be used to solve this problem (Wannar, 2000).

Biological nitrogen removal processes involve both nitrification and denitrification. Nitrification, the two-step oxidation of ammonia into nitrate by autotrophic bacteria, is as follows:

$$NH_4^+ + 3/2O_2 \rightarrow NO_2^- + H_2O + 2H^+ \tag{4-9}$$
$$NO_2^- + 1/2O_2 \rightarrow NO_3^- \tag{4-10}$$

The oxygen required for the complete oxidation of ammonia is 4.57 g O_2/gN. The amount of alkalinity required for the complete oxidation of ammonia is 7.14 g $CaCO_3$/gN, and the pH may decrease. Nitrite is oxidized easier, and the accumulation of nitrite is usually not observed. Biological denitrification involves the biological oxidation of many organic substrates using nitrate or nitrite as the electron accepter instead of oxygen. Some anoxic volume or time must be included to provide biological denitrification to complete the total nitrogen removal by nitrate, and nitrite reduction into nitrogen gas. Denitrification requires an electron donor, which can be supplied in the form of influent BOD, by endogenous respiration, or by an external carbon source. Reaction stoichiometry for methanol as an electron donor is as follows:

$$6NO_3^- + 5CH_3OH \rightarrow 3N_2\uparrow + 5CO_2 + 7H_2O + 6OH^- \tag{4-11}$$

Methanol is often used owing to its inexpensive cost, but it takes a few weeks for activated sludge to adapt to methanol (Hidaka *et al.*, 2003). 3.57 g $CaCO_3$/gN of alkalinity is produced and the pH increases. The growth of nitrifying bacteria is slower, and it is necessary to keep the growth condition for the aerobic treatment of wastewater containing organics and ammonia. Biological nitrogen removal with aerobic and anoxic tanks in series requires the addition of electron donors and alkalinity, as organics and alkalinity are consumed in an aerobic tank. To avoid the addition of electron donors and alkalinity, an anoxic/aerobic process is developed. In this process, nitrate is fed to the anoxic reactor from nitrate in the return activated-sludge flow and by pumping mixed liquor from the aerobic zone (**Figure 4-7a**). Organics in the influent can be used for denitrification, and the decrease in pH by nitrification can be recovered by alkalinity generated during denitrification. The nitrogen removal efficiency is limited depending upon the recirculation of nitrified liquid from the aerobic zone to the anoxic zone including sludge return, R (–). The theoretical maximum nitrogen removal efficiency is R/(1 + R), excluding nitrogen removal

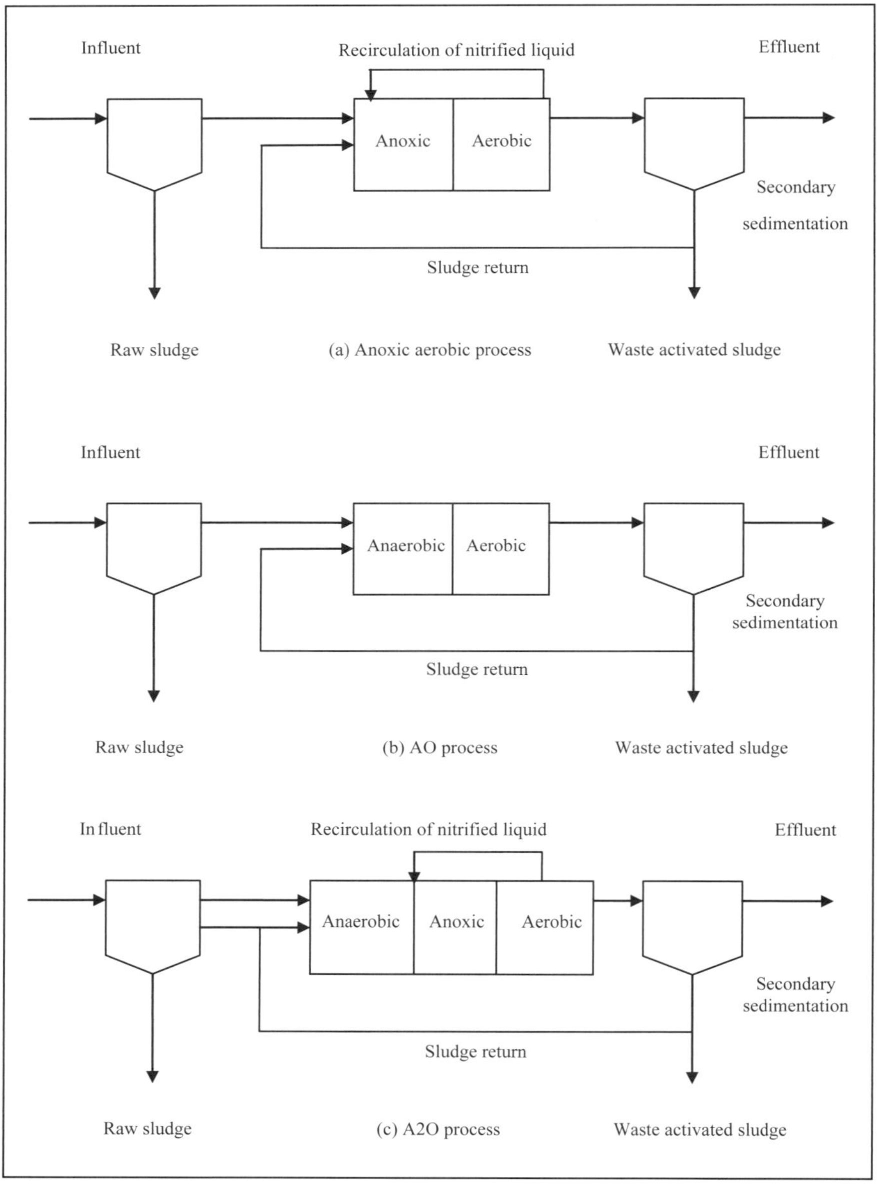

FIGURE 4-7. Schematic diagram of biological nitrogen and removal processes.

by sludge wasting. Typically, R is set to 2, and the maximum nitrogen removal is only around 0.7. Another anoxic tank with additional electron donors can be added to improve nitrogen removal.

Biological phosphorus removal processes include the use of an anaerobic tank prior to an aerobic tank. Under these conditions, a group of heterotrophic bacteria, called polyphosphate-accumulating organisms (PAO), are selectively enriched in activated sludge. Generally, all bacteria contain a fraction (1.5 to 2.5%) of phosphorus in their biomass, and some portion of phosphorus is removed by sludge wasting. When PAOs accumulating large quantities of polyphosphate within their cells grow, the fraction of phosphorus accumulating biomass increases to 3 to 5%, and the removal of phosphorus is enhanced. The further suitable treatment of waste activated sludge is essential to complete the phosphorus removal. The basic process configuration consists of an anaerobic zone followed by an aerobic zone (A/O process) (**Figure 4-7b**). An A_2O process is a modification of the A/O process and provides an anoxic zone for denitrification (**Figure 4-7c**). Balance between nitrogen removal and phosphorus removal conditions should be controlled considering that nitrifying bacteria require a long SRT, phosphorus removal requires sludge wasting, and electron donors are required in both anaerobic and anoxic tanks.

Recently, the anaerobic ammonium oxidation (anammox) process has been focused upon as a new ammonia removal technology. This process was reported by the Delft University of Technology (Mulder *et al.*, 1995). Reaction stoichiometry for anammox is as follows:

$$NH_4^+ + 1.32\ NO_2^- + 0.066HCO_3^- + 0.13H^+$$
$$\rightarrow 1.02\ N_2 + 0.26\ NO_3^- + 0.066CH_2O_{0.5}\ N_{0.15} + 2.03H_2O \qquad (4\text{-}12)$$

Theoretically, this is an advanced technology in that the 'electricity requirement for aeration is low', 'organics are not required as an electron donor,' and 'sludge generation is low.' This is expected to treat wastewater containing especially low carbon/nitrogen (C/N) such as concentrate from sludge dewatering processes. However, there are challenges such as 'nitrite accumulation by stopping further nitrification into nitrate,' and 'maintaining annamox bacteria with low growth rates in the reactors'; pH control and heat shock methods, and a nonwoven fabric carrier for immobilizing the anammox bacteria (Date *et al.*, 2009), were developed.

For these treatments, the aerobic, anoxic, and anaerobic conditions need to be controlled. Under anoxic conditions, there is no dissolved oxygen (DO), although NO_2 and NO_3 are present. Under anaerobic conditions, there is no DO, NO_2, or NO_3 present. Occasionally, the term 'anaerobic' is used to describe an anoxic condition, but this is not the case when considering this definition.

The principal biological processes used for wastewater treatment can be divided into two main categories: suspended growth and biofilm processes. In most wastewater treatment plants, suspended growth processes are adopted. In some wastewater treatment plants, biofilm processes such as cells immobilized in polyethylene glycol for nitrification (Tanaka *et al.*, 1996) are adopted. Biological filtration processes, which can work as both filtration and biofilm processes, are proposed (Hidaka *et al.*, 2003). Biofilm processes are suitable to maintain nitrifying bacteria with low growth rates. For denitrification, preventing the media from floating from the generated nitrogen gas is important. On the other hand, for phosphorus removal, biofilm processes are not very popular, as sludge wasting is required.

Table 4-3 summarizes the process types used in wastewater treatment plants in Japan (JSWA, 2009b). In larger plants, the conventional activated sludge process is the most popular, and in smaller plants, the oxidation ditch process is the most common. Other processes have also been adopted, but not many. A novel oxidation ditch system for biological nitrogen and phosphorus removal is currently under development (Chen *et al.*, 2009).

Tertiary treatment is defined as the additional treatment of a secondary effluent, and advanced treatment is defined as the treatment of any wastewater to obtain a higher water quality than from secondary treatment. Such treatment includes micro- and ultra-filtration, coagulation, and rapid filtration for SS and BOD removal; biological phosphorus removal, coagulation, and chemical precipitation for phosphorus removal, biological nitrification and denitrification, breakpoint chlorination, and ammonia stripping for nitrogen removal; and activated carbon adsorption, ozonation, and advanced oxidation for the removal of trace and non-biodegradable organics. As an example, a wastewater treatment plant in Shiga prefecture in Japan has adopted coagulation, multi-stage nitrogen removal, and sand filtration to preserve the water quality in Lake Biwa. A wastewater plant in Kyoto city in Japan has adopted ozonation to treat dye wastewater. However, wastewater from only 17% of the population in Japan is treated by advanced processes (JSWA, 2009b), which is lower than that of other developed countries.

4.2.2 *Sludge treatment*

Sludge generated from water treatment contains 98–99% water and organics, and further treatment is required to make it suitable for final disposal. Unit processes are classified as follows (JSWA, 2009a):

Decrease in quantity: thickener, dewatering, drying
Decrease in solids: digestion, incineration, sintering
Stabilization: anaerobic digestion, composting, incineration, sintering, carbonization, fuel making

TABLE 4-3. Number of wastewater treatment plants and treatment methods (JSWA, 2009b).

	<5	5–10	10–50	50–100	100–500	>500	Sum
Primary sedimentation	1		1				2
A2O	1	3	6	5	9	1	25
Anoxic-aerobic	3	3	8		4		18
Step feed nitrification and dentrification	1	1	6	3	6		17
AO	12		6	3	10		31
Conventional activated sludge	46	51	331	117	130	12	687
Conventional extended aeration	37	6	2				45
Sequencing batch reactor	61	8	2				71
Aerobic filtration	22	6					28
Anaerobic-aerobic filtration	41	2					43
Advanced oxidation ditch	48	10					58
Oxidation ditch	780	96	33				909
Others	87	18	30	11	19		165
Total	1,140	204	425	139	178	13	2,099
Advanced treatment	99	28	63	29	84	8	311

If one plant uses two or more methods, the one with the highest flow rate is counted. 'Primary sedimentation' means only sedimentation without secondary treatment.

Two of the most important parameters during these processes are an organic containing ratio (VSS/SS) and a water containing ratio.

$$[\text{Water containing ratio}] = [\text{Mass of water}]/([\text{Dry weight}] + [\text{Mass of water}]) \quad (4\text{-}13)$$

The former is an index of sludge stabilization, and the latter is an index of sludge volume to be treated. For example, when the water containing ratio decreases from 0.99 to 0.97, the sludge volume decreases by two-thirds. Water from sludge treatment needs further treatment, and is returned to the wastewater treatment process, or is treated separately.

Thickening is a procedure used to increase the solid content of sludge by removing a portion of the liquid fraction. A volume reduction is beneficial for the subsequent treatment processes. Thickening is generally accomplished by physical means, including gravity, floatation, centrifugation, a gravity belt, and a rotary drum. Gravity and mechanical thickening can produce concentrated sludge with water containing ratios of 0.96 to 0.98, and 0.96, respectively.

Anaerobic digestion is a bacterial process carried out in the absence of oxygen. Through the solubilization of sludge, and acid and methane fermentation, anaerobic digestion decreases the organic containing ratio and generates a biogas with a high proportion of methane that can be used to both heat the tank and run engines or micro-turbines. Anaerobic digestion has been focused upon for global environmental protection and recycling-oriented societies (Noike 2009). Reaction stoichiometry for the anaerobic digestion of glucose is as follows:

$$C_6H_{12}O_6 = 6CH_4 + 6CO_2 \tag{4-14}$$

Here, the sum of COD before and after digestion remains constant. Under an anaerobic condition, no oxygen molecules are consumed, and during the production of organic acids, hydrogen, and methane, the sum of COD expressed as gO_2 is conservative. Therefore, the mass balance of COD can be easily checked. 50% of the carbon is converted into methane, and the other half is converted into CO_2. On the other hand, the COD of CO_2 is 0, and therefore 100% of the COD is converted into methane. The principal anaerobic digestion processes can be divided into some main categories such as single-phase digestion and two-phased digestion. In two-phased digestion, solubilization and acid fermentation, and methane fermentation, are divided. The temperature is generally set at around 35°C (mesophilic), or 55°C (thermophilic), and occasionally the first and second stages have different temperature conditions. It was reported that for the anaerobic digestion of waste activated sludge, mesophilic methane fermentation is preferable (Kobayashi *et al.*, 2009). Recently, innovative technologies such as hyper thermophilic treatment at around 70–80°C (Lee *et al.*, 2008), subcritical water hydrolysis treatment (Sato *et al.*, 2010), ozonation (Kobayashi *et al.*, 2009), and attached growth media have been proposed. Among the 2000 sewage treatment plants in Japan, around 300 have adopted the anaerobic digestion of sludge, and this number has remained almost constant over the past twenty years. The larger plants usually adopt incineration treatment. In addition, only thirty out of the 200 plants with anaerobic digestion are utilizing produced methane gas for energy recovery. Gravity thickening used to be the most popular, and the solid concentration of sludge fed into digesters is 1.5–2.5%. Therefore, a lot of energy is required for heating the reactors. Mechanical thickening has recently become more popular, and sludge concentrations can be as high as 4–5%, which reduces the required energy for heating, and the remaining gas can be utilized as an energy source.

By dewatering the thickened or digested sludge, easy-handling cake solids are generated, and the water containing ratios are decreased from 96–98% to around 80%. Coagulants or lime are sometimes added to improve the dewaterability. Dewatering processes include a solid bowl centrifuge, belt-filter press, and recessed-plate filter press. Dewaterability can be measured through a leaf test or capillary suction time (CST) (JSWA, 1997).

Heat drying involves the application of heat to evaporate water and reduce the moisture content of biosolids below that achieved by conventional dewatering methods. Heat drying is used as a pre-treatment for the incineration, sintering, and carbonization processes, and the water containing ratios reach around 0.2–0.5. The incineration of sludge involves the total conversion of organic solids into oxidized end products, primarily CO_2, water, and ash. The major advantages of incineration are a maximum volume reduction, the destruction of pathogens and toxic compounds, and potential energy recovery.

4.2.3 *Water reclamation technologies*

4.2.3.1 Reclaimed water quality level and treatment level

Water is an essential resource for life, social activities, and environmental protection. In addition, industries are unable to develop without access to water. The amount of fresh water in ground water, rivers, and lakes is only 0.8% out of about 1.4 billion km^3 water existing on Earth, and its availability is very limited. Owing to the pollution of water environments and water shortages derived from an increase in urban populations, the necessity of water reuse has greatly increased. In Japan and other countries, the technologies for the reclamation of wastewater and rain water, and the desalination of seawater, have therefore been actively studied. In Japan, the need for the restoration of water environments lost through the increase in urbanization and the enhancement of living environments such as through fire protection capabilities, are increasing with the development of wastewater treatment technologies and an improvement in effluent qualities. Based on these phenomena, water reuse has attracted greater attention. The quality of reclaimed water should be controlled based on each purpose, and should maintain safe levels for people and ecosystems. The state of California in the USA has a regulation called Title 22, which is one of the strictest standards ever implemented, and has been referred to in the determination of water reuse guidelines in other countries. The U.S. Environmental Protection Agency (USEPA) has created guidelines for water reclamation since the 1980s, and has promoted water reuse for individual purposes. In addition, Australia established guidelines in 2006 and 2008, and is trying to mitigate damage to public health and ecosystems through water reuse. Regulations related to BOD, TSS, turbidity, and bacteria (total or fecal coliforms) have been set up for the reuse of urban water. Secondary treatment, filtration, and disinfection processes have been proposed to meet these standards. Reclaimed water is also used for irrigation, recreation, industry, and even drinking water. The parameters of the standards for these uses are almost same as those for municipal water in the Australian guidelines, although the levels are different. The Californian guidelines include a regulation parameter for TOC for groundwater recharge.

In Japan, the Ministry of Land, Infrastructure, Transport, and Tourism (MLIT) and the National Institute for Land and Infrastructure Management (NILIM) established a committee for water quality standards for recycling wastewater as a means to review the guidelines for recycling wastewater (March, 1981) and a manual for reclaimed water quality for landscape and recreational use (March, 1990). This committee discussed standards for water reclamation technologies to secure the sanitary safety, and created a manual for these standards in April, 2005. This manual determined the standards for water quality and facilities for flushing, sprinkling, and landscape and recreational use (**Table 4-4**). *E.coli*, turbidity, color, odor, pH, appearance, and residual chlorine were selected as the standard parameters. The minimum water quality and water reclamation facilities required, and the target values in the operation and maintenance of the facilities, are regulated as the water quality and facility standards, and as control targets, respectively.

4.2.3.2 Disinfection technologies

In developed countries including Japan, it is often mistakenly thought that water-borne diseases cannot occur owing to the establishment of environmental sanitary facilities such as water supply and sewerage systems. In fact, while typical water-borne diseases such as cholera and dysentery have almost been eradicated, newly emerging and reemerging water-borne diseases may still occur. After 1998, for example, outbreaks of water-borne diseases in Japan were reported by the Risk Management Procedure for Water Supply and Health from the Ministry of Health, Labor, and Welfare (MHLW). Up to six outbreaks of these diseases occur each year from drinking water from springs and wells. In addition, small facilities such as private water networks have been contaminated with microorganisms, particularly *Cryptosporidium*. Cases of pathogenic *E.coli*, viruses, and *Campylobacter* have also been reported. Moreover, some people have been infected by the norovirus accumulated in fish and shellfish. Although an understanding of these conditions is insufficient owing to a lack of epidemiological information in Japan, infection through recreational use such as bathing has occurred in many developed countries. Pollutions by pathogens are caused by a lack of proper treatment and countermeasures of fecal matter from humans and other animals. In a combined sewer system, untreated wastewater is discharged into water environments on rainy days, which is considered a social problem. Thus, combined sewer systems are being improved from the viewpoint of public sanitation. It was reported that if untreated wastewater is discharged into an enclosed water area, the total number of coliforms will increase dramatically during a period of rain, decrease gradually after the rain stops, and after two days grow to a higher number than during dry weather. Recreational, bathing, and drinking water source areas are located downstream of about 600 out of 3,000 combined sewer system outlets in Japan, and the effects of combined sewer overflow (CSO) on the receiving areas are of great concern. In water reuse for irrigation in developing countries, it was reported that water-borne diseases occur by pathogens remaining in insufficiently treated reclaimed water. Therefore, it is necessary to prevent water-borne diseases by pathogens during water reuse.

Water-borne diseases are mainly caused by enteric pathogens. These pathogens are generally classified into three categories: bacteria, protozoa, and viruses. Disinfection by chlorine after proper wastewater treatment is effective

TABLE 4-4. Guidelines for water reclamation by Ministry of Land, Infrastructure, Transport and Tourism (2005).

	Where are the standards applied?	**For flushing use**	**For sprinkling use**	**For landscape use**	**For recreational use**
E.coli		Not detect (in 100 mL)[1]	Not detect (in 100 mL)[1]	Refer to remarks[2]	Not detect (in 100 mL)[1]
Turbidity		(Control target) ≤2	(Control target) ≤2	(Control target) ≤2	≤2
pH	Exit of treatment facilities for reuse	5.8–8.6	5.8–8.6	5.8–8.6	5.8–8.6
Appearance		Shall not be unpleasant	Shall not be unpleasant	Shall not be unpleasant	Shall not be unpleasant
Chromaticity		–[3]	–[3]	40 degree[3]	10 degree[4]
Odor		Shall not be unpleasant[3]	Shall not be unpleasant[3]	Shall not be unpleasant[3]	Shall not be unpleasant[3]
Residual Chlorine	Responsibility demarcation point	(Control target) Free 0.1 mg/L Combined 0.4 mg/L[4]	(Control target[5]) Free 0.1 mg/L Combined 0.4 mg/L[4]	Refer to remarks[6]	(Control target[5]) Free 0.1 mg/L Combined 0.4 mg/L[4]
Treatment Level		Sand filtration	Sand filtration	Sand filtration	Coagulation-sedimentation + sand filtration

1: Measured using specific enzyme substrate culture medium methods.
2: The present standard (coliform groups count: 1000 CFU/100 mL) is adopted *pro tempore*.
3: Depending on the users of reclaimed water, standard values and more stringent standards are determined.
4: An additional dose of chlorine is based on an individual agreement with the supply destination.
5: This value is not applied if a residual effect of chlorine is not needed.
6: This value shall not be stipulated, as treatment other than chlorine disinfection is carried out case by case from ecological correctness as the water may be used according to the prerequisite that humans shall not touch it.

for bacteria such as *Shigella*. Because protozoa including *Cryptosporidium* have resistance to chlorination, disinfection by ozone and UV exposure is effective. For the control of viruses, disinfection by ozone and UV exposure, coagulation, and filtration are effective. In the following section, disinfection processes using chlorine, ozone, and UV treatment are explained.

(1) *Chlorination*

Chlorine is one of the most prevalent disinfectants, and the main chlorine compounds used in water reclamation are Cl_2, sodium hypochlorite (NaClO), and chlorine dioxide (ClO_2). Calcium hypochlorite ($Ca(OCl_2)$) is used in small treatment plants for convenience. NaClO has been adopted in many cities owing to the risks and regulations regarding the use and storage of liquid chlorine. In the disinfection process for the sanitary safety of reclaimed water, suspended solids (SS) affect the efficiency of disinfection. In particular, chlorine is ineffective for the inactivation of microorganisms associated with SS (Dietrich *et al.*, 2003, Gideon *et al.*, 2008) (**Figure 4-8**). Gideon *et al.*, (2008) investigated the effect of organic compounds and SS on the disinfection efficiency using chlorine. They found that SS inhibits the inactivation of total coliform, while the existence of organic compounds is not very relevant to the inactivation efficiency. Microorganisms associated with SS are a very important factor in the disinfection process for water reuse. In some cases, the number of microorganisms associated with SS cannot be measured through conventional methods such as membrane filtration or MPN. Moreover, the

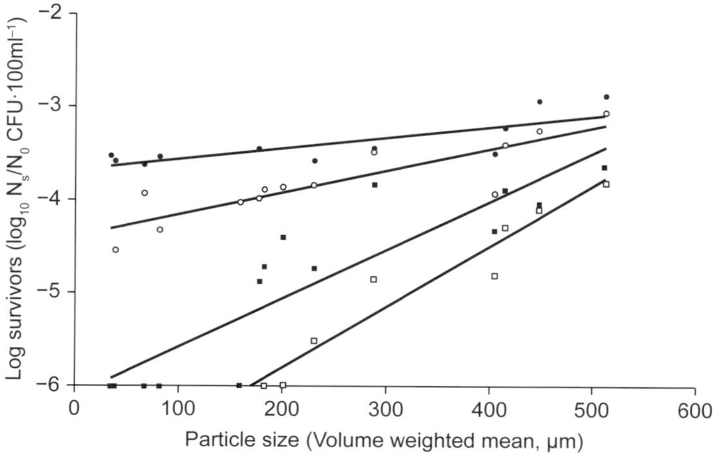

FIGURE 4-8. Proportion of total surviving coliforms against particle size in manipulated grey water, following disinfection at initial chlorine concentrations of 10 (closed circle), 20 (open circle), 40 (closed square) and 80 (open square) mg/L. Cases of zero coliform survival have a value of –6 (Gideon *et al.*, 2008).

disinfection of such microorganisms may be insufficient, because SS protect them from disinfectants such as chlorine. To attain sufficient inactivation, it is therefore important to remove SS before the disinfection process.

(2) *Ozonation*

Ozone consists of three oxygen atoms and has strong oxidation ability after fluorine and OH radicals. In addition to the oxidation of many organic and inorganic compounds, ozone can inactivate pathogens by damaging the cell membrane and nucleic acid, and through an alternation of enzymes and proteins. The mechanisms of disinfection by chlorine and ozone differ in that chlorine affects the enzyme reaction. Along with chlorine, ozone is recognized as an efficient disinfectant, and has been adopted in water treatment plants, especially in Europe. In the U.S.A., wastewater treatment plants have adopted ozonation as a disinfection process since the 1970s. However, in those days, operation and maintenance problems often occurred, and disinfection by ozonation was recognized has being worse than UV as an alternative to chlorination. The application of ozonation to wastewater treatment and water reclamation requires know-how on a proper design and operation. Pei *et al.* (2002) studied parameters for the design of a disinfection process for wastewater using ozone exposure. It was found that, depending on the quality of the effluent, a 2–15 mg/L dose of ozone is required to meet the WHO standards for irrigation (1,000 FC/100 mL). Such a dose can inactivate enteroviruses sufficiently. They proposed that *Clostridium*, which has resistance to ozone, is the more proper candidate as a resistant microorganism indicator than bacteriophages, because some bacteriophages are rapidly inactivated by ozone exposure. The inactivation of *E. coli* in secondary effluent by ozonation can achieve 2 to 5 logs with 1.5 to 4.5 mg/L of ozone consumption. Based on a batch experiment, the range of ozone consumption for the inactivation of bacteriophages Qβ and T4 for a similar inactivation was determined to be higher (Ab. Wahid, 2011). **Figure 4-9** shows the inactivation of *E. coli* and bacteriophages Qβ and T4 in a secondary effluent by ozonation.

Other researchers have also reported the significant inactivation of viruses, phage, and *E. coli* through ozonation treatment (Harata *et al.*, 2004; Liberti *et al.*, 1999; Lim *et al.*, 2010; Shin *et al.*, 2003; Paraskeva *et al.*, 2005). Therefore, the applicability of ozonation for water reclamation and reuse should be considered. However, optimization of the process design is important prior to the production of a disinfection by-product.

Compared with UV exposure, ozonation has a high removal ability of UV absorbance and color, which can be an advantage for certain reuse applications. From an operational viewpoint, the transfer of ozone from a gas phase to a water phase was found to be a very critical factor, because *E. coli* and total coliforms can be inactivated rapidly through ozonation. From a semi-batch experiment, Pei *et al.*, (2002) found no differences between hydraulic retention times of 2 and 10 minutes as shown in **Table 4-5** and **Figure 4-10**. A short contact time and an enhanced mass transfer of ozone are required for an ozone reactor for wastewater treatment.

(3) *UV*

UV disinfection inactivates microorganisms or bacteria by inhibiting cellular respiratory activity and proliferation through the destruction of their DNA inhabiting within the mitochondria and nucleus of the microorganism. However, the microorganisms have repair mechanisms such as dark reactivation, which does not require light, and photo reactivation, in which an inactivated microorganism receives light and actives it through UV sterilization. With regard to photo reactivation, photo reactivating enzymes from inside the cell use and activate near-ultraviolet light, which cleaves the dimer caused by the UV and turns it back into original base. Dark reactivation does not require light since it repairs damaged DNA caused by exposure to a mutagen, turns it into an enzymatic reaction, and causes DNA breakage and a DNA break repair metabolism. This is not a problem when examining UV irradiation showing a disinfection effect in microorganisms, but a problem does remain in how to maintain the same UV irradiation under the same conditions. If it irradiates at a specific level of energy continuously, the irradiation is influenced by the permeability and temperature. UV has a short wavelength such that its permeability is limited. Its permeability also differs depending upon the material. Moreover, if the temperature is too

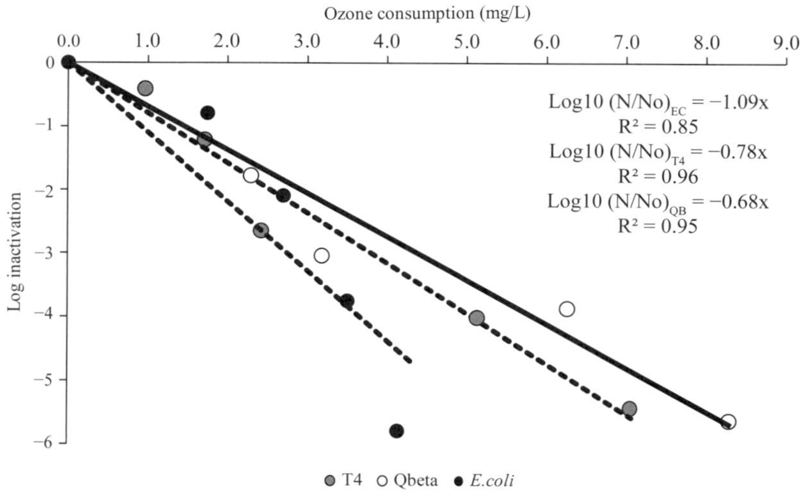

FIGURE 4-9. *E.coli* and bacteriophages Qβ and T4 inactivation in secondary effluent (Ab. Wahid, 2011).

high or low, the irradiance level decreases, which is a necessary consideration when using a standard lamp. The factors influencing the effects of disinfection include the type of microorganism, water permeability, deterioration in the UV output lamp, amount of pollutants on the surface of the lamp, and flow of treatment.

Based on a batch experiment, UV treatment was shown to achieve an inactivation of 2–5logs for *E.coli* in a secondary effluent at 45 to 130 mJ/cm², and for bacteriophage Qβ and T4 at approximately 100 to 200 mJ/cm² or more

TABLE 4-5. Wastewater characteristics, average (min–max) (Pei *et al.*, 2002).

Parameter	Tertiary effluent	Secondary effluent	
	Indianapolis (USA)	Evry (France)	Washington (UK)
Suspended solids (mg/L)	2.3 (1–4)	5 (3–6)	18 (7–33)
COD (mg/L)	30 (24–38)	36 (26–56)	71 (41–150)
TOC (mg/L)	8 (5.5–10.2)	< 10 (< 10–14)	26 (< 11–30)
UV254 abs (/m)	15.5 (12.5–20.8)	22.2 (17.4–20.8)	34.9 (26–50.9)
pH	7 (6.9–7.2)	7.3 (7.3–7.4)	7.5 (7.4–8.0)
Fecal coliforms (log CFU per 100 mL)	–	3.6–4.5	4.3–6.5

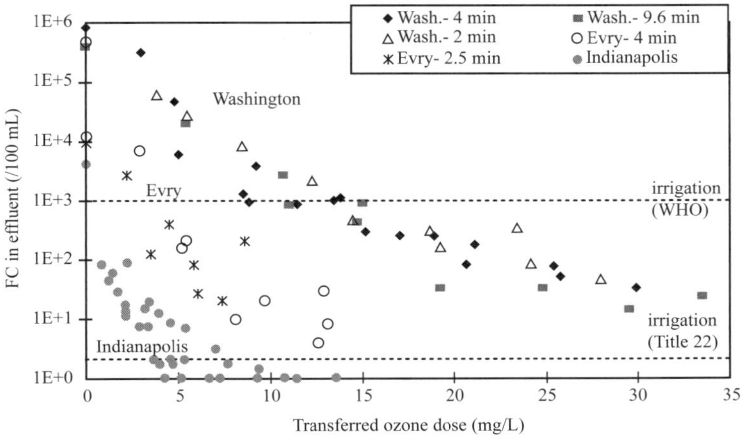

FIGURE 4-10. Ozone exposure performance for FC inactivation on three different effluents: comparison of concentration level after ozonation with reuse standards (Pei *et al.*, 2002).

(Ab. Wahid, 2011). **Figure 4-11** shows the inactivation efficiency of the UV treatment process for *E.coli* and bacteriophages Qβ and T4.

Several water reuse studies have shown that a series of coagulation, high-rate filtration, and UV disinfection require a UV dose of 80 mJ/cm² to achieve 5-logs rotavirus and MS2, while 20 mJ/cm² is needed for *E.coli* (Nasser *et al.*, 2006). In other studies, 3-log reductions of MS2 at a UV dose of 140 mJ/cm² were attained by coagulation and rapid sand filtration, followed by UV treatment (Rajala *et al.*, 2003), and only 35 mJ/cm² was needed for *E.coli* and the total phage treated by sand-ultra filtration and UV treatment (Gómez *et al.*, 2006). In comparison with human viruses, a few laboratory studies found that poliovirus (PV1) is less resistant than MS2 and GA (FRNA Group II), while Qβ phage was found to have an intermediate sensitivity when exposed to a UV dose of 0 to 150 mJ/cm² (Gantzer *et al.*, 2006). Previous studies showed that, for a 4-log inactivation, Ade40 and Ade41 required a UV dose of 124–112 mJ/cm² (Meng and Gerba, 1996), 226–203 mJ/cm² (Thurston-Enriquez *et al.*, 2003) and 222 mJ/cm² (Ko *et al.*, 2005). In addition, for MS2 to achieved 4-log inactivation, a UV dose of 65 mJ/cm² (Meng and Gerba, 1996), 119 mJ/cm² (Thurston-Enriquez *et al.*, 2003), and 118 mJ/cm² was needed (Ko *et al.*, 2005). In the long-term 2 Enhance Surface Water Treatment Rule (LT2ESWTR), the USEPA established in 2006 that a delivered UV dose of 186 mJ/cm² is required for a 4-log inactivation of all viruses for drinking water.

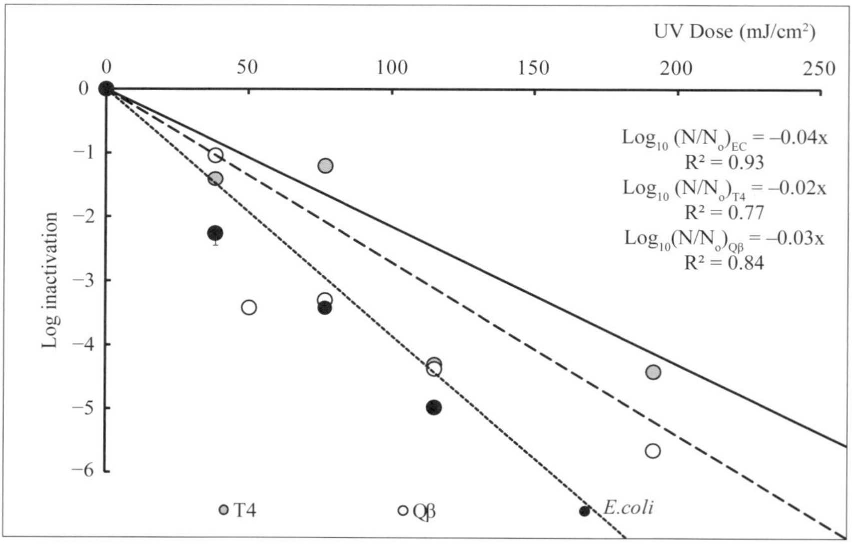

FIGURE 4-11. *E.coli* and bacteriophages Qβ and T4 inactivation in secondary effluent (Ab. Wahid, 2011).

4.2.3.3 Physico-chemical treatment

To meet the water standards for reclamation water, it is necessary to remove contaminants that are difficult to remove through biological treatment. For the reuse of treated waste water, in addition to BOD, SS, and chromaticity, organic inorganic micro contaminants have been recently highlighted. Organic micro contaminants in treated wastewater may be toxic to the human body and ecosystems, and there has been recent interest in measures to remove such contaminants efficiently. It is known that RO membranes remove various organic micro contaminants, but in some cases, chemical oxidation, adsorption, and ion exchange treatments are more efficient. In this chapter, in addition to treatment using a membrane process, other techniques for removing organic micro contaminants, i.e., absorption, chemical oxidation, and UV treatment, are examined below.

(1) *Adsorption*

Adsorption is used as the final stage after biological treatment, and is usually used to remove residue materials as a post-treatment of another process. Activated carbon is a very significant adsorbent used in the adsorption process, and is also used to decompose the precursors of the disinfection byproduct produced during the disinfection. It was reported that activated carbon has a low adsorption affinity with regard to low molecular polar organics (Metcalf and Eddy, 2006). When the biological activity of a carbon contact tank or other biological treatment processes is decreased, it is likely to be difficult to remove low molecular polar organics using activated carbon. Metals from domestic sewage and/or industrial wastewater can be removed through an adsorption treatment (Demirbas, 2008). As a reuse technology, the adsorption process has certain constraints in terms of being able to secure a physical distribution and transportation system for a large amount of adsorbents, obtaining an activated carbon contact tank site, and a propensity to cause non-reproducible waste adsorbents. The adsorption ability of activated carbon is easily affected by the pH, temperature, and flow rate, and therefore the monitoring and control of the process are important.

(2) *Chemical oxidation*

To decompose organic materials containing toxicity, a chemical oxidation treatment is generally used as a water reuse technology. For this process, an oxidizing agent such as added chlorine or ozone decomposes the target substances as it reacts directly with the organic materials in the water. During wastewater treatment, a more advanced water treatment that takes the effects of reuse water or a water circulation system into account is required, and therefore the application of ozone treatment seems to be an effective oxidation process. The removal of chromaticity or odor through ozone treatment has been regarded as a necessary process to ensure stability and proper aesthetics. In addition, ozone with strong oxidation power shows disinfecting effects against bacteria, viruses, and protozoa, and therefore it is a highly recommended reuse technology effectively used in the removal of pigmentation, deodorization, biodegradability, an improvement in flocculation efficiency, and the stabilization or mineralization of poisonous and toxic materials in persistent substances. An advanced oxidation process produces a large amount of hydroxyl radicals according to the reaction of water with various oxidizing agents, including ozone, UV, and hydrogen peroxide, and it aims to facilitate oxidation decomposition. After the final reaction during oxidation, it turns into a low-molecular substance or inorganic gas, and unlike other treatments, no residue is discovered with a few exceptions. Chemical oxidation for water reuse can be utilized to control the presence of odors, hydrogen sulfide, chromaticity, iron, and manganese, as well as for disinfection, biofilm growth inhibition, and so on. Oxidizing agents that are most commonly used in reuse include hydroxyl radicals, chlorine, ozone, chlorine dioxide, and hydrogen peroxide. These oxidizing agents are added during specific time periods (odor control, prevention of membrane fouling) and during the final treatment stage (disinfection) right before the water supply. Meanwhile, chlorine or ozone treatment can produce carcinogenic disinfection byproducts such as trihalomethane, nitrosamines, and bromate, and a solution for the inhibition of such byproducts is an urgent matter.

(3) *UV*

For the removal of organic materials in water, interest in the UV process is growing owing to its disinfection capability during water treatment; a combined application of UV exposure with ozone or hydrogen peroxide produces hydroxyl radicals, which efficiently decompose persistent substances, and the UV process can decomposes various compounds such as *N*-nitrosodimethylamine (NDMA) or other organic micro contaminants. Since most compounds are not decomposed during UV treatment alone, a method combining this process with hydrogen peroxide has been generally used. For the decomposition of pharmaceuticals included in secondary effluent from wastewater treatment plants, a concomitantly used process that uses UV with hydrogen peroxide was reported to be useful in producing by-products and removing pharmaceuticals. However, because UV treatment requires a high amount of energy, further solutions for energy reduction are needed.

4.2.3.4 Membrane technologies

(1) *Membrane separation processes*

Membrane separation processes aim at recovering target components from a mixture through a perm-selective membrane, which allows some components to pass through more easily than others. A marked difference in the penetrability between the target components and other materials in a mixture allows for a separation to take place. For water treatment, water containing as few contaminants as possible is the target product. A driving force is necessary in the membrane separation processes; otherwise, no components will penetrate through the membrane. Examples of driving forces are pressure, electrical potential, and concentration gradient. Membrane technologies applied to municipal wastewater treatment are predominantly pressure driven. Based on the membrane pore size, pressure-driven membrane separation processes are classified into microfiltration (MF), ultrafiltration (UF), nanofiltration (NF), and reverse osmosis (RO). Membranes are often made from organic polymers, usually including cellulose acetate, polysulfone, polyamide, polyurea, and polycarbonate.

Among all pressure-driven membranes, RO is the densest with a pore diameter of less than 1 nm, which renders it capable of removing components with a very low molecular weight of less than 200 Da. It can remove high levels of ions such as Na^+, Cl^-, NO_3^-, Ca^{2+}, and Mg^{2+} for desalination when sea water is used to enlarge the water supply. Owing to the marked concentrated polarization of ions, high osmotic pressure across an RO membrane is inevitable. Therefore, a higher operation pressure is required to overcome the osmosis pressure, which makes RO energy intensive. The operating pressure generally ranges from 1,550 to 3,200 kPa for new RO systems. Small organics such as herbicides and pesticides, probably present in micro pollutant surface water, can also be effectively retained by an RO membrane.

NF membranes fall in between ultrafiltration and reverse osmosis membranes, and are characterized by their pore size, on the order of nanometers, and a negatively charged surface in a neutral pH solution. Despite larger pores than in an RO membrane, a negatively charged surface assists in efficiently rejecting multivalent ions. Compared with RO, NF is advantageous in terms of osmotic pressure owing to the poor retention of monovalent ions, and is more suitable for water softening. A lower operation pressure is required, typically ranging from 500 to 1,500 kPa. Moreover, NF has been regarded as an advanced treatment technology for secondary effluent for reclaiming water. The molecular weight cut-off range for NF membranes is 200 to 1,000 Da. Attractive organic micro pollutants such as endocrine disrupting compounds (EDCs) and pharmaceutical and personal care products (PPCPs), which are generally slowly biodegraded by activated sludge, or are non-biodegradable, display moderate to high retention efficiencies in NF systems.

UF membranes have pores ranging from 1 to 100 nm, and effectively remove substances with a molecular weight in excess of 5,000 Da. Applications of UF include the removal of oil, turbidity, viruses, and bacteria, and the recovery of valuable substances such as proteins, and dyes. Since UF membranes have no ability to retain either multivalent or monovalent ions, osmotic pressure can be neglected and the operating pressure is low (50 to 500 kPa).

MF membranes are characterized by a larger pore size. They can remove particles on the order of microns (10^{-6} m) in size, including *Giardia*,

Cryptosporidium, coliforms, and particles shielding pathogens from disinfectants. Despite the difference in pore size between MF and UF, their operating pressure is similar.

Over the past two decades, novel membrane materials and configurations with lower cost, higher permeate flux, higher selectivity and retention efficiency, and higher resistance against fouling have been well developed. Given their excellent effluent quality and small footprint, membrane separation processes will be increasingly used in the future for water reclamation to alleviate water shortages.

(2) *Membrane bioreactor for municipal wastewater reclamation*

A membrane bioreactor (MBR) is regarded a successful membrane technology in the area of water treatment. Since the concept of an MBR was first established about half a century ago, a number of researches have been dedicated to promoting its practical application in the real world. This goal has now been achieved, as confirmed by the rising number of plants adopting MBR technology globally. As it increasingly matures, MBR technology will certainly enlarge its market share among the various water treatment technologies.

A classical membrane bioreactor integrates both the biological degradation and the separation of membrane technology, namely, it realizes dual functions in one unit. MF or UF membranes are commonly used in an MBR. Activated sludge growing in an MBR consistently fulfills the mission of decomposing biodegradable organic contaminants, as in any individual conventional activated sludge process, while the membrane removes particles, colloids, and/or non-degradable organics, as in any individual membrane separation process. Above all, sludge suspending in the reactor, which should have settled in the secondary sedimentation to maintain its concentration, can be efficiently retained by membranes and reliably stopped from loss. Therefore, secondary sedimentation with a large footprint is reasonably omitted and substituted by an MF or UF membrane in an MBR, diminishing the footprint. Additionally, the sludge concentration is evidently elevated with the help of the membrane, as compared to a conventional activated sludge process. Hence, the volume of MBR is substantially saved, and the organic volumetric load subsequently increases by several times.

It has been definitively proved through extensive researches that a variety of pollutants show relatively high or even complete removal efficiencies in an MBR. The removal performance can be enhanced and/or expanded using a flexible combination with other units, namely, integrated processes. The removal performance is reviewed in the next sections.

1) Removal of regular pollutants

The typical removal efficiency of turbidity, SS, and COD is >99%, >99%, and >90% (Du *et al.*, 2008; Liu *et al.*, 2000; Yang *et al.*, 2009a; Wei *et al.*, 2006), respectively, according to data from pilot-scale MBRs. Suspended particles and most colloids bigger than 0.05–0.5 μm are removed predominantly by the size exclusion of the membrane. In addition to bio-degradation, membranes have also been found to contribute to COD removal. Wang and Wu (2009) found that the typical bimodal distribution of molecular weight (MW) of dissolved organic matter (DOM) in an MBR supernatant was 6.1–382,000 kDa, whereas that in an effluent was 10.9–1870 kDa; the absence of DOM with an MW higher than 1,870 kDa demonstrated the retention of a large DOM by the membrane itself. Furthermore, it was observed that all organic carbon was removed with a gradually increasing efficiency (Wang and Waite, 2009), for which membrane pore narrowing or gel formation might be a possible explanation.

2) NH_4^+–N and TN removal

In an MBR, intensive aeration creates an aqueous environment rich in oxygen; the sludge retention time is also easily regulated at high levels. Under such conditions favorable for nitrobacteria, the nitrification of NH_4^+–N is easily achieved with a removal efficiency of higher than 90% (Du *et al.*, 2008; Yang *et al.*, 2006; Yang *et al.*, 2009a). However, to completely remove nitrogen, an anoxic environment is required for denitrification, which is a focus of recent studies. An anoxic environment can be created through either time or spatial alternation. For time alternation, a sequencing batch MBR (SBMBR) is employed. Zhang *et al.*, (2006) observed a TN removal of > 65% in such a reactor at a COD/TN ratio of 6.3 and TN load of 0.22 kg/(m^3•day). For spatial alternation, the most popular process is anoxic/oxic-MBR (A_2/O-MBR). In Cao *et al.*'s pilot scale study using an anaerobic/anoxic/oxic-MBR (A_1/A_2/O-MBR) (Cao *et al.*, 2007), where the recirculation ratio from membrane zone to anoxic zone is 300%, the TN removal was found to be 74.4%, which is close to the theoretical value. Other processes have also been proposed, such as an up-flow anaerobic sludge blanket (UASB) and an aerobic MBR (i.e., UASB-MBR) (An *et al.*, 2008), enabling simultaneous methanogenesis and nitrogen removal. The TN removal reached 82.8% with a sludge recirculation ratio (from MBR to UASB) of 800%. It was found that the realization of a shortcut biological nitrogen removal (SBNR) process can accomplish such a high TN removal efficiency.

It is interesting to note that spatial alternation can also take place in a sludge floc, which explains the fact that a single-tank MBR is additionally effective in removing TN. The heterogeneous environment (for instance, DO, substrate concentration, and microbiology) within a sludge floc along its thickness direction enables the simultaneous nitrification and denitrification (SND), namely, nitrification in the outer sphere, and denitrification in the inner sphere. He *et al.*, (2009) achieved a TN removal of around 86.6% in a single-tank MBR at DO = 0.8 mg/L. The effectiveness of SND may influenced by the food/microorganism (F/M) ratio, COD/TN, DO, pH, and steric factors such as the size and compactness of the sludge floc (He *et al.*, 2009). A high COD/TN is advantageous to SND (He *et al.*, 2009; Meng *et al.*, 2008), whereas for DO and pH, the optimal values lay within a moderate range (around DO = 1 mg/L and pH = 7). Additionally, SND can be facilitated through the addition of carriers (Wang *et al.*, 2008b; Yang *et al.*, 2009).

3) Phosphorus removal

Phosphorus can be removed in three ways: biological accumulation, membrane retention, and precipitation with additional chemical agents. Biological phosphorus removal by phosphorus accumulating bacteria requires an alternation of anaerobic and aerobic environments. Both time alternation and spatial alternation are feasible.

For time alternation, the sequencing batch membrane bioreactor (SBMBR) developed by Zhang *et al.*, (2006) exhibited a stable TP removal of around 90%. Dai *et al.*, (2007) investigated the denitrifying phosphorus removal (DPR) progress in an SBMBR, and found that the number of denitrifying phosphorus removing bacteria (DPB) can reach as high as 69.6% of the total number phosphate accumulating organisms. DPR is a process during which NO_3^-–N in an anoxic environment acts as an electro-acceptor for phosphorus uptake. Micro-organisms exhibiting such talent are referred to as denitrifying phosphorus removing bacteria (DPB).

For spatial alternation, Zhang *et al.*, (2007) constructed an MBR system featuring an enhanced biological phosphorus removal (EBPR) system composed of one anaerobic zone, two anoxic zones, and a membrane (aerobic) zone. The authors studied the effect of SRT (20, 30, 40, and 50 d) on phosphorus removal. It is generally accepted that a short SRT (high excess sludge discharge) is advantageous to phosphorous removal, which is achieved by means of discharging excess sludge abundant in phosphorous. However, a high total phosphorus (TP) removal was surprisingly achieved even when SRT was prolonged to 40 d in the EBPR-MBR system (**Figure 4-12**). The researchers ascribed this unanticipated performance to the increased phosphorus concentration in the excess sludge (expressed as kg-P/kg-MLSS) with an increased SRT. This phenomenon is exciting, simultaneously benefiting both the reduction of excess sludge discharge and sufficient phosphorus removal (Zhang *et al.*, 2007). Denitrifying phosphorus removal (DPR) was also observed to take place in this system, contributing to 52% of the biological phosphorus removal (Zhang *et al.*, 2008). It should be noted that, not only microorganisms, but also extracellular polymeric substances (EPS) surrounding them, can play a role in phosphorus removal. Compared to EPS poor in phosphorus from conventional activated sludge (CAS) process, EPS from EPBR-MBR contains much more polyphosphate (characterized by the 'end poly-P' and 'middle poly-P' peaks in **Figure 4-13**) (Zhang *et al.*, 2009). Polyphosphate in EPS is discharged along with excess sludge.

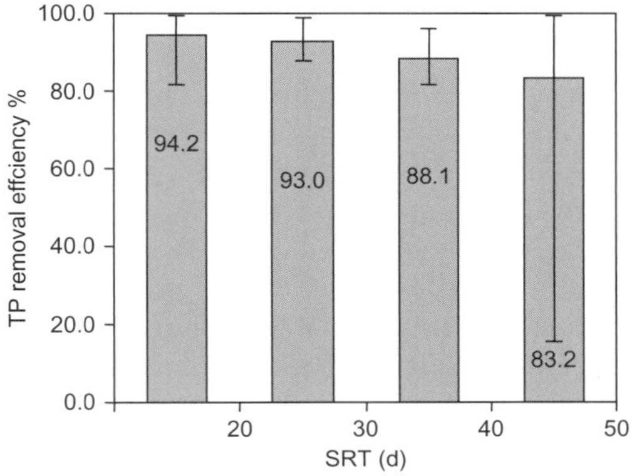

FIGURE 4-12. Average phosphorus removal in the EBPR-MBR at different SRTs.

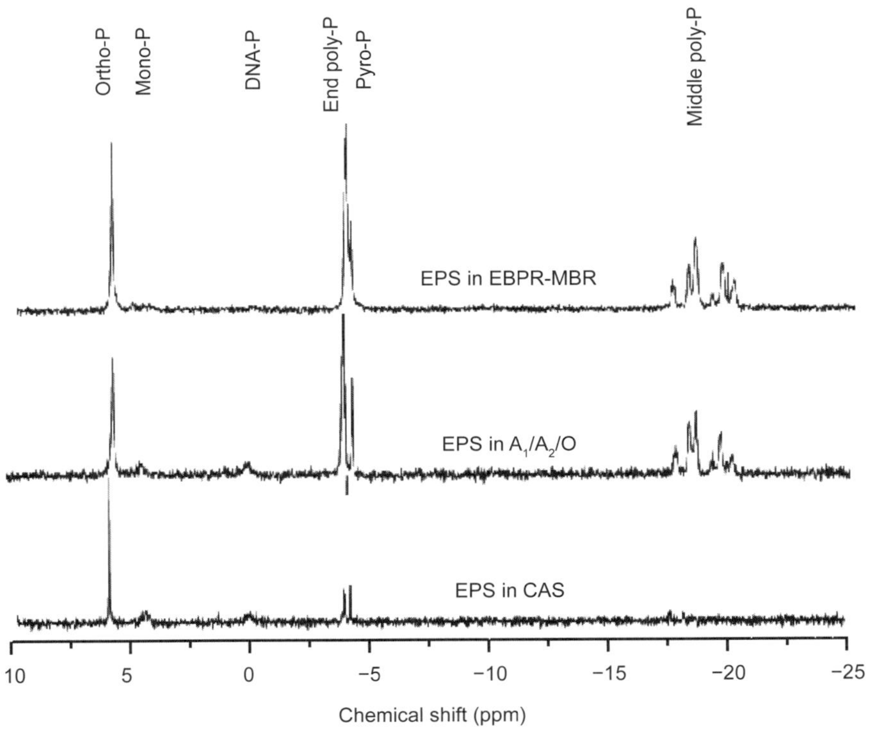

FIGURE 4-13. Typical ^{31}P-NMR spectra from the EPS investigated from three sludge samples (modified from that provided by Zhang *et al.* (2009)).

Membranes themselves can also retain phosphorus, as demonstrated by the lower phosphorus concentration in the effluent than in the supernatant. The phosphorus contained in a supernatant includes inorganic phosphates of small molecular size, as well as colloidal phosphorus typically larger than 0.025 μm (Hens and Merckx, 2002). Colloidal phosphorus is believed to chemically bond with macromolecules such as protein-like substances (according to a three-dimensional excitation-emission matrix (EEM) fluorescence spectroscopic analysis (Zhang, 2008)) from soluble microbial products (SMP). **Figure 4-14** shows the obvious existence of colloidal phosphorus (accounting for ca. 50% of TP in a supernatant); this fraction is well intercepted during membrane filtration. It should be noted, however, that a new microfiltration membrane (generally with a pore size within the range of 0.05–0.5 μm) is never effective in rejecting colloidal phosphorus. Only after the membrane fouling layer grows to a certain degree can colloidal phosphorus be well retained. This may be supported by Wang and Waite's finding (2008) that after the formation of a well-grown Ca-alginate gel layer (alginate is a model foulant for polysaccharide), the apparent diameter of the equivalent channels increased to within the range of 10–30 nm according to modeling evaluation. At this time, colloidal

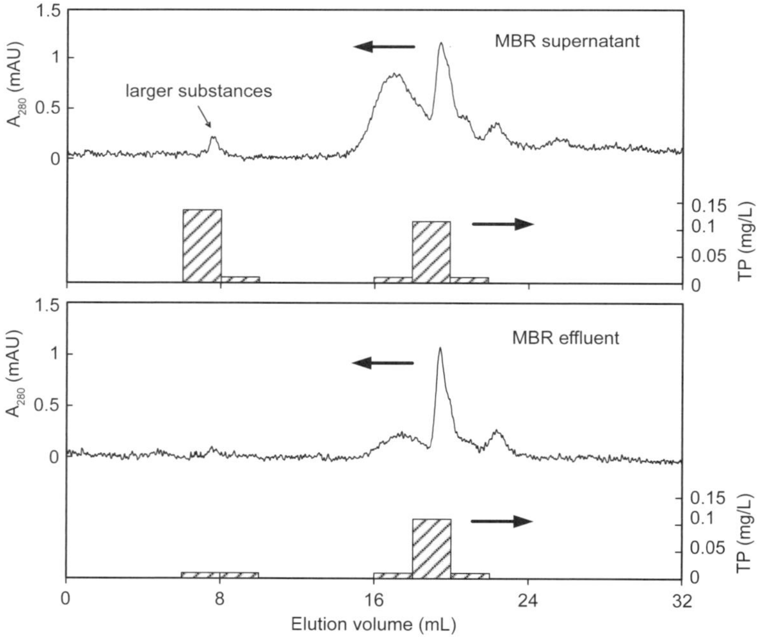

FIGURE 4-14. Molecular weight distribution and TP distribution in MBR supernatant and effluent (modified from that provided by Zhang (2008)). The molecular weight distribution was measured using a gel filtration chromatography (GFC) instrument.

phosphorus larger than 0.025 μm will be reasonably rejected by the gel layer. Zhang further confirmed this phenomenon, namely, fouling-layer enhanced phosphorus retention (Zhang, 2008). The retention efficiency of colloidal phosphorus increased rapidly from 2% to 60–70% in accord with the trans-membrane pressure (TMP) elevation during one month of operation (2%, corresponding to a freshly cleaned 0.4-μm membrane), and the retention efficiency reached as high as 87.3% when the membrane was severely fouled (TMP >50 kPa at a constant flux of ~12 L/(m^2•h)).

Phosphorus removal through precipitation with additional chemical agents is another way to ensure a low phosphorus concentration in an effluent, which is also common in conventional activated sludge processes. For instance, Cui *et al.*, (2009) developed an electrocoagulation-MBR (EC-MBR) process by installing an iron-anode plate and a titanium-cathode mesh into the down-flow zone of a submerged MBR (SMBR). By controlling the electric current intensity and current-on time to adjust the iron ion release, the TP removal could be maintained at over 95% with an effluent TP of < 0.5 mg/L.

4) Virus removal

Although reclaimed water provides an alternate way to enlarge a water supply, the inevitable exposure of humans to reclaimed water has recently raised concerns regarding health safety during utilization. The presence of pathogens is a big challenge to human health, and can induce epidemic diseases if appropriate control is not used. An MBR has almost no difficulty in retaining protozoa, helminths (as well as their eggs), fungal spores, and bacteria owing to their larger sizes than membrane pores. However, viruses small in size can easily penetrate microfiltration membranes commonly used in an MBR, and have therefore attracted a great deal of attention lately. The virus removal performance by an SMBR fed with real municipal wastewater was investigated by Wu *et al.*, (2010), who used indigenous somatic coliphages (SC) as a virus indicator. The feed SC concentration was $(2.81 \pm 1.51) \times 10^4$ PFU/ml. It was found that in the SMBR system, biomass played the most important role in SC removal, achieving a high removal efficiency of over 98%. Membrane rejection was an essential supplement to biomass, which is particularly the case when the membrane is fouled, allowing the effluent SC to be less than 10 PFU/ml. A fouling-layer enhanced virus retention has also been reported by a few other researchers, using phage T4 (Lv *et al.*, 2006; Zheng *et al.*, 2005; Zheng and Liu, 2007), phage f2 (Zheng and Liu, 2007), and others as indicators. Further study by Wu *et al.*, (2010) revealed that a gel layer plays a more important role in SC rejection than a cake layer, especially at higher filtration fluxes (**Figure 4-15**), whereas Lv *et al.*, (2006) found that a cake layer is the main contributing factor in phage T4 retention. A possible explanation for these opposite findings is the different fluxes used in these two studies. Lv *et al.*, operated the system inconstant-pressure mode with a TMP of only 8.5 kPa and a flux of merely

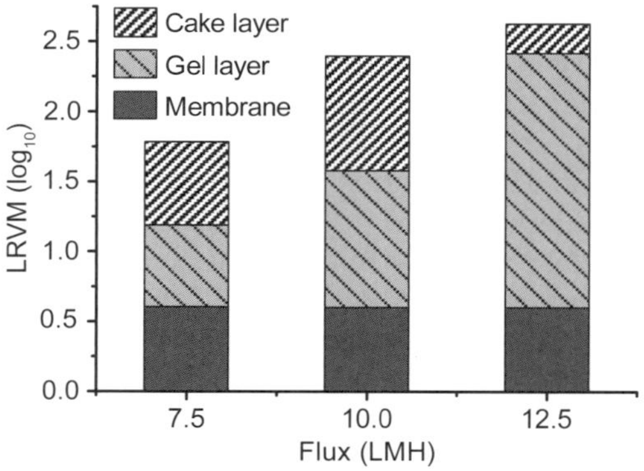

FIGURE 4-15. SC rejection contribution of different fouling layers at three fluxes when the membrane is severely fouled (Wu *et al.*, 2010). LRVM denotes the $\log_{10}$ removal value by the membrane (LRVM = $\log_{10}(C_s/C_e)$, where C_s and C_e indicate the SC concentrations in the supernatant and effluent, respectively).

around 3 LMH (estimated from Lv *et al.*, (2006)), which is much smaller than those in **Figure 4-15**; the gel layer should be much looser with wider channels (Wang and Waite, 2008) and is ineffective in retaining viruses.

5) Removal of micropollutants

In addition to microbial risk, chemical risk has come into the field of vision of many researchers owing to the ubiquitous traces of micropollutants in reclaimed water, including EDCs and PPCPs. Such micropollutants have been found to potentially harm humans and other living creatures. EDCs may be grouped into the following types: 1) phenolic compounds including bisphenol A (BPA), alkylphenols (such as 4-n-octylphenol (4-n-OP), 4-n-nonylphenol (4-n-NP), and 4-tert-octylphenol (4-tert-OP)) and alkylphenolpolyethoxylates (such as nonylphenolpolyethoxylates (NPnEO)), and 2) steroidal estrogens including estrone (E1), 17β-estradiol (E2), 17α-estradiol (17α-E2), estriol (E3), and 17α-ethynylestradiol (EE2). PPCPs may be grouped as follows: 1) anti-inflammatory and analgesic drugs such as aspirin (ASA), diclofenac (DCF), ibuprofen (IBP), and ketoprofen (KTP), 2) antibiotics such as sulfamethoxazole (SMZ), norfloxacin (NOR), and cefmetazole (CMZ), 3) lipid regulating drugs such as clofibric acid (CFA), 4) developers such as diatrizoate (DTZ), 5) musks such as tonalide (AHTN) and gallaxolide (HHCB), and 6) other drugs such as caffeine (CAF), carbamazepine (CBZ), and sulpiride (SLP).

The typical concentrations of EDCs and PPCPs in municipal wastewater, and their residual concentrations after conventional activated sludge (CAS) treatment, are shown in **Table 4-6**. The effluent concentrations are in the range of dozens to hundreds of nanograms per liter, which is much lower than in regular pollutants on the order of micrograms per liter. However, toxicity can be induced by micropollutants of such low concentration. For instance, it has been reported that the hormone system of adult trout and/or roach, as representative fish types, is clearly disrupted when the concentration of 4-*tert*-OP reaches 10 μg/L, or when that of E1 or E2 exceeds 10 ng/L (Routledge *et al.*, 1998). Moreover, for EE2, which is highly endocrine disruptive, a concentration of as low as 0.1 ng/L may be high enough to induce estrogenic abnormalities (Routledge *et al.*, 1998). It is clear that CAS, yielding EDC concentrations much higher than the minimum toxicity-inducing values, cannot satisfy the requirements in ensuring safe water reuse. In this regard, the utilization of an MBR should be worth trying owing to its potential advantages.

Chen *et al.*, (2008a) made a comparison of the BPA removal between an MBR and CAS system. The influent concentration of BPA ranged from 0.05 to 20 mg/L. A stable removal of over 93.7% was obtained in the MBR, while that in the CAS reactor was less stable in spite of a similar average efficiency. Moreover, the MBR could bear much higher volume loadings than the CAS reactor without affecting the BPA removal efficiency. With regard to NP*n*EO removal, similar results were obtained in these two systems (Chen *et al.*, 2008b). Zhou *et al.*, (2011) compared the removal of eight typical EDCs in an MBR system with a sequencing batch reactor (SBR). The eight target EDCs selected were 4-*n*-OP, 4-*n*-NP, NP*n*EO, BPA, E1, E2, E3, and EE2. For substances that are relatively easily biodegradable such as BPA, E2 and E3, both reactors exhibited excellent removal efficiency of over 95% (**Figure 4-16**). However, with regard to more refractory compounds such as E1 and EE2, the MBR exhibited a better performance than the SBR.

The concentration itself has a quantifiable relationship with the existence of EDCs and PPCPs, but is not directly related to toxicity. An estrogenicity assay has been developed to build a bridge between concentration and toxicity. This was determined through a yeast estrogen screen (YES) assay and expressed as an E2 equivalent quantity (EEQ) (cf. Zhou *et al.*, (2009b) for details). In Zhou *et al.*'s study, it was found that the EEQ removal efficiency of around 90% in the MBR was much higher than the around 50% efficiency in the SBR (Zhou *et al.*, 2011). The EDC removal behavior of a full-scale MBR combined with an anaerobic-anoxic-oxic process (i.e. $A_1/A_2/O$-MBR) was also investigated (Wu *et al.*, 2011). It was found that more than 97% of BPA, E1, E2, 17α-E2, and E3, 87% of EE2, and over 70% of 4-*n*-OP and 4-*n*-NP can be removed through this combined process. The estrogenicity was markedly reduced from 72.1 ng-EEQ/L in the influent to 4.9 ng-EEQ/L in the effluent. An analysis of the performance profile of this process revealed that the front zones, i.e., the anaerobic zone and the anoxic zone, accounting for about one-half of the total

TABLE 4-6. Removal of typical EDCs and PPCPs at conventional municipal wastewater treatment plants in Beijing, China.

	Name	Abbreviation	Solubility in water (mg/L)	log K_{ow} [a]	Concentration (ng/L) Influent	Concentration (ng/L) Effluent
EDCs	4-*n*-octylphenol	4-*n*-OP	–	5.00[b]	104.3[c]	70.1[c]
	4-*n*-nonylphenol	4-*n*-NP	–	5.40[b]	71.3[c]	42.0[c]
	Bisphenol A	BPA	120[d]	3.32[d]	974.3[c]	85.3[c]
	Estrone	E1	30[d]	3.13[d]	606.3[c]	145.4[c]
	17α-estradiol	17α-E2	3.9[d]	3.94[d]	77.6[c]	64.5[c]
	17β-estradiol	E2	3.6[d]	4.01[d]	83.4[c]	37.3[c]
	Estriol	E3	441[d]	2.45[d]	368.3[c]	20.4[c]
	17α-ethynylestradiol	EE2	11.3[d]	3.67[d]	430.8[c]	92.6[c]
PPCPs	Tonalide	AHTN	–	5.7[e]	218.2[f]	109.2[f]
	Gallaxolide	HHCB	–	5.9[e]	1968.1[f]	827.0[f]

a: Common logarithm of *n*-octanol/water partition coefficient.
b: Estimated using chemical modeling software, HyperChem 7.1.
c: Averaged from data provided by Zhou *et al.*, (2010) analyzing samples from three conventional municipal wastewater treatment plants.
d: Obtained from Comerton *et al.*, (2008), where the log K_{ow} values were estimated using chemical modeling software, HyperChem 7.5.
e: According to Zhou *et al.*, (2009a).
f: Averaged from data provided by Zhou *et al.*, (2009a) analyzing samples from three conventional municipal wastewater treatment plants.

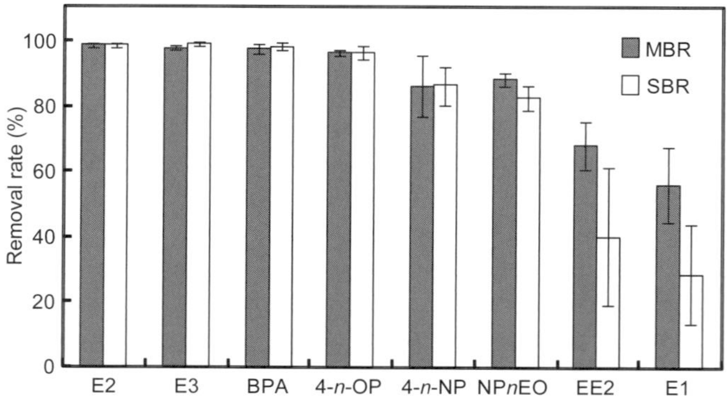

FIGURE 4-16. Removal rates of target EDCs using an MBR and SBR (Zhou *et al.*, 2011).

volume, play a significant role, with a local EEQ removal efficiency of 37% and 80%, respectively (without taking into account the sludge recirculation), therefore contributing to nearly 90% of the overall EEQ removal.

The capacity to effectively remove EDCs and PPCPs has been widely validated in MBR-based systems. The next work is to depict their fate for a deep understand of the removal mechanisms. Improvements can be made based on the contributions of the different mechanisms, for the purpose of enhancing the removal performance. The major fate is usually considered to be physical adsorption by sludge or biodegradation. According to Wu *et al.*'s observations (2011), adsorption by sludge plays a prominent role in the removal of most of the selected hydrophobic EDCs, the solid-water distribution coefficients (the ratio of solid-phase concentration to aqueous-phase concentration) of which were found to be mostly above 10,000 L/kg-SS. In some cases, biodegradation is predominant. This was observed by Chen *et al.*, who used an MBR to remove BPA (Chen *et al.*, 2008a) and NP*n*EO (n = 1–4) (Chen *et al.*, 2008b). Zhou *et al.*, (2011) investigated the effect of SRT on EDC removal in an MBR (**Figure 4-17**) As can be seen, the removal of 4-*n*-NP, NP*n*EO, E1, and EE2, as well as the total estrogenicity expressed as EEQ, was effectively improved by an increase in the SRT. From the viewpoint of physical adsorption, an increased SRT is adverse to EDC removal owing to a decrease in excess sludge discharge. The opposite trend observed by Zhou might suggest an enhancement in the biodegradation. Further research is needed to reveal the detailed mechanisms of this, however.

6) A comparison of MBR effluent quality with reclaimed water regulation

Table 4-7 summarizes the results obtained from several studies with respect to the removal efficiency and effluent concentrations of the turbidity, SS, COD, NH_4^+–N, TN, and TP. The corresponding water quality standards for municipal reclaimed water for different uses, including miscellaneous urban use, scenic purposes,

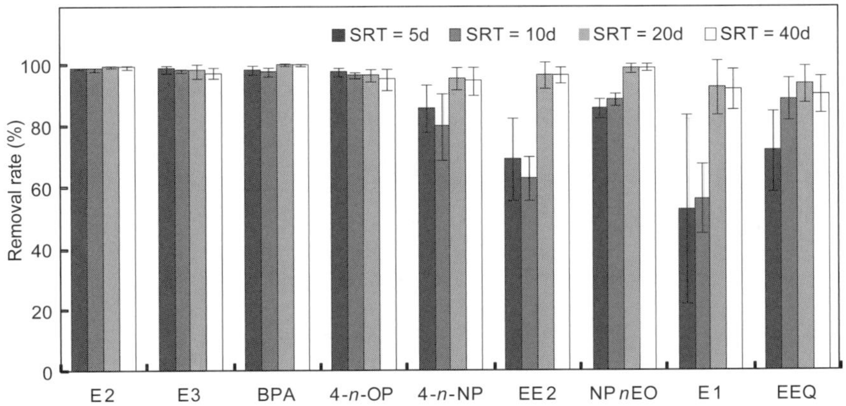

FIGURE 4-17. Effect of SRT on EDC removal using an MBR (Zhou *et al.*, 2011).

TABLE 4-7. The performance of several MBR-based processes as reported in the literature.

Number	MBR-based process	Removal efficiency	Effluent concentration	Reference
1	MBR	COD: 86% NH_4^+–N: 58.2% Turbidity: 99.4%	COD: 38.5 mg/L NH_4^+–N: 19.5 mg/L Turbidity: 0.96 NTU	Wei *at al.*, 2006
2	A/O-MBR	COD: 93.9 ± 5.1% NH_4^+–N: 95.2 ± 2.7% TN: 77.3 ± 8.4% TP: 68.0 ± 14.1% SS: 100%	COD: 22.0 ± 16.0 mg/L NH_4^+–N: 1.9 ± 1.1 mg/L TN: 15.0 ± 5.6 mg/L TP: 4.0 ± 1.8 mg/L SS: 0	Wang and Wu, 2009
3	MBR	COD: >90% NH_4^+–N: >95%	COD: 16.85 ± 12.75 mg/L NH_4^+–N: <1 mg/L	Liu *et al.*, 2000
4	MBR	COD: 84.8% NH_4^+–N: 91.2% TN: 45.6% TP: 43.1%	COD: 49.6 mg/L NH_4^+–N: 3.9 mg/L TN: 30.9 mg/L TP: 3.4 mg/L	Yang *et al.*, 2006
5	A^2/O-MBR	E: 97.7% COD: 91.3% NH_4^+–N: 98.6% TN: 74.4% TP: 88.4%	BOD_5: 2.54 mg/L COD: 24.9 mg/L NH_4^+–N: 0.68 mg/L TN: 14.3 mg/L TP: 0.98 mg/L	Cao *et al.*, 2007
6	UASB-MBR	TOC: ~100% NH_4^+–N: 98.2% TN: 82.8%	NH_4^+–N: <3 mg/L TN: <10 mg/L	An *et al.*, 2008
7	SBMBR	COD: 97.7% NH_4^+–N: 93.1% TN: 67.6% TP: 91.4%	COD: 9.2 mg/L NH_4^+–N: <10 mg/L TN: 20 mg/L TP: 2 mg/L	Zhang *et al.*, 2006
8	MBR	COD: 95.6% NH_4^+–N: 80% TN: 70%	COD: <35.6 mg/L NH_4^+–N: <10 mg/L TN: 20 mg/L	Yang *et al.*, 2009b
9	SBMBR	COD: 95.2% NH_4^+–N: 95% TN: 50.5% TP: 96.4%	COD: 13.6 mg/L NH_4^+–N: 0.7 mg/L TN: 9.22 mg/L TP: 0.07	Xia *et al.*, 2009
10	ICAS-MBR[a]	COD: 95.4 ± 3.55% NH_4^+–N: 93.1 ± 5.95% TN: 69.0 ± 6.00% TP: 82.6 ± 14.22%	COD: 16.9 ± 10.13 mg/L NH_4^+–N: 2.4 ± 2.22 mg/L TN: 14.3 ± 2.94 mg/L TP: 0.85 ± 0.64 mg/L	Wang and Wu, 2009

(Continued)

TABLE 4-7. Continued.

11	A/O-MBR	COD: 94.8 ± 3.96% NH_4^+–N: 92.5 ± 7.10% TN: 78.0 ± 6.94% TP: 64.6 ± 9.93%	COD: 17.3 ± 10.98 mg/L NH_4^+–N: 2.7 ± 2.66 mg/L TN: 10.2 ± 3.69 mg/L TP: 1.7 ± 0.42 mg/L	Wang and Wu, 2009
12	SBMBR	COD: 57.34–88.59% NH_4^+–N: 56.55–96.72% TN: 52.22–83.41% TP: 34.74–82.86%	COD: 11.34–33.19 mg/L NH_4^+–N: 0.68–6.14 mg/L TN: 3.93–11.14 mg/L TP: 0.36–0.89 mg/L	Zhang *et al.*, 2007

a: Intermittently cyclic-activated sludge-membrane bioreactor

groundwater recharge, farmland irrigation, and industrial use, as established by the Chinese Ministry of Construction, are shown in **Table 4-8~Table 4-12**. The feasibility of an MBR effluent as reclaimed water can be evaluated through a comparison of the MBR effluent quality with the water quality standards. It can be seen that an MBR effluent can meet the COD standards most easily, but has some difficulties in meeting the TN and TP standards. MBR-based processes are relatively qualified to supply reclaimed water and mitigate water shortages.

(3) Advanced membrane filtration of secondary effluent

Under the pressure of water scarcity and the subsequent increase in stringent water quality standards, many municipal wastewater plants face the challenge of renovation. An advanced membrane filtration of secondary effluent is one preferable option for upgrading such plants. The membranes employed in advanced filtration are generally ultrafiltration or nanofiltration. In view of the safe utilization of reclaimed water, an advanced membrane filtration should mainly target possibly toxic substances such as pathogens and micropollutants. Microfiltration membranes, with a relatively large pore size, are believed to have trouble successfully finishing such tasks, while an RO is too effective in retaining pollutants, and thus applying high pressure is a huge waste of energy and cost.

Table 4-13 shows several advanced membrane-based treatments used to further purify municipal secondary effluent. To mitigate membrane fouling and enhance the pollutant removal, pretreatments are adopted before membrane filtration, among which coagulation is a popular option. The blocking

TABLE 4-8. Water quality standard for miscellaneous urban water uses (GB/T 18920-2002).

	Miscellaneous urban water uses				
	Toilet flushing	Road cleaning/ fire-fighting	Urban irrigation	Car washing	Architectural construction
Turbidity (NTU)	5	10	10	5	20
SS (mg/L)	–	–	–	–	–
COD_{Cr}(mg/L)	–	–	–	–	–
NH_4^+–N (mg/L)	10	10	20	10	20
TN (mg/L)	–	–	–	–	–
TP (mg/L)	–	–	–	–	–
Total coliform (/L)	3	3	3	3	3
Faecal coliform (/L)	–	–	–	–	–

TABLE 4-9. Water quality standard for scenic purposes (GB/T 189215–2002).

	Scenic uses					
	Aesthetic use			Recreational use		
	Water-course	Impoundment	Water-scaping	Water-course	Impoundment	Water-scaping
Turbidity (NTU)	–	–	–	5	5	5
SS (mg/L)	20	10	10	–	–	-
COD_{Cr}(mg/L)	–	–	–	–	–	-
NH_4^+–N (mg/L)	5	5	5	5	5	5
TN (mg/L)	15	15	15	15	15	15
TP (mg/L)	1	0.5	0.5	1	0.5	0.5
Total coliform (/L)	–	–	–	–	–	-
Faecal coliform (/L)	10,000	10,000	2,000	500	500	ND

TABLE 4-10. Water quality standards for groundwater recharge (GB/T 19772-2005).

	Groundwater recharge	
	Surface recharge	Injection recharge
Turbidity (NTU)	10	5
SS (mg/L)	–	–
COD_{Cr}(mg/L)	40	15
N (mg/L)	NH_4^+–N : 1 NO_3^-–N: 15 NO_2^-–N: 0.02	NH_4–N: 0.2 NO_3^-–N: 15 NO_2^-–N: 0.02
TP (mg/L)	1	1
Total coliform (/L)	–	–
Faecal coliform (/L)	1,000	3

TABLE 4-11. Water quality standard for farmland irrigation (GB 20922-2007).

	Farmland irrigation			
	Fiber crops	**Dry grain**	**Wet grain**	**Open-air vegetables**
Turbidity (NTU)	–	–	–	–
SS (mg/L)	–	–	–	–
COD_{Cr}(mg/L)	200	180		150
NH_4^+–N (mg/L)	–	–	–	–
TN (mg/L)	–	–	–	–
TP (mg/L)	–	–	–	–
Total coliform (/L)	–	–	–	–
Faecal coliform (/L)	40,000	40,000		40,000

TABLE 4-12. Water quality standards for industrial uses (GB/T 19923-2005).

	Industrial uses				
	Cooling water				
	Uniflow cooling water	**Recirculated cooling water**	**Rinsing water**	**Boiler feedwater**	**Process water**
Turbidity (NTU)	–	5	–	5	5
SS (mg/L)	30	–	30	–	–
COD_{Cr}(mg/L)	–	60	–	60	60
NH_4^+–N (mg/L)	–	10	–	10	10
TN (mg/L)	–	–	–	–	–
TP (mg/L)	–	1	–	1	1
Total coliform (/L)	–	–	–	–	–
Faecal coliform (/L)	2,000	2,000	2,000	2,000	2,000

of membrane pores can be reduced owing to the formation of larger flocs with the help of coagulants.

With regard to regular pollutants, the removal and effluent concentration of COD was most frequently reported in the studies listed in **Table 4-13**. Similar to MBR-based processes, advanced membrane filtration concerns not only regular pollutants but also micropollutants and microbiological indexes. Although water quality standards with respect to micropollutants have yet to be established, the awareness of the levels of such substances in an effluent is of paramount importance. Much work should be conducted to promote the establishment of standards for micropollutants. Until then, the effluent quality of advanced membrane filtration procedures can be comprehensively evaluated.

TABLE 4-13. The performance of several advanced membrane-based treatments reported in the literature.

Advanced membrane-based treatment	Removal efficiency	Effluent concentration	Reference
1. NF	Nonylphenol: 74%; Bisphenol A: 95% Estrone: >76% 17α-Ethynyl estradiol: >71% Genistein: >97%	Nonylphenol: 0.54 μg/L Bisphenol A: 4.08 ng/L Estrone: ND 17α-Ethynyl estradiol: ND Genistein: ND	Lee *et al.*, 2008
2. NF	COD: 66% NH_4^+–N: 31% TN: 88% TP: 86% Bisphenol A: 54.5% Nonylphenoleth oxylate: 55.4% Estradiol: 16.7% Estriol: 61.5% 17α-Ethynyl estradiol: 46.3%	COD: 12.8 mg/L NH_4^+–N: 0.09 mg/L TN: 3.9 mg/L TP: 0.5 mg/L Bisphenol A: 0.25 μg/L Nonylphenolethoxylate: ND Estradiol: 0.0049 μg/L Estriol: 0.002 μg/L 17α-Ethynyl estradiol: 0.029 μg/L	Chen, 2008
3. NF	COD: 90% NO_2^- + NO_3^-: 80% Total soluble P: 70%	COD: 10.2 mg/L NO_2^- + NO_3^-: 3.5 mg/L Total soluble P: 1.4 mg/L	Wang *et al.*, 2008a
4. Flotation + BAC + NF	BOD: 53% SS: 80% Conductivity: 88% HPC[a]: 100%	BOD: 5.08 mg/L SS: <1 mg/L Conductivity: 286 μS/cm HPC : ND	Ran *et al.*, 2004
5. MF+NF	DOC: 85–95% Estrone: 30–50%		Jin *et al.*, 2010
6. Coagulation + UF	P: 94%	P: 0.3 mg/L	Citulski *et al.*, 2009
7. UF	COD: 19% DOC: 25% Turbidity: 90% Faecal coliform: 99.94% Total coliform: 99.96%	COD: 12–20 mg/L DOC: 5–8 mg/L Turbidity: 0.2–1 NTU Faecal coliform: 6 CFU/ 100 ml Total coliform: 17 CFU/100 ml	Dialynas *et al.*, 2008
8. Coagulation + Sedimentation + UF	Turbidity: >79% COD_{Mn}: 34–45% TOC: 20%	Turbidity: <0.1 NTU COD_{Mn}: 4.71–7.25 mg/L TOC: 5.05–5.35 mg/L	Li *et al.*, 2008
9. UF	Turbidity: 95.1–100% COD: 10.3–59.1	Turbidity: <0.45 NTU COD: 13.73–40.36 mg/L	Yang *et al.*, 2007
10. Sand filtration + UF	Turbidity: 96% COD: 27% Total N: 76% Total coliform: 100% HPC: 100%	Turbidity: 0.07 NTU COD: 44 mg/L TN: 4.9 mg/L Total coliform: 0 HPC: 0	Wu *et al.*, 2004

a: heterotrophic plate counts

4.2.3.5 Confidence in a system

Water reuse facilities also require a high level of reliability similar to water treatment plants. There is a possibility of damage caused by use of inappropriately treated reuse water, and therefore strict standards should be applied in terms of the overall water quality parameters available. For the continual production of highly reliable and safe reuse water, it is necessary to carefully focus on reliability during the design, construction, and operation. In particular, as pathogenic organisms may cause human infections, the reliability of treatment is very significant. Therefore, the monitoring of certain water quality items such as turbidity is necessary to confirm the reliability and analyze the functions of treatment. To maintain highly reliable reuse water, the following aspects are required: 1) operation of the reuse and water supply systems by qualified personnel, 2) Performance of treatment process and Alarm system in trouble, 3) a comprehensive program for water quality protection that guarantees an accurate sampling and analysis protocols, and 4) an adequate emergency facility for the retention of reuse water requiring retreatment.

4.3 SYSTEM AND TECHNOLOGIES FOR MUNICIPAL WASTEWATER UNDER SUSTAINABLE DEVELOPMENT

4.3.1 *Advances and perspectives of treatment level in municipal wastewater systems*

4.3.1.1 Perspective toward water quality preservation

A number of artificial chemical substances used in human activities are released into sewage water. Some chemical substances can be removed by sewage water treatment when an activated sludge process is applied. However, persistent chemical substances are not adequately removed through a biological sewage treatment process and are released into aquatic environments. Recently, several pharmaceuticals, personal care products, and endocrine disrupting compounds have been detected in sewage effluents and surface waters in several countries. Although their effects on aquatic ecosystems are still unknown, the presence of such chemicals is of concern owing to their adverse effects on aquatic organisms living in the receiving waters. Most chemical substances are of anthropogenic origin, and sewage treatment plant effluents are important points of discharge for the chemicals present in rivers, streams, and surface waters. Therefore, the elimination of chemical substances within sewage treatment plants is considered a very important.

Some water treatment processes are clearly more effective for reducing the concentration of a broad range of chemical substances. Conventional biological treatments are not always efficient for a reduction of microcontaminants, as shown in **Figure 4-18**, where pharmaceuticals and personal care products demonstrate a wide range of removal (Narumiya *et al.*, 2009). Coagulation, flocculation, and precipitation processes are mostly ineffective for removing dissolved chemical substances (Ternes *et al.*, 2002; Westerhoff *et al.*, 2005). Biological sewage treatment processes, such as the activated sludge process, have been shown to greatly reduce the concentration of compounds that are biodegradable or readily bind to particles (Alcock *et al.*, 1999; Ternes *et al.*, 1999, b; Drewes *et al.*, 2002; Snyder *et al.*, 2004; Joss *et al.*, 2005). Reverse osmosis (RO) and nanofiltration (NF) membranes provide effective barriers for the rejection of contaminants, while microfiltration (MF) and ultrafiltration (UF) membranes provide a selective removal of contaminants with specific properties (Snyder *et al.*, 2006).

Oxidative processes such as chlorination and ozonation are effective for reducing the concentration of several classes of microcontaminants. However, the removal efficacy is a function of the contaminant structure and oxidant dose (Zwiener *et al.*, 2000; Adams *et al.*, 2002; Huber *et al.*, 2003; Snyder *et al.*, 2003; Ternes *et al.*, 2003; Pinkston *et al.*, 2004). The removal efficiency of pharmaceuticals and personal care products (PPCPs) through the ozonation process was studied, as shown in **Figure 4-19** (Kim *et al.*, 2009). The ozonation process can be used to effectively remove a variety of PPCPs, although some PPCPs such as 2-QCA, DEET, and cyclophosphamide showed a relatively low degradability compared with other PPCPs. However, since the DOC concentration is not decreased at lowered concentrations of PPCPs, the formation of intermediate products through the degradation of the parent PPCPs is of concern.

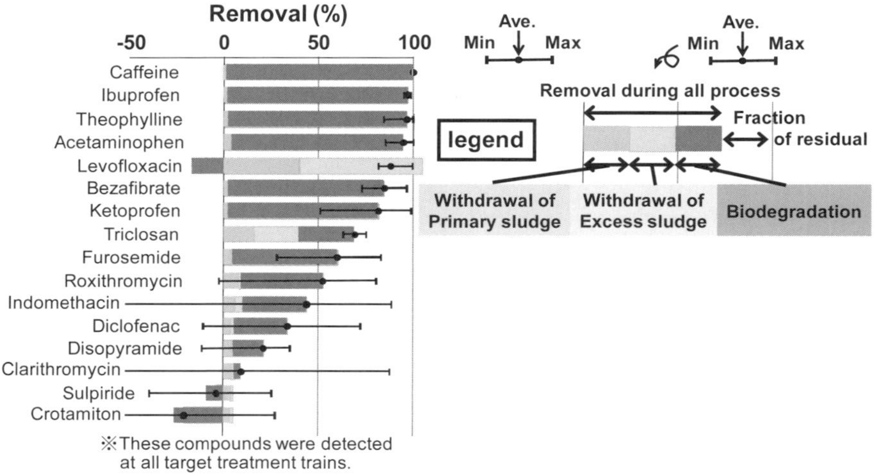

FIGURE 4-18. Removal efficiency of pharmaceuticals and personal care products in conventional biological treatment (Narumiya *et al.*, 2009).

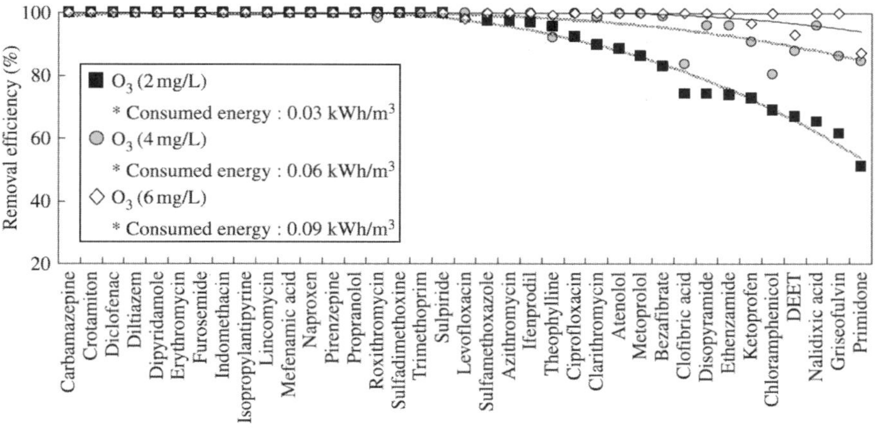

FIGURE 4-19. Removal efficiency of pharmaceuticals and personal care products (PPCPs) through the ozonation process (Kim *et al.*, 2009).

4.3.1.2 Perspective toward water resources

Under the increasing demand of water resources, people around the world will face a shortage of water. Wastewater reclamation and reuse are attracting attention as environmentally sound and stable water resource techniques. Therefore, reclaimed waste water is considered a sustainable and attractive alternative in water-threatened areas around the world. However, only 1.3% of treated wastewater is recycled in Japan, and most reclaimed waste water is used for urban areas, with little applied to agricultural irrigation fields. In urban areas, reclaimed waste water is used for toilet flushing, landscape irrigation, and recreational uses. Reclaimed waste water is also sprinkled on roads to decrease the atmospheric temperature. A total of 14 million m^3/year of reclaimed waste water is utilized in the agricultural sector in Japan. The major application is in Kumamoto city, which is located in the western part of Japan, where the reclaimed waste water is used as an irrigation water source for rice cultivation. Since 1985, the rice fields in Kumamoto city have been irrigated using reclaimed waste water (25,000 to 30,000 m^3/day) blended with river water during only the summer months. Very little reclaimed waste water is applied to dry-farming irrigation in Japan.

In 2005, guidelines for the urban use of reclaimed waste water were promulgated by the Ministry of Land, Infrastructure, and Transport (MLIT) in Japan. To ensure human health standards, these guidelines regulate the values of *E. coli* on total coliform, turbidity, pH, and residual chlorine concentration. However, there are no criteria for the agricultural reuse of reclaimed waste water for vegetables in Japan. Application of the state of California's reuse criterion, Title 22 (CDHS, 1978), which is one of the most stringent standards of wastewater reclamation in the world, has been planned for protecting public health. The use of the Title 22 system is preferable owing to its treatment process, which guarantees sufficient safety against pathogenic infections by viruses (Tanaka *et al.*, 1998).

Wastewater contains various kinds of pathogens and hazardous chemicals, and may pose a risk to human health and adversely affect the environment, such as soil and groundwater, as a result of long-term irrigation. Therefore, careful attention to the treatment and application of reclaimed waste water is essential. In particular, pathogens are considered to be important risks from reclaimed wastewater usage. Among the different types of pathogens, viruses have a high potential of infectivity and are difficult to reduce during conventional wastewater treatment. Therefore great attention should be paid to viruses existing in reclaimed wastewater.

Membrane treatment

Reclamation treatment for wastewater using membrane filtration has attracted greater attention, as membrane treatments are advantages for the removal of particulate matter, pathogens, organic compounds, and so on. Microfiltration (MF) is generally used to reduce turbidity and certain kinds of colloidal suspension. Microfiltration is also applicable for removing bacteria and spores from water, and is applied in water purification plants. Ultrafiltration (UF) is the process for separating extremely small particles and dissolved molecules. The primary basis for separation is the molecular size, and materials ranging in size from 1 to 1,000 K molecular weight are retained by ultrafiltration membranes, while salts and water will pass through. Ultrafiltration offers higher removal than MF, especially in viruses. Nanofiltration (NF) is a relatively recent membrane filtration process, which ranges somewhere between ultrafiltration (UF) and reverse osmosis (RO). The most different speciality of nanofiltration membranes is the higher rejection of multivalent ions. Nanofiltration membranes are used in softening water, industrial wastewater treatment, wastewater reuse, product separation in the industrial fields, and recently, desalination as a two-pass nanofiltration system. RO is a filtration process that removes many types of molecules and ions from water, and is most commonly used for drinking water purification from seawater, removing the salt and other substances from the water. It is also used to clean wastewater effluent and brackish groundwater. The effluent is firstly treated at an effluent treatment plant with other type of membrane filtration, and the clear effluent is then subjected to a reverse osmosis system.

Virus removal through ultrafiltration

For waste water reclamation, virus removal is considered important. There are different treatment processes for removing viruses from water, and the ultrafiltration process is an attractive option owing to a relatively high throughput compared with nanofiltration and the RO process. Ultrafiltration offers a higher removal of viruses and is often applied to the reclamation treatment process. To achieve a higher removal, the coagulation process along with ultrafiltration is considered a viable alternative in terms of higher virus removal and a reduction in coagulant and chlorine dosages.

Qβ phage is often used as a model virus, and the removal efficiency of the model virus and cumulative probability at the dosage of coagulant PAC 0 and 20 mg/L were examined (**Figure 4-20**). As shown in **Figure 4-20**, the removal efficiency increased to 4 to 5 logs based on the coagulant dosage. If the guaranteed amount of virus removal is considered to be 95%, removal efficiencies of 1.4 and 5.8 log can be achieved at dosages of PAC 0 and 20 mg/L, respectively. The average efficiencies were 3.2 to 6.9 log at dosages of PAC 0 and 20 mg/L. The effect of the pore size distribution of UF membranes on virus rejection was also investigated (Urase *et al.*, 1994). Contrary to the expectations based on the nominal molecular weight cutoff size, none of the five membranes tested provided a complete barrier for the model virus used in this study. This may hold true in this experiment, especially at a dosage of PAC 0 mg/L. Viruses are usually negatively charged and can be attached easily to coagulants, which are positively charged. Furthermore, spiked viruses are easily attached to the particles in feed water. More than 60% of spiked Qβ phage was attached to the particles. However, nearly 100% of Qβ phage was detected from a residue at a dosage of PAC 20 mg/L. The coagulants make the particle size larger, which results in a higher virus removal through the UF process.

4.3.1.3 Perspective of water quality control of a combined sewer overflow (CSO)

Many of the cities in Japan that constructed sewage collection systems during the early 20th century applied single-pipe systems that collect both sewage and storm water runoff. This type of collection system is called a combined sewer system, and is cheaper and quicker to construct. However, combined sewers can cause serious water pollution problems from combined sewer overflows (CSOs), which are caused by large variations inflow between dry and wet weather conditions. This type is no longer used when building new sewer

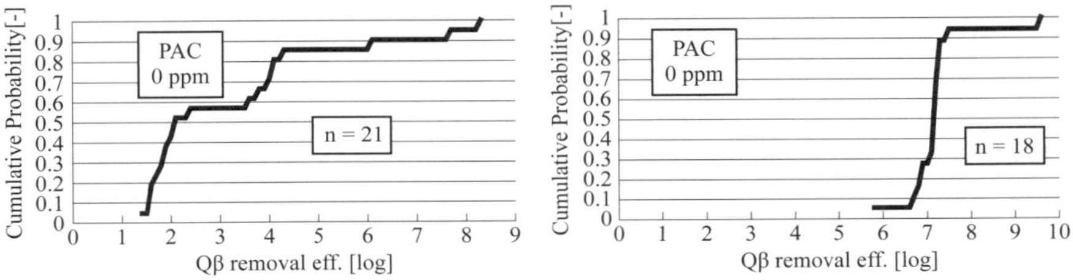

FIGURE 4-20. Qβ phage removal efficiency and cumulative probability (Shigematsu *et al.*, 2007).

TABLE 4-14. Pollutants of concern and consequences of CSOs (US EPA 2001).

Pollutants	Principal Consequences
Bacteria (e.g., FC, *E. coli*, enterococci) Viruses Protozoa (e.g., *Giardia, Cryptosporidium*)	Beach closures Odors Shellfish bed closures Drinking water contamination Adverse public health effects
Trash and floatables	Aesthetic impairment Devaluation of property Odors Beach closures
Organic compounds Metals Oil and grease Toxic pollutants	Aquatic life impairment Adverse public health effects Fishing and shellfishing restrictions
Biochemical oxygen demand (BOD)	Reduced oxygen (O_2) levels and fish kills
Solids deposits (sediments)	Aquatic habitat impairment Shellfish bed closures
Nutrients (e.g., nitrogen (N), phosphorus (P))	Eutrophication. algal blooms Aesthetic impairment
Flow shear stress	Stream erosion

systems, but many older cities continue to operate combined sewers. The CSO is the discharge of wastewater and storm water from a combined sewer system directly into rivers, lakes, or oceans when the volume of wastewater exceeds the system capacity after periods of heavy rainfall or snow melt.

The untreated sewage from these overflows can contaminate waters, causing serious water quality problems and threatening drinking water supplies as shown in **Table 4-14**. It can also create threats to public health for those who make contact with the discharged water. The CSO can carry bacteria, fungi, viruses, and protozoa, and can cause diseases such as cholera, dysentery, infectious hepatitis, and diarrheal syndrome. People can be exposed to the contaminants from sewage in drinking water sources and through direct contact in areas of high public access such as basements, streets, and water used for recreational purposes. The CSO can also contain nutrients and pollutants such soil, grease, pesticides, and chemicals such as pharmaceuticals and personal care products. These pollutants will be flushed out into aquatic environments after heavy rains and result in water contamination.

(1) *Measures to address CSOs*

Administrative agencies have been undertaking projects to address CSOs since the 1990s. In Japan, an oil ball, which is made by oil released into sewage, was found at Odaiba Seaside Park in Tokyo Bay. It was revealed that the oil ball came from CSOs after a heavy rain, and CSOs came to be considered important problems in maintaining water environments. The Ministry of Land, Infrastructure, Transport, and Tourism of Japan decided that municipalities must address the CSOs problem by 2023 by introducing newly developed technologies or facility improvements.

CSOs must be controlled because storm flows in combined sewerage systems are intermittent and highly variable in both pollutant concentration and flow rate. There are several kinds of measures to address CSOs, such as CSO sewer separation, CSO storage, treatment facilities, and wastewater treatment plant improvements. CSO sewer separation has been undertaken in the United States and Japan, which means building a second piping system. However, since high costs and/or physical limitations may preclude the building of a separation system, CSO separation is not actual and reasonable in many cases.

The use of a CSO storage facility, such as a tunnel that can store the sewage water when heavy rains occur, is one solution. Storage tunnels store combined sewage but do not treat it. When the storm is over, the flows are pumped out of the tunnel and sent to a wastewater treatment plant. However, a strategy to optimize a CSO control system should be considered before constructing a storage facility. An optimized system maximizes the use of an existing system before a new construction, and sizes the storage volume in accord with the wastewater treatment plant capacity to obtain the cheapest storage and treatment system possible. Reducing storm water flows is a soft and low impact way for the environment. Administrative agencies may implement low-impact development techniques to reduce storm water flows into a collection system. This includes renovating streets, parking lots, and walkways with permeable pavement, and using pervious concrete in green building roofs. Screening and disinfection facilities apply simple treatment processes for CSOs without storing sewage water. These facilities use fine screens to remove solids from combined sewage. Sodium hypochlorite is injected and mixed into sewage water for disinfection as it travels through a series of fine screens to remove debris. These fine screens have openings that range in size from 4 to 6 mm. The flow is sent through the facility at a rate that provides sufficient time for the sodium hypochlorite to inactivate existing bacteria.

(2) *CSO control in Japan*

1) National policy of CSO control

In 2001, the Ministry of Land, Infrastructure, Transport and Tourism (MLIT) set up the Committee for Countermeasures against Combined Sewer Overflows (CSOs), and carried out a nationwide investigation on their actual condition and ordered municipal governments to take action to improve the CSO situation within ten years according to the following temporary goals:

- The discharge of pollutant loads from a CSO should be reduced to less than that from separate sewer systems when each municipality is applied.
- The frequency of CSOs including untreated storm sewage at each outlet should be halved to protect the public health.
- As many floating substances as possible should be removed at CSO outlets through a screening process.

The MLIT promulgated the Sewerage Law in 2004, and has requested mid- and small-sized municipalities applying combined sewer systems to complete their CSO control by achieving the above goals by 2013, and large cities to do so by 2023.

The promulgation of the Sewerage Law includes reducing as much CSO as possible from CSO outlets by elevating weir height etc. in structural improvement. This promulgation also includes the prevention of floatables, including oil balls, by placing screens at each CSO outlet. In addition, to make a combined sewer system equivalent or greater than a separate sewer system in terms of water quality management, the water quality of CSOs should be below 40 mg/L of BOD for total rainfall events of 10 to 30 mm when the total pollutant loads from the CSO outlet are estimated by dividing into the total water amount of the CSO.

2) SPIRIT 21 for CSO control

Further, to develop appropriate technologies for sewage systems, MLIT initiated the Sewage Project, Integrated and Revolutionary Technology for the 21st Century (SPIRIT 21), in 2002. As the first theme, the project conducted research during 2002 through 2005 in the following areas of CSO treatment and instrumentation: (a) floatables/debris removal (mostly by screening), (b) high rate filtration, (c) coagulation and separation, (d) disinfection, and (e) measurement and control instrumentation. A total of twenty-four technologies were proposed in all by private industry and field tested in thirteen cities. All technologies were successfully evaluated and proposed for use in practical applications. For example, up-flow filters employing special filter media and operated with or without an additional coagulant, achieved a suspended solid removal rate of up to 70% at surface load rates of up to 1,000 m/day. In addition, bromine disinfection was found feasible for use in CSO disinfection, with the main advantage being short reaction times (JIWET, 2011).

3) Wet-weather wastewater treatment process (3 W treatment process)

The wet-weather wastewater treatment process (3 W treatment process) was innovated by the city of Osaka to treat storm water combined with sewage

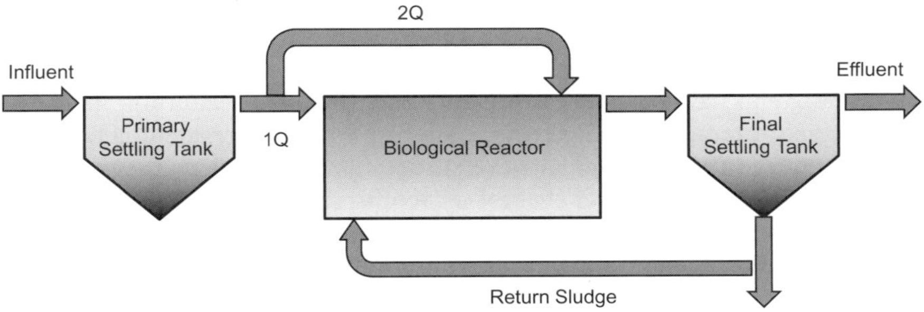

FIGURE 4-21. Wet weather wastewater treatment process (3 W treatment process) applied in Osaka, Japan.

after heavy rains by adapting an existing conventional activated sludge process. After primary sedimentation, stormy sewage exceeding 1 Q (designed for fine-weather treatment capacity of the rector) bypasses the latter part of the reactor during the activated sludge process though a step channel along the rector, as shown in **Figure 4-21**.

The initial sorption of suspended matters and soluble contaminants is expected to occur in the reactor within 30 min when sewage is introduced into the reactor. The contaminants sorped into the activated sludge then gradually degrade in the rector. The 3 W treatment process sorps the contaminants in stormy sewage into the activated sludge at the latter half of the reactor for a relatively short retention time until the storm sewage exits the reactor. The contaminants sorped into the activated sludge will be degraded while returning to the head of the reactor, or while staying at the beginning or middle part of the reactor. The activated sludge reaching the latter part then recovers the sorpability of the contaminants. The activated sludge sorping the contaminants in the storm sewage is sent to a final settling tank, where it settles if the sludge retains sufficient settleability.

According Osaka's application of this process to actual storm sewage, 60% and 70% reductions in BOD and SS were respectively demonstrated in comparison with conventional operation. BOD in discharge after 3 W treatment was reported to be 12 mg/L, which is similar to that under dry weather conditions. Among the CSO control technologies, the 3 W treatment process is quite economic allowing to a small restructuring of existing facilities such as modification of the step channel. However, when advanced treatment such as nitrogen removal is applied and a long-term operation of wet weather treatment is necessary, the water quality will deteriorate for a certain period of time.

4.3.2 *Technology for energy saving and resource recycling in municipal wastewater systems*

Sewage systems have been developed and constructed for waterborne disease prevention, water quality conservation, improvements in residential environments, and flood prevention. However, the resources included in sewage have been recognized, and a higher level of requirements beyond conventional use is now expected owing to the characteristics of the sewage system itself: the collection of dispersed materials from residences, a complete sludge and wastewater treatment system, the collection and storage of data, and so on. In addition, low carbon strategies should be included in the system.

Energy recovery from sewage using a heat-pump for areal air conditioning, and small amounts of hydraulic power generation, are conducted. Treated water is used for urban water usage after filtration, ozonation, and/or membrane filtration.

The energy included in sludge is recovered through methane fermentation (including the combined fermentation of sewage sludge and organic solid waste), a sludge-drying particulate processing, and low-temperature carbonization. Methane gas is used for the generation of heat and electric power, vehicle fuels, and commercial municipal gases after refining. Dried sludge particles and carbonized sludge are used as alternatives to coal in electric power generation plants and paper mills. Gasification technologies have also been developed and used. These technologies contribute to a low carbon society by decreasing N_2O emissions and the use of fossil fuels.

Phosphorus is not a renewable resource, and there is a concern regarding its potential depletion (see **Figure 4-22** (Steem, L., 1998)). Recovery technologies for phosphorus have also been developed and applied. Examples of these technologies are shown in **Figure 4-23** and **Figure 4-24**.

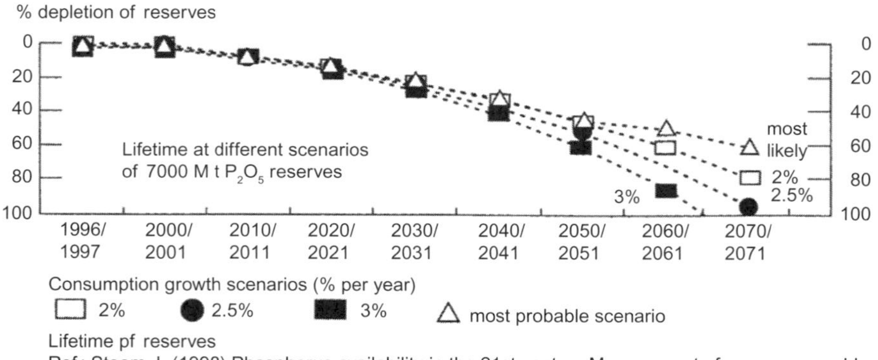

Ref.: Steem, L (1998) Phosphorus availability in the 21st century Management of a non-renewable resource, Phosphorus & Potassium. British Sulphur Publishing, No. 217, 25-31.

FIGURE 4-22. Trend of depletion of available phosphorus resources.

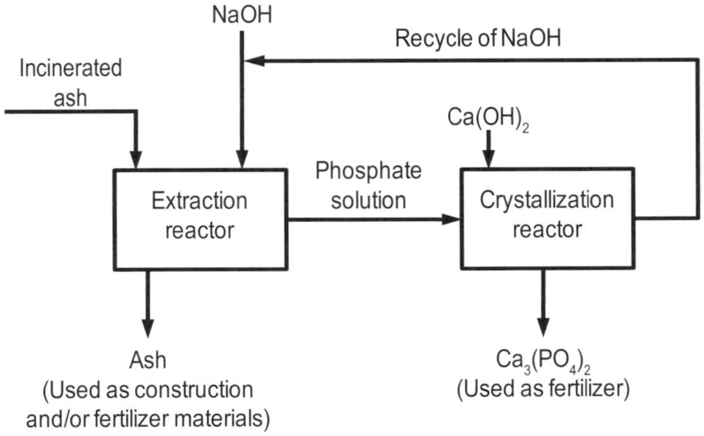

FIGURE. 4-23. Recovery technology of phosphorus as $Ca_3(PO_4)_2$.

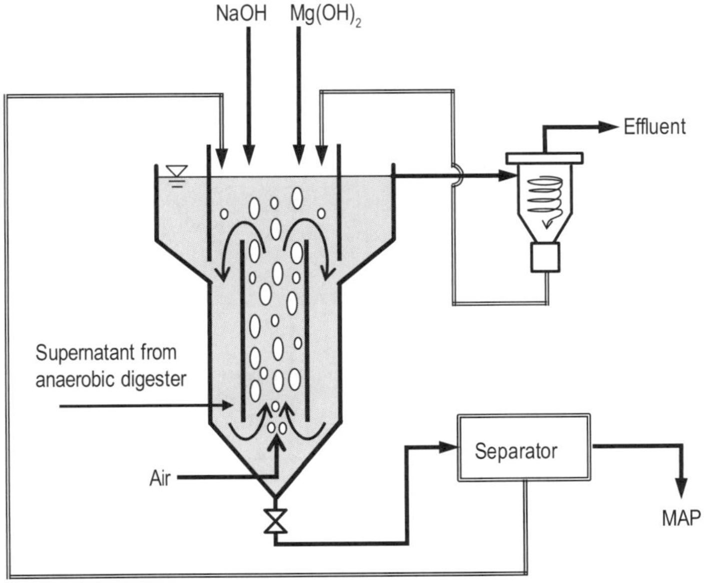

FIGURE 4-24. Recovery technology of phosphorus as MAP.

4.3.3 *Development of combined municipal wastewater and solid waste treatment system toward sustainable development*

A higher level sewage system has been recognized for satisfying the following conditions:

1. to be effective against even unknown pollutants and pathogenic microorganisms to secure a safe water environment;
2. to support sound water recycling;
3. to contribute to an enhancement of welfare;
4. to have energy saving and resource recycling capabilities, and contribute to sustainable development; and
5. to be responsive to urgent disasters.

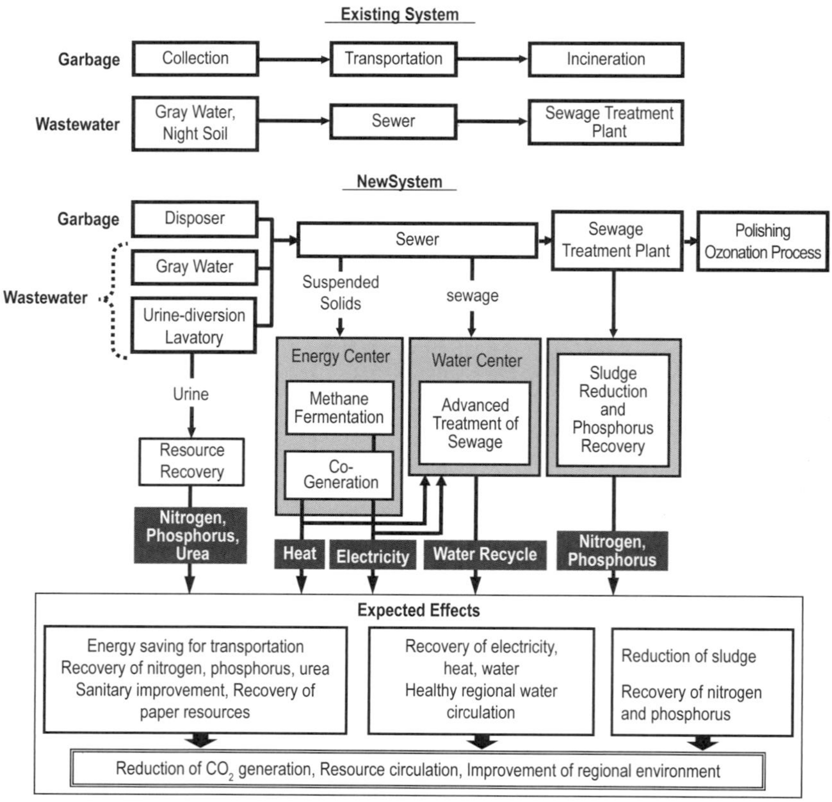

FIGURE 4-25. Outline of combined municipal wastewater and solid waste treatment system (Tsuno, 2006).

Although several alternative and innovative systems may exist, herein, we propose the Combined Municipal Wastewater and Solid Waste Treatment System (Tsuno, 2006) as an example of a higher level sewage system. A schematic of this system is shown in **Figure 4-25**.

As the concept of this system, the materials included in municipal wastewater and solid waste are considered resources that should be recovered and recycled. Kitchen garbage, as well as domestic wastewater, is taken into a sewage system, and recovered from the sewage pipe on the way to the final treatment plant. Energy (electricity and heat) is generated from the recovered organics through an effective use of methane fermentation and a power generation facility. A portion of the wastewater is also picked up at the same point and treated with generated electricity and heat by using an advanced coagulation-biofiltration process and ozonation to obtain water sustainable for a sound water circulation in the area. Methane fermentation is also applied for raw sludge. Phosphorus is recovered from urine in each house, and from an innovative and advanced excess sludge reduction and phosphorus recovery treatment process at the final treatment plant. Effluent is discharged from the final treatment plant after ozonation. An outline of the developed innovative technologies supporting this system is summarized in **Table 4-15**.

TABLE 4-15. Supporting technologies developed.

Technologies	Outline
Phosphorous recovery process from urine	Automatic recovery process of phosphorous as MAP by 90%
Collection process of solids from pipes	High concentration SS recovery process and its concentration process to 100,000 mg/L
Effective methane fermentation process	Process of 435 L of methane gas/1 kg of dried garbage under 10–20 kg COD/(m^3•d) loading rate and 55)°C
Power generation process	Micro-turbine with combined municipal gas and biogas burning process and heat recovery
Effective advanced treatment process	Automatically operated coagulation-biofiltration system to accomplish BOD of 5 mg/L, TN of 2 mg/L, TP of 0.2 mg/L and SS of 3 mg/L under HRT of 3 hours
Excess sludge reduction and advanced phosphorous recovery advanced process	A2O process with phosphorous crystallization (recovery of 80%) unit and excess sludge reduction (90%) unit by ozonation
Polishing ozonation process	High concentration ozonation for disinfection of chlorine tolerant pathogenic organisms, destruction of EDCs, PPCPs, POPs and genotoxic substances

4.3.4 *Asset management of sewerage system toward sustainability*

4.3.4.1 Stock management for sustainable municipal wastewater system

(1) *Asset and stock management of sewage works*

The infrastructure in Japan has rapidly improved during a period of high economic growth. The supply of infrastructural services is available only when stocks are soundly kept to a certain extent. These vast amounts of stocks deteriorate from day to day, and possess the risk of functional failures in addition to causing increases in the costs of maintenance, repair, and reconstruction. The important tasks under stable economic growth during the 21st century are to maintain and manage infrastructural stocks properly: to plan their reconstruction, maintain their functionality, and ensure safety while managing costs.

Asset and stock management have received a lot of attention as a means of properly managing the infrastructure for many years to come. When applying the general notion of asset management to sewage works, the objects are the properties, organizations, and human resources, such as funding and accounting methods based on the present value of the facilities, and utilizing the top sewage treatment plants. To carry out asset management, we need to comprehend and consider the present state of enormous amounts of stocks, how they will change, and what kind of management is needed at which time beforehand. The aim of stock management is to equalize businesses and minimize the life cycle cost (LCC) by integrally considering new establishments, maintenance, management, and reconstruction.

Sewerage systems, which play various roles, have a duty to provide continual and steady services that are essential for our daily lives and social activities. When sewerage systems stop operating, no alternative systems are available. Therefore, proper stock management is necessary to maintain and improve the function of a sewerage system for future use. The number of cities that have established sewerage systems has increased rapidly in Japan since the 1960s. The total amount of sewer pipelines reaches about 400,000 km, and the total number of sewage treatment plants is about 2,000, including those currently under construction. Therefore, not only is this a high number of facilities to manage, the depreciation of these facilities owing to their extended use is becoming clear. In particular, about 4,700 road sagging events caused by the breaking of old pipes occurred in 2007 in large cities, which began establishing such pipelines in earlier years. This is a serious problem, and it is inevitable that additional problems accompanying the depreciation of older sewerage systems will continue to expand nationwide. Therefore, appropriate maintenance and renewal are required to secure their continual functionality and avoid serious social problems such as road sagging. To achieve these, it is necessary to maintain and improve the functionality of existing sewerage systems as an infrastructure and social foundation. At the same time, techniques to properly and effectively maintain and manage entire facilities have to be established in the near future.

(2) *A stock management scheme*

Rebuilding means a complete or partial reconstruction or renewal of the target facilities, owing to the expansion in the amount of water discharge, and can be separated into two areas: □ renewal, which is a complete reconstruction or replacement of a target facility, and □ life extension, which is a partial reconstruction or replacement. Repairing old or broken facilities to maintain their sewerage functionality is called mending. Maintenance is defined as an action to retain the functionality of a sewerage system, such as its operation, maintenance, inspection, and cleaning of disposal facilities, which do not require additional construction.

To execute stock management without fail, establishing a plan that sets definite and specific objectives, and is regularly inspected and reviewed, is required (**Figure 4-26**). In addition, establishing and reviewing such plans should be conducted with the participation of the citizenry. Upon the implementation of stock management, managers should set the goals of the sewage works (Goal A), and the individual aims of each facility to achieve Goal A, i.e., Goal B, by considering what role the sewage works are required to take and the characteristics of the sewerage facilities as managed by local authorities. These aims are set to realize sustainable sewage works through the appropriate management of facilities, while considering the changes in social needs and a heightening of their functions. Citizens should also participate in setting these aims through public comments.

Upon the management of a facilities plan, Goal B needs to be set by each management plan for three types of facilities: pipelines, civil structures such as concrete sewage treatment plant structures, and the machinery and equipment of the sewage treatment plants. To achieve these goals, an individual plan of building, inspection, repair, and rebuilding are then made. Among these, an 'inspection and maintenance plan' is intended to determine the state of a facility, and a 'building plan' and 'repair and rebuilding plan' are intended to determine the amount of business necessary for countermeasures, both of which will be regularly inspected and reviewed. In addition, an information system based on a database will be created and developed as a tool to optimize the stock management.

As the building method, the best financial measure based on Goal A has to be chosen by each facility (pipeline, civil structure, and machinery and equipment facilities). We need to consider the social conditions and what kind of role people expect from the sewerage system when establishing such a plan.

A check-up and investigation consist of the comprehension of the state of a facility and a diagnosis based on this state. We should comprehend the condition of the facilities efficiently and systematically through clearing up which part of the check-up and investigation needs to be prioritized by considering how important the facility is and how long it has been used. In addition to the target facilities themselves, we need to clarify which items need to be checked and investigated and how often. Upon a diagnosis, an evaluation will be made regarding which stage of deterioration the facilities are in based on their present state. We have to quantify the state of the facilities precisely using numerical values and indexes, and make an effort to evaluate the probability of accidents and other problems in the future. We then need to examine whether facilities that have a high probability of accidents and other problems require urgent provisioning. At the same time, the soundness of the facility changes and the probability of future accidents and problems need to be predicted. The social influence when functional disorders occur also needs to be considered. Additionally, the soundness of facilities without data on check-ups and investigations needs to be estimated.

(3) *Implementation of stock management*

We choose the appropriate method of maintaining, repairing, and rebuilding facilities based on estimating how well the facilities will recover and how long their lives can be extended if a particular method is adopted, considering the

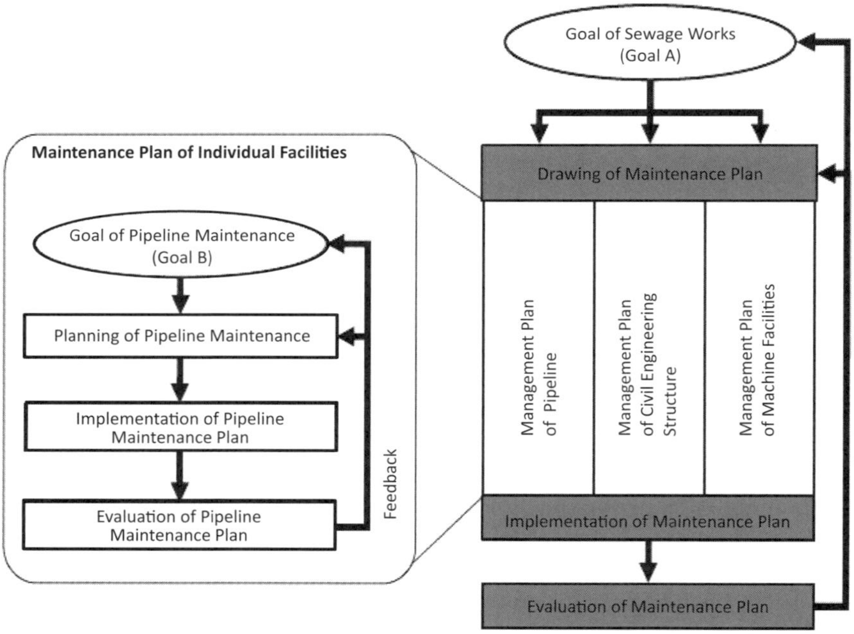

FIGURE 4-26. Maintenance plan of facilities at a sewerage system, and a PDCA cycle (MLIT, 2008).

life-cycle cost, and if possible, including the probability of accidents or other problems.

To establish a building plan, as well as a maintenance, repair, and rebuilding plan, the annual project expenses when minimizing the life-cycle cost of each type of facility (pipeline facilities, civil structures, and facilities such as sewage treatment plants) need to be first determined (**Figure 4-27**). Second, the probability of future accidents and problems, and the influence such a plan has on the overall operational costs, are evaluated synthetically, and the costs are distributed properly considering the importance of each facility.

The facility maintenance project is evaluated based on the degree of attainment of Goal A on the sewage works, and reviewed if necessary. Similarly, pipeline management projects, civil engineering structural management projects, and facility management projects are also evaluated based on the degree of attainment of Goal B, and reviewed if necessary.

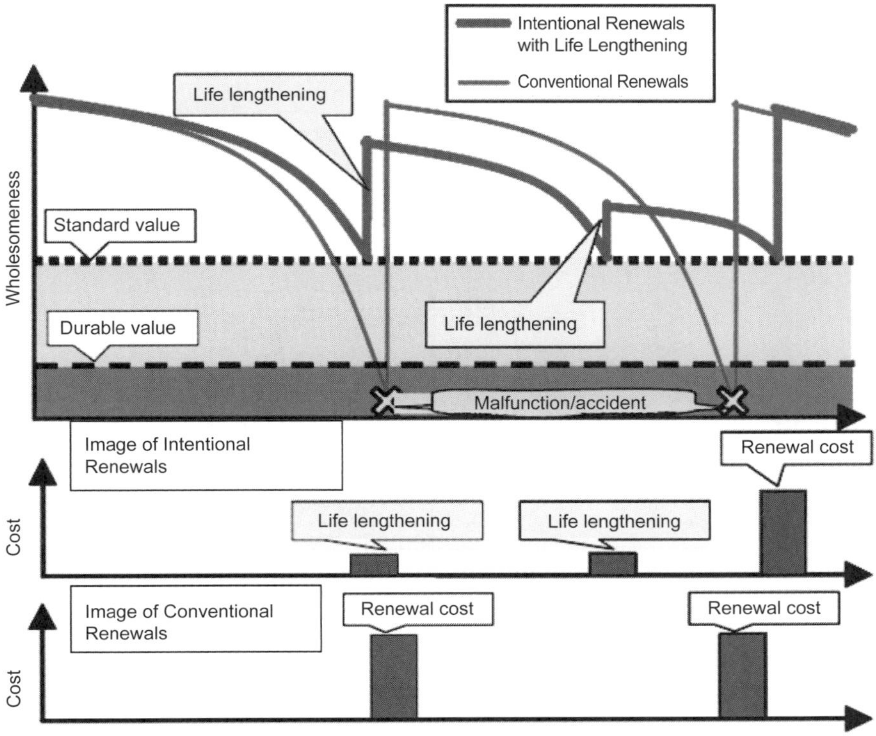

FIGURE 4-27. Cost reductions from a life extension of existing sewerage facilities through stock management (MLIT, 2008).

4.3.4.2 Sewerage Financial Management for Sustainable Municipal Wastewater System

Money transfers from the general account of the local authorities cover the expenses for the maintenance of a sewerage system in Japan, i.e., the basic interest redemption expenses for a bond issuance of the facilities, and the maintenance and operational costs of sewerage systems (**Figure 4-28**). For example, since the expense for constructing main sanitary sewers among separate sewer system were the aid object by the national government, they are subsidized by the national government and public organizations issue bonds under the permission of the national government. Public organizations also secure the expenses for constructing sewers that connect to individual residences as independent businesses. In addition, user fees depending on the housing area are collected when each house connects to the sewerage system. When redeeming issued bonds, the national government has appropriated a local tax grant for 100% of the rainwater channels and 50% of the sanitary sewer and sewage treatment facilities. In contrast, the maintenance costs for sanitary sewers, the treatment of drainage from rainwater and advanced sewage treatment are paid from the general revenue of the local authorities, which means the expenditures are public expenses. However, concerning most

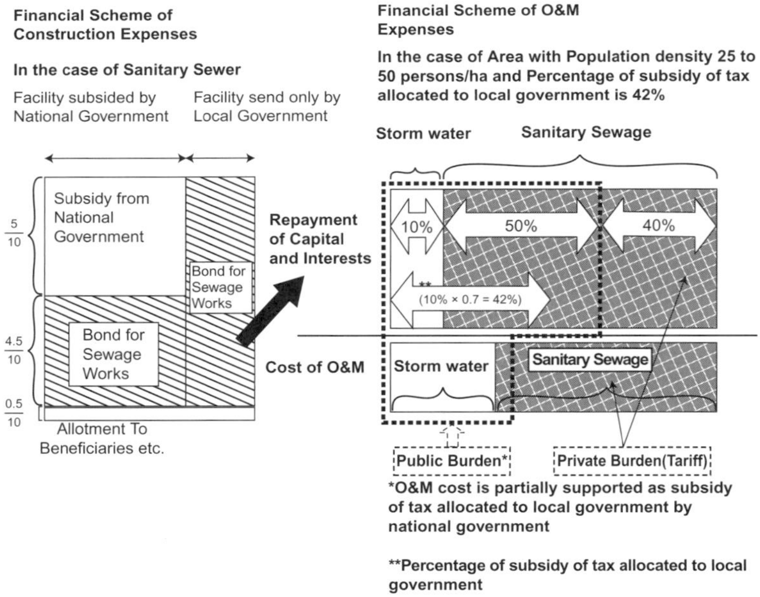

FIGURE 4-28. Example of a financial scheme for construction, and the O&M costs of Japanese sewage systems (MLIT, 2007).

expenditures for drainage processing, users pay a charge; in other words, it is in principle a private expense.

In reality, however, many municipalities do not receive sufficient charge amounts owing to a low connection rate and an inappropriate charge amount. For instance, the fees they collected in 2006 correspond to only 70% of the total expenses necessary to maintain the sewerage system. Therefore, public organizations run sewage works by transferring money from a general account. This tendency will become more conspicuous in medium-sized municipalities, which are under the influence of an aging and declining population. Since fee incomes are expected to decline from now on, reinforcement of a management base is essential.

For sewage works, the burden of expense has to be divided between the public and private enterprises in an appropriate proportion considering the difference between the redemption period of the debt issuance and the useful life of the facilities. Developing an outlook to reinforce the management base from not a short-term point of view, but from long-term point of view, stock management such as improvements and a renewal of the facilities, and a consideration of both revenue and expenditures mentioned in the previous section, is necessary. Furthermore, it is essential to create suitable sewage works by drawing up a mid-term management plan that considers every stage of sewerage management from planning and construction to maintenance.

REFERENCES

Ab.Wahid, M. (2011), Occurrence and reduction of pathogens in wastewater treatment and consideration for wastewater reclamation and reuse (Doctoral Dissertation), Retrieved from Dissertations and Theses database Kyoto University. (repository.kulib.kyoto-u.ac.jp).

Adams, C., *et al.*, (2002), Removal of antibiotics from surface and distilled water in conventional water treatment processes, *Journal of Environmental Engineering,* 128, pp. 253–260.

Alcock, R.E., *et al.*, (1999), Assessment of organic contaminant fate in waste water treatment plants, *Chemosphere*, 38(10), pp. 2247–2262.

An, Y.Y., *et al.*, (2008), The integration of methanogenesis with shortcut nitrification and denitrification in a combined UASB with MBR, *Bioresource Technology*, 99, pp. 3714–3720.

Ayhan Demirbas, (2008), Heavy metal adsorption onto agro-based waste materials: A review, *Journal of Hazardous Materials*, 157, pp. 220–229.

Cao, B., *et al.*, (2007), Pilot test on enhanced biological nitrogen and phosphorus removal by using A^2/O-MBR, *China Water and Wastewater*, 23, pp. 22–26 (in Chinese).

CDHS, (1978), California Code of Regulations. In Wastewater Reclamation Criteria; Title 22; California Department of Health Services, Sanitary Engineering Section, Berkeley.

Chen, J.H., *et al.*, (2008a), Bisphenol A removal by a membrane bioreactor, *Process Biochemistry*, 43, pp. 451–456.

Chen, J.H., *et al.*, (2008b), Comparison of bisphenol A (BPA) and nonylphenolethoxylate (NP*n*EO) removal with a membrane bioreactor versus a conventional activated sludge reactor, *Acta Scientiae Circumstantiae*, 28, pp. 433–439 (in Chinese).

Chen, J.H. (2008), Removal of endocrine disrupting chemicals in wastewater by a membrane reactor and enhanced process with nanofiltration membrane, Dissertation for the Doctoral Degree, Beijing: Tsinghua University (in Chinese).

Chen, X., *et al.*, (2009), Evaluation of a novel oxidation ditch system for biological nitrogen and phosphorus removal from domestic sewage, *Proceedings of the 3rd IWA-ASPIRE Conference*. Taipei, Taiwan. 18–22 October.

Citulski, J., *et al.*, (2009), Optimization of phosphorus removal in secondary effluent using immersed ultrafiltration membranes with in-line coagulant pretreatment—implications for advanced water treatment and reuse applications, *Canadian Journal of Civil Engineering*, 36(7), pp. 1272–1283.

Comerton, A.M., *et al.*, (2008), The rejection of endocrine disrupting and pharmaceutically active compounds by NF and RO membranes as a function of compound and water matrix properties, *Journal of Membrane Science*, 313, pp. 323–335.

Cui, Z.G., *et al.*, (2009), Enhanced phosphorus removal in membrane bioreactor combined with electrocoagulation, *China Water and Wastewater*, 25, pp. 1–4 (in Chinese).

Dai, W.C., *et al.*, (2007), Enrichment of denitrifying phosphate accumulating organisms in sequencing batch membrane bioreactors, *Environmental Science*, 28, pp. 517–521 (in Chinese).

Date, Y., *et al.*, (2009), Microbial diversity of anammox bacteria enriched from different types of seed sludge in an anaerobic continuous-feeding cultivation reactor, *Journal of Bioscience and Bioengineering*, 107(3), pp. 281–286.

Dialynas, E., *et al.*, (2008), Integration of immersed membrane ultrafiltration with coagulation and activated carbon adsorption for advanced treatment of municipal wastewater, *Desalination*, 230(1–3), pp. 113–127.

Dietrich, J.P., *et al.*, (2003), Preliminary assessment of transport processes influencing the penetration of chlorine into wastewater particles and the subsequent inactivation of particle-associated organisms, *Water Research,* 37, pp. 139–149.

Drewes, J.E., *et al.*, (2002), Fate of pharmaceuticals during indirect potable reuse, *Water Science and Technology*, 46(3), pp. 73–80.

Du, C., *et al.*, (2008), Bacterial diversity in activated sludge from a consecutively aerated submerged membrane bioreactor treating domestic wastewater, *Journal of Environmental Sciences*, 20, pp. 1210–1217.

Gideon P. Winwarda, Lisa M. Averyb, Tom Stephensona, Bruce Jeffersona (2008), Chlorine disinfection of grey water for reuse: Effect of organics and particles, *Water Research,* 42, 483–491.

Gómez., Plaza, F., Garralon, G., Pérez, Jand Gómez, M.A. (2006), A comparative research of tertiary wastewater treatment by physico-chemical-UV process and macrofiltration-ultrafiltration technologies, *Desalination,* 202, pp. 369–376.

Harata, N., *et al.*, (2004), Kinetics and mechanisms of virus inactivation by ozone (In Japanese), *Japan Society on Water Environment,* 27(1), pp. 33–40.

He, S.B., *et al.*, (2009), Factors affecting simultaneous nitrification and de-nitrification (SND) and its kinetics model in membrane bioreactor, *Journal of Hazardous Materials,* 168, pp. 704–710.

Hens, M., *et al.*, (2002), The role of colloidal particles in the speciation and analysis of "dissolved" phosphorus, *Water Research*, 36, pp. 1483–1492.

Hidaka, T., *et al.*, (2003), Advanced treatment of sewage by pre-coagulation and biological filtration process, *Water Research*, 37(17), pp. 4259–4269.

Huber, M.M., *et al.*, (2003), Oxidation of pharmaceuticals during ozonation and advanced oxidation processes, *Environmental Science and Technology,* 37(5), pp. 1016–1024.

Japan sewage works association (JSWA) (1997) Standard methods for sewage analysis.

Japan sewage works association (JSWA) (2009a) Sewage facility planning and design manual.

Japan sewage works association (JSWA) (2009b) Japanese sewage white book.

Japanese Society of Water Environment (JSWE) (ed.) (2009), *Water Pollution Control and Management in Japan*. Gyosei.

Jin, X., *et al.*, (2010), Removal of natural hormone estrone from secondary effluents using nanofiltration and reverse osmosis, *Water Research*, 44(2), pp. 638–648.

Joss, A., *et al.*, (2005), Removal of pharmaceuticals and fragrances in biological wastewater treatment, *Water Research,* 39(14), pp. 3139–3152.

Kim, I., *et al.*, (2009), Discussion on the application of UV/H_2O_2, O_3 and O_3/UV processes as technologies for sewage reuse considering the removal of pharmaceuticals and personal care products, *Water Science and Technology*, 59(5), pp. 945–955.

Ko, G., *et al.*, (2005), UV inactivation of adenovirus type 41 measured by cell culture mRNA RT-PCR. *Water Research*, *39*(15), pp. 3643–3649.

Kobayashi, T., *et al.*, (2009), Upgrading of the anaerobic digestion of waste activated sludge by combining temperature-phased anaerobic digestion and intermediate ozonation, *Water Science and Technology*, 59(1), pp. 185–193.

Lee, J., *et al.*, (2008), Comparison of the removal efficiency of endocrine disrupting compounds in pilot scale sewage treatment processes. *Chemosphere*, 71, pp. 1582–1592.

Lee, M., *et al.*, (2009), Comparative performance and microbial diversity of hyperthermophilic and thermophilic co-digestion of kitchen garbage and excess sludge, *Bioresource Technology*, 100(2), pp. 578–585.

Li, Q.X., *et al.*, (2008), Experimental Study on Advanced Treatment of Municipal Wastewater by Coagulation-Sedimentation-UF, *Environmental Science and Technology*, 31(6), pp. 119–120 (in Chinese).

Liberti, L., *et al.*, (1999), Advanced treatment and disinfection for municipal wastewater reuse in agriculture, *Water Science and Technology*, 40(4–5), pp. 235–245. IWA Publishing.

Lim, M.Y., *et al.*, (2010), Characterization of ozone disinfection of murine norovirus, *Applied and Environmental Microbiology*, 76(4), pp. 1120–1124. American Society for Microbiology.

Liu, R., *et al.*, (2000), A pilot study on a submerged membrane bioreactor for domestic wastewater treatment, *Journal of Environmental Science and Health Part A*, 35, pp. 1761–1772.

Lv, W., *et al.*, (2006), Virus removal performance and mechanism of a submerged membrane bioreactor, *Process Biochemistry*, 41, pp. 299–304.

Meng, Q.J., *et al.*, (2008), Effects of COD/N ratio and DO concentration on simultaneous nitrification and denitrification in an airlift internal circulation membrane bioreactor, *Journal of Environmental Sciences*, 20, pp. 933–939.

Meng, Q.S. *et al.*, (1996), Comparative inactivation of enteric adenoviruses, polioviruses and Coliphages by ultraviolet irradition, *Water Research*, 30, pp. 2665–2668.

Ministry of Land Infrastructure and Transport, MLIT (2008a) Sewage Works in Japan, FY 2008, Japan Sewage Works Association.

Ministry of Land Infrastructure and Transport, MLIT (2008b) Fundamental Concept of Stock Management in Sewage Works, http://www.mlit.go.jp/common/000056589.pdf

Ministry of Land Infrastructure and Transport, MLIT: http://www.mlit.go.jp/crd/sewerage/keikaku/01-1-1.html (last access: January 10, 2014).

Mulder, A., *et al.*, (1995), Anaerobic ammonium oxidation discovered in a denitrifying fluidized bed reactor, *FEMS Microbiology Ecology*, 16, pp. 177–184.

Narumiya, M., *et al.*, (2009), Occurrence and fate of pharmaceuticals and personal care products during wastewater treatments, *JSCE Journal of Environmental Engineering*, 46, pp. 175–186.

Nasser, A.M., *et al.*, (2006), UV disinfection of wastewater effluents for unrestricted irrigation. *Water Science and Technology*, 54(3), pp. 83–88.

Paraskeva, P. *et al.*, (2005), Treatment of a secondary municipal effluent by ozone, UV and microfiltration: Microbial reduction and effect on effluent quality, *Desalination*, 186, pp. 47–56.

Pei Xua, Marie-Laure Janexb, Philippe Savoyeb, Arnaud Cockxc, Valentina Lazarova, 2002, Wastewater disinfection by ozone: main parameters for process design, *Water Research*, 36 1043–1055.

Pinkston, K.E., *et al.*, (2004), Transformation of aromatic ether- and amine-containing pharmaceuticals during chlorine disinfection, *Environmental Science and Technology*, 38, pp. 4019–4025.

Rajala, R.L., *et al.*, (2003), Removal of microbes from municipal wastewater effluent by rapid sand filtration and subsequent UV irradiation, *Water Science and Technology*, 47(3), pp. 157–162.

Ran, X.J., *et al.*, (2004), Application of "flotation + BAC + NF" process for reclaimed waste water reuse project, *Environment Protection in Petrochemical Industry*, 27(4), pp. 23–26 (in Chinese).

Routledge, E.J., *et al.*, (1998), Identification of estrogenic chemicals in STW effluent. 2. In vivo responses in trout and roach, *Environmental Science and Technology*, 32, pp. 1559–1565.

Sato, H., *et al.*, (2010), Subcritical Water Hydrolysis Treatment for Sewage Sludge Reduction and Energy Recovery by High-rate Digestion, WEFTEC2010, New Orleans, USA, pp. 3948–3960.

Shigematsu, T., *et al.*, (2007), The Removal Efficiency of Coliphage under the Ultrafiltration Process in Wastewater Reclamation for Agricultural Use, *Proceedings of IWA 14th Conference on Health-related Water Microbiology (Water Micro)*, pp. 202–203.

Shin, G.-A., *et al.*, (2003), Reduction of Norwalk virus, poliovirus 1, and bacteriophage MS2 by ozone disinfection of water, *Applied and Environmental Microbiology*, 69(7), pp. 3975–3978. American Society for Microbiology.

Snyder, S.A., *et al.*, (2003), Pharmaceuticals, personal care products, and endocrine disruptors in water: Implications for the water industry, *Environmental Engineering Science,* 20(5), pp. 449–469.

Snyder, S.A., *et al.*, (2004), Biological attenuation of EDCs and PPCPs: implications for water reuse, *Ground Water Monitoring and Remediation*, 24(2), pp. 108–118.

Snyder, S.A., *et al.*, (2007), Role of membranes and activated carbon in the removal of endocrine disruptors and pharmaceuticals, *Desalinization*, 202(1–3), pp. 156–181.

Steem, L. (1998), Phosphorus availability in the 21st century, Management of non-renewable resources, *Phosphorus & Potassium*, British Sulphur Publishing, 217, pp. 25–31.

Tanaka, H., *et al.*, (1998), Estimating the safety of wastewater reclamation and reuse using enteric virus monitoring data, *Water Environmental Research*, 70, pp. 39–51.

Tanaka, K., *et al.*, (1996), Application of nitrification by cells immobilized in polyethylene glycol, *Progress in Biotechnology*, 11, pp. 622–632.

Tatsuya, N. (2009), Methane fermentation, Gihodo, Tokyo, Japan.

Ternes, T.A., *et al.*, (1999) Behaviour and occurrence of estrogens in municipal sewage treatment plants—II. Aerobic batch experiments with activated sludge, *Science of the Total Environment*, 225, pp. 91–99.

Ternes, T.A., *et al.*, (2002), Removal of pharmaceuticals during drinking water treatment, *Environmental Science and Technology,* 36, pp. 3855–3863.

Ternes, T.A., *et al.*, (2003), Ozonation: A tool for removal of pharmaceuticals, contrast media and musk fragrances from wastewater? *Water Research,* 37 (8), pp. 1976–1982.

Tsuno, H. (2006), Final Report of Research on Development of resource recycling type urban wastewater and solid waste processing system, *CREST Program* by Japan Science and Technology Agency.

Tsuno, H., *et al.*, (2001), Advanced Treatment of Sewage by Pre-Coagulation and Biofilm Process, *Water Science and Technology*, 43(1), pp. 327–334.

U.S. Environmental Protection Agency (2001), Report to Congress: Implementation and Enforcement of the CSO Control Policy, EPA 833-R-01–003, Washington, DC.

Urase, T., *et al.*, (1994) Effect of pore size distribution of ultrafiltration membranes in virus rejection in crossflow conditions, *Water Science and Technology*, 30(9), pp. 199–208.

Wang, J., *et al.*, (2008a), Advanced treatment of produced water with a novel composite nanofiltration membrane, *Journal of Southwest Petroleum University* (Science & Technology Edition), 30(1), pp. 112–116 (in Chinese).

Wang, J.F., *et al.*, (2008b), Organics and nitrogen removal and sludge stability in aerobic granular sludge membrane bioreactor, *Applied Microbiology and Biotechnology*, 79, pp. 679–685.

Wang, X.M., *et al.*, (2008), Impact of gel layer formation on colloid retention in membrane filtration processes, *Journal of Membrane Science*, 325, pp. 486–494.

Wang, X.M., *et al.*, (2009), Role of gelling soluble and colloidal microbial products in membrane fouling, *Environmental Science and Technology*, 43, pp. 9341–9347.

Wang, Z.W., *et al.*, (2009), Distribution and transformation of molecular weight of organic matters in membrane bioreactor and conventional activated sludge process, *Chemical Engineering Journal*, 150, pp. 396–402.

Wannar, J. (1994), Activated Sludge: Bulking and Foaming Control, CRC Press.

Water Reuse, Metcalf & Eddy, 2006.

Wei, C.H., *et al.*, (2006), Pilot study on municipal wastewater treatment by a modified submerged membrane bioreactor, *Water Science and Technology*, 53, pp. 103–110.

Westerhoff, P., *et al.*, (2005), Fate of endocrine-disruptor, pharmaceutical, and personal care product chemicals during simulated drinking water treatment processes, *Environmental Science and Technology,* 39(17), pp. 6649–6663.

Winwarda, G.P., *et al.*, (2008), Chlorine disinfection of grey water for reuse: Effect of organics and particles, *Water Research,* 42, pp. 483–491.

Wu, C.Y., *et al.*, (2011), Removal of endocrine disrupting chemicals (EDCs) in a large scale membrane bioreactor plant combined with anaerobic-anoxic-oxic process for municipal wastewater reclamation, *Water Science and Technology*, 64, pp. 1511–1518.

Wu, G., *et al.*, (2004), A pilot study on municipal sewage treatment and reuse via UF membrane process, *Membrane Science and Technology*, 24(1), pp. 38–41 (in Chinese).

Wu, J., *et al.*, (2010), Indigenous somatic coliphage removal from a real municipal wastewater by a submerged membrane bioreactor, *Water Research*, 44, pp. 1853–1862.

Xia, S.B., *et al.*, (2009), Enhanced biological phosphorus removal in a novel sequencing membrane bioreactor with gravitational filtration (GFS-MBR), *Desalination and Water Treatment*, 9, pp. 259–262.

Xua, P., *et al.*, (2002), Wastewater disinfection by ozone: Main parameters for process design, *Water Research,* 36, pp. 1043–1055.

Yang, L., *et al.*, (2006), Pilot study on simultaneous nitrogen and phosphorus removal in submerged membrane bioreactor treating municipal wastewater, *China Water and Wastewater*, 22, pp. 13–17 (in Chinese).

Yang, Q., *et al.*, (2009a), Treatment of municipal wastewater by membrane bioreactor: a pilot study*,* *International Journal of Environment and Pollution*, 38, pp. 280–288.

Yang, S., *et al.*, (2009b), Comparison between a moving bed membrane bioreactor and a conventional membrane bioreactor on organic carbon and nitrogen removal, *Bioresource Technology*, 100, pp. 2369–2374.

Yang, X.H., *et al.*, (2007), Advanced Treatment of Wastewater by GAC and Super-membrane, *Liaoning Chemical Industry*, 36(9), pp. 629–631 (in Chinese).

Zhang, H.M., *et al.*, (2006), Comparison between a sequencing batch membrane bioreactor and a conventional membrane bioreactor, *Process Biochemistry,* 41, pp. 87–95.

Zhang, Z.C. (2008), Characteristics of biological phosphorus removal process using membrane bioreactor, Dissertation for the Doctoral Degree, Beijing: Tsinghua University (in Chinese).

Zhang, Z.C., *et al.*, (2007), Impact of sludge retention time on enhanced biological phosphorus removal using membrane bioreactor, *In 2nd IWA-ASPIRE Conference & Exhibition 2007*, Perth.

Zhang, Z.C., *et al.*, (2009), Study on P forms in extracellular polymeric substances in enhanced biological phosphorus removal sludge by ^{31}P-NMR spectroscopy, *Spectroscopy and Spectral Analysis*, 29, pp. 236–239 (in Chinese).

Zhang, Z., *et al.*, (2008), Enhanced phosphorus removal in biological nitrogen and phosphorus removal process using membrane bioreactor, *Journal of Tsinghua University*, 48, pp. 1472–1474 (in Chinese).

Zheng, X., *et al.*, (2005), Evaluation of virus removal in MBR using coliphages T4, *Chinese Science Bulletin*, 50, pp. 862–867.

Zheng, X., *et al.*, (2007), Virus rejection with two model human enteric viruses in membrane bioreactor system, *Science in China Series B*, 50, pp. 397–404.

Zhou, H.D., *et al.*, (2009b), Evaluation of estrogenicity of sewage samples from Beijing, China, *Environmental Science*, 30, pp. 3590–3595 (in Chinese).

Zhou, H., *et al.*, (2009a), Distribution and elimination of polycyclic musks in three sewage treatment plants of Beijing, China, *Journal of Environmental Sciences*, 21, pp. 561–567.

Zhou, H., *et al.*, (2010), Behaviour of selected endocrine-disrupting chemicals in three sewage treatment plants of Beijing, China, *Environmental Monitoring and Assessment*, 161, pp. 107–121.

Zhou, Y.J., *et al.*, (2011), Removal of typical endocrine disrupting chemicals by membrane bioreactor: In comparison with sequencing batch reactor, *Water Science and Technology*, 64, pp. 2096–2102.

Zwiener, C., *et al.*, (2000), Oxidative treatment of pharmaceuticals in water, *Water Research,* 34(6), pp. 1881–1885.

5

Integrated Management of Lake Watersheds

5.1 UNIQUE CHARACTERISTICS OF LAKES

Lakes are among the most vulnerable of all water resource systems. Although a source of essential resources necessary for the well-being of humans and ecosystems, lake basins are easily impacted by complex land and water relationships. As primary freshwater storage systems, they receive water, sediments, pollutants, nutrients, and biota from inflowing rivers, land runoff, groundwater aquifers, and the atmosphere. Used for a wider range of purposes than any other type of water body, they are much more vulnerable to stresses, and are more difficult to manage, than other water systems.

Lakes can be created through natural processes, such as glacial scouring, plate tectonics, and volcanic activities. Humans also construct lakes by building dams across flowing water systems, pooling the water behind these structures, and creating reservoirs (also called dams in some countries). Whatever their origin, lakes and reservoirs can be considered 'standing water' or 'lentic' systems. This is in contrast to the flowing waters of rivers, which are known as 'lotic' systems. Because lakes have both influent and outflowing rivers, a lake basin represents a complex combination of both lentic and lotic waters, which is an important consideration for lake-basin management. In fact, the occurrence and management of lake-basin problems is a function of the distinguishing characteristics of lentic water systems, including a long residence time, complex internal response dynamics, and an integrated nature.

5.1.1 *Characteristics of lakes*

5.1.1.1 Long retention time

The water residence time refers to the average amount of time water spends in a lake. Large lakes are typically characterized by large water volumes and resulting long retention times, giving it a buffer capacity in that it is able to tolerate large inputs of water, as well as the pollutants and sediment loads carried by this water, without exhibiting immediate negative changes. This incremental response can make it difficult to notice degradation problems until they have become serious lake-wide problems. Long water retention times allow suspended materials in the water column, including pollutants, to settle to the bottom of the lake, thereby ensuring their role as a sink for many types of materials. A long water retention time also ensures that, even when remedial programs are implemented to restore a degraded lake, it can take a very long time, if ever, for the lake to recover. It also leads to lags in ecosystem responses that poorly match the human management time scale.

5.1.1.2 Complex response dynamics

In contrast with lotic water systems, lakes do not necessarily respond to perturbations or pollution in a linear fashion, largely owing to their lake volume and long retention times, which provide a degree of buffering capacity to such disturbances. The results can be a non-linear response (hysteresis) to increasing pollutant loads. For example, a lake can receive a considerable nutrient load without exhibiting a significant amount of degradation until the nutrient concentration reaches a critical level resulting in a shift from the existing trophic state. A lake can exhibit a rapid degradation after this critical level is reached. The same buffering capacity also hinders an achievement of the positive goals of the nutrient remediation programs. Even after pollutant loads have been reduced, a lake will not necessarily exhibit a positive response to such remediation efforts for some period of time until the lake has been flushed or otherwise its previous high nutrient content is neutralized. Further, experience also suggests that only a certain degree of recovery may be possible, and that the original good condition may not be achievable.

5.1.1.3 Integrating nature

Because of their location at the terminus of a drainage basin, lakes receive pollutant inputs through rivers and other inflowing channels from all sources within their surrounding drainage basin, as well as beyond (through long-term atmospheric deposition). These inputs are mixed within a lake, thereby ensuring that both the resources and problems associated with them are disseminated and integrated throughout the water volume of the lake. In fact, this integrating nature often provides a 'trigger' for the development and implementation of remediation programs, in that many symptoms of pollutant loadings only become visible after they have reached a lake and have had sufficient time to become visible. Thus, the condition of a lake can be viewed as a type of 'barometer' of human activities within the lake basin. This observation has an enhanced significance in that lakes are used for a greater range of human uses than any other type of water system, thereby ensuring that such degradation can affect a greater number and range of these uses. Algal blooms are an example of this type of phenomenon, noting that algal cells require the same nutrients, temperature, and light requirements in both rivers and lakes. However, excessive nutrient loads result in algal blooms in lakes because the algae have sufficient time to accumulate to nuisance levels in a lentic environment, whereas this is not possible in a lotic river environment.

5.1.2 *Value and uses of lakes*

Lakes are among the most dramatic and pleasing features of the planetary landscape, and the most variable type of inland water system. Although rivers represent flowing water systems, lakes are primarily water storage bodies. Their sizes, shapes, and depths vary considerably based on their specific mode of origin. They are dynamic aquatic ecosystems, being at the same time storehouses for large quantities of water, sources of food, and recreational areas for humans. Lakes are home to an amazing range of biodiversity, in some cases containing organisms found nowhere else on Earth. For many indigenous lakeshore communities, lakes also provide the very foundation of their livelihood.

Lakes also are significant repositories of natural and human history, with ancient local political centers often arising at or near lakeshores. Specific lifestyles based entirely on lakes and their resources have developed in some locations, an example being indigenous cultures. lakes also have a fundamental religious and spiritual significance in many countries.

The large quantities of water stored in lakes are especially useful for meeting the water requirements of humans and ecosystems when natural climatic conditions cannot otherwise supply water when it is most needed. On the other hand, the ability of lakes to store water also helps protect the lives and property of downstream communities during flooding events. At the same time, it may significantly raise lake water levels, thereby affecting people living along shorelines. Because water can also absorb a large amount of heat, lakes with large water volumes can also moderate the local climate by reducing the range of atmospheric temperature fluctuations.

Lakes are also among the most vulnerable and fragile aquatic ecosystems. They are sinks for inflowing substances, including sediment, minerals, aquatic plant nutrients, and organic materials from their drainage basin. Such materials tend to accumulate in the water column or on the lake bottom. In sparsely populated drainage basins, this typically leads to a relatively slow aging process. In densely populated or industrialized drainage basins, however, human activities can significantly accelerate this natural aging process, and degrade the water quality and lake bottom environment. Because of this property, lakes serve as sensitive indicators and unique records of the effects of human and natural activities inside their drainage basins, and even sometimes from activities occurring outside their basins.

Regardless of their size, lakes are the primary repositories of rich aquatic biodiversity, with many having a range of native and endemic species. However, their biodiversity is very sensitive to hydrological disturbance, water quality degradation, and the introduction of non-native species without sufficient scientific knowledge of their impact. Lakes can suffer significant losses of their endemic and native species when, in the absence of natural predators or other control mechanisms, invasive species proliferate and replace them. There are many lakes around the world with relatively saline or salty water caused by the weathering of minerals from the bedrock of their drainage basins, and typically exist in closed drainage basins. Others have become saline over time because of an excessive diversion or withdrawal of water.

What cannot to be forgotten is the intrinsic beauty of lakes, many of which exhibit breathtaking aesthetic features. Lakes evoke a pleasing range of emotional, spiritual, and intellectual responses. Lakes have been described as 'pearls on a river string,' and 'islands of water in an ocean of land.' Although they are extremely important features, the aesthetic value of lakes is the most difficult to quantify, as compared to their other uses.

5.1.3 *Lake and lake watershed issues*

The withdrawal or diversion of excessive quantities of water from lakes can reduce their water levels and volumes to the extent that the water quality, and the biological communities they support, is seriously threatened and the shoreline characteristics altered. A dramatic example is the demise of the Aral Sea, a large lake located in a closed drainage basin in south central Asia. Owing to a significant diversion of the lake's inflowing tributaries for irrigation purposes during the last half century, the Aral Sea has shrunk significantly in surface area and volume, experienced a major increase in salinity, and undergone fundamental changes to its biological communities. Water withdrawal for land reclamation can also have a profound impact. In addition, hydraulic structures used for water withdrawal or diversion can change the flow patterns in lakes. They can also change the relationship between upstream and downstream communities and water use potential. The actions in an upstream lake drainage basin, for example, can significantly affect the downstream drainage basins in regard to flood risks, water supply, the ecosystem, services, and so on. In contrast, water withdrawal in a downstream drainage basin may limit or otherwise impact the potential water use in an upstream drainage basin.

A variety of pollutants can degrade the lake water quality. Excessive nutrient loads (primarily phosphorus and nitrogen) can cause an accelerated eutrophication, the accelerated growth of algae and aquatic plants to nuisance levels (e.g., algal blooms and floating weeds), with an accompanying degradation of water quality and significant imbalances to lake ecosystems and their biological communities. Excessive nutrient inputs can stimulate the growth of toxic blue-green algal species detrimental to both livestock and human health. They can also interfere with beneficial human water uses, cause taste and odor problems in drinking water, and provide precursors of trihalomethanes, which have been identified as carcinogenic chemical compounds.

5.2 THE HISTORY AND AN EXAMPLE OF ENVIRONMENTAL CONSERVATION IN LAKE BIWA, JAPAN

5.2.1 *Profile of Lake Biwa*

Lake Biwa is the largest lake in Japan with an area of 674 km^2 and a maximum depth of more than 100 meters. It can be divided into two parts at its narrowest point, where it is spanned by the Lake Biwa Bridge. These two parts are called the Northern and Southern Lakes (**Figure 5-1**).

Lake Biwa is also an ancient lake along with Lake Baikal and Lake Tanganyika, and was formed about four million years ago. Owing to this long history, Lake Biwa has a high bio-diversity, with approximately 600 kinds of animals, including many indigenous species, and approximately 500 kinds of

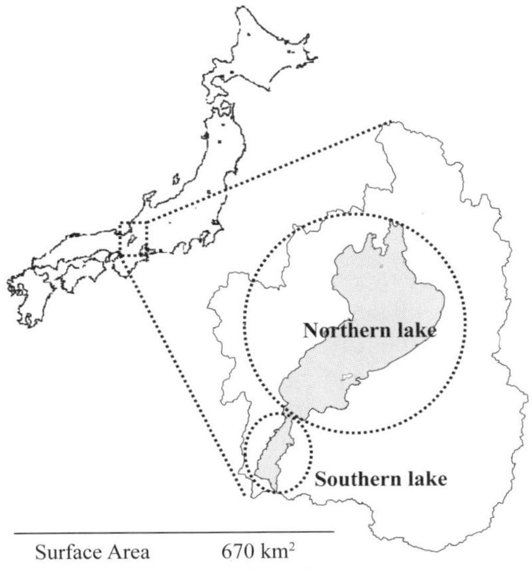

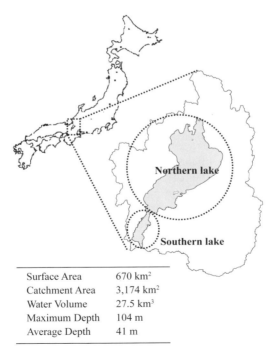

Surface Area	670 km^2
Catchment Area	3,174 km^2
Water Volume	27.5 km^3
Maximum Depth	104 m
Average Depth	41 m

FIGURE 5-1. Outline of Lake Biwa and its watershed.

plants. Lake Biwa, noted for its abundant and steady supply of water, has been a valuable resource for people, industry, agriculture, and other uses downstream of the region for a very long time. Today, it plays an extremely important role as a water source for fourteen million people.

Most of the fisheries in Shiga prefecture, whose boundary is very similar to that of the Lake Biwa watershed, are concentrated in Lake Biwa. Lake Biwa is Japan's leader in fresh water fisheries. In particular, about 70% of all ayu fish (sweetfish) hatched are from Lake Biwa. Lake Biwa has also recently attracted a lot of attention as a recreation area owing to its scenic beauty. There are more than sixty popular swimming locations around the shores of Lake Biwa. Each year the lake receives approximately forty-three million visitors.

5.2.2 *Recent changes in Lake Biwa and its surrounding social situation*

5.2.2.1 Water quality

Several institutes in Shiga prefecture have been monitoring the water quality of Lake Biwa. There are forty-seven fixed monitoring points around the lake where observations are carried out, and observations are made once every month. The following graphs show the year-to-year changes in the mean surface water quality as monitored monthly, and an estimate of the inflow load to the lake (**Figure 5-2**). The trends in each water quality parameter show that water pollution peaked in the 1970s and 1980s, and since then, the deterioration of the lake has stopped owing to the use of various pollution prevention measures. Indeed, the inflow load since 1990 has been decreasing for all water quality indices. Contrary to these general trends, however, COD(Mn) has shown a reversal (worsening or not improving) since 1985.

5.2.2.2 Red tides and blue-green algal blooms

A moderate level of eutrophication stimulates the propagation of phytoplankton, and increases the number of plankton-eating fish. However, excessive levels of eutrophication cause phenomena such as red tides and blue-green algal blooms from the excessive growth of phytoplankton.

From late April to early June, when the water temperature is between 15 to 20°C, an abnormal growth of the phytoplankton *Uroglena americana* turns the water surface into a reddish-brown color, or into malodorous water, producing a red tide. This was first observed in 1977, but the frequency of occurrence has been decreasing year by year (**Figure 5-3**).

On the other hand, a similar abnormal growth of blue-green algal blooms, including *Microcystis* spp. and *Anabaena* spp., has occurred annually over the past twenty years, turning the water surface green. In the Southern lake of Lake Biwa, blue-green algal blooms have occurred from August to September since the phenomenon was first observed in 1983 (**Figure 5-3**).

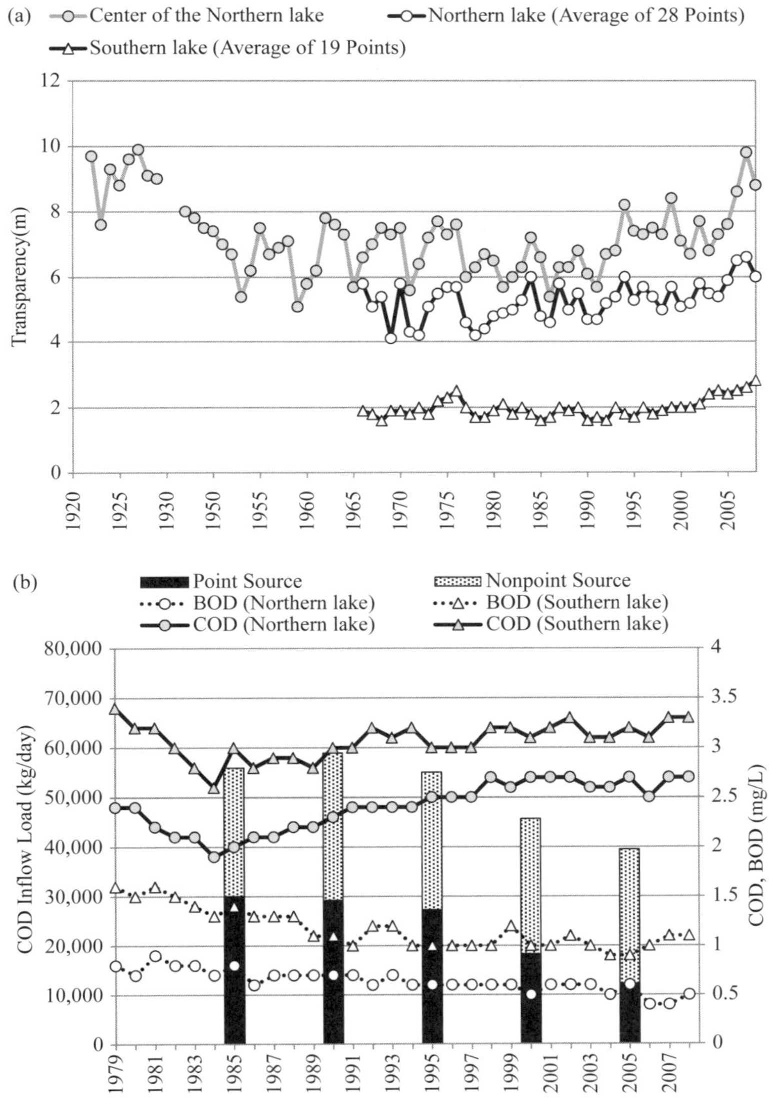

FIGURE 5-2. Continued.

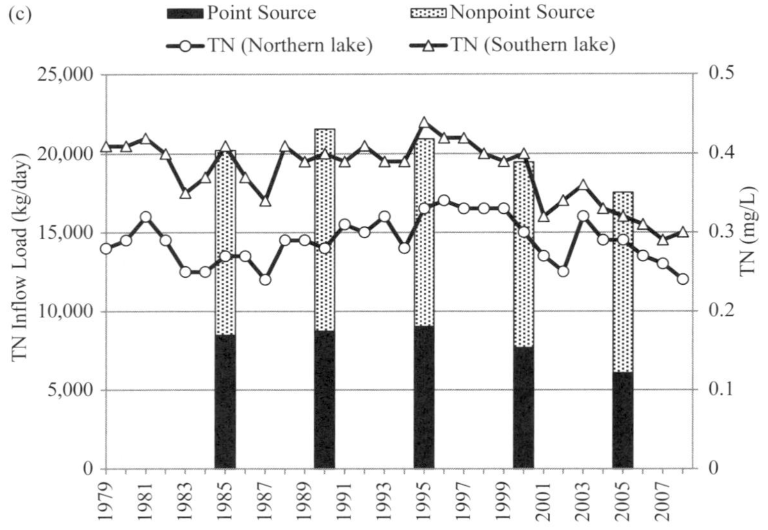

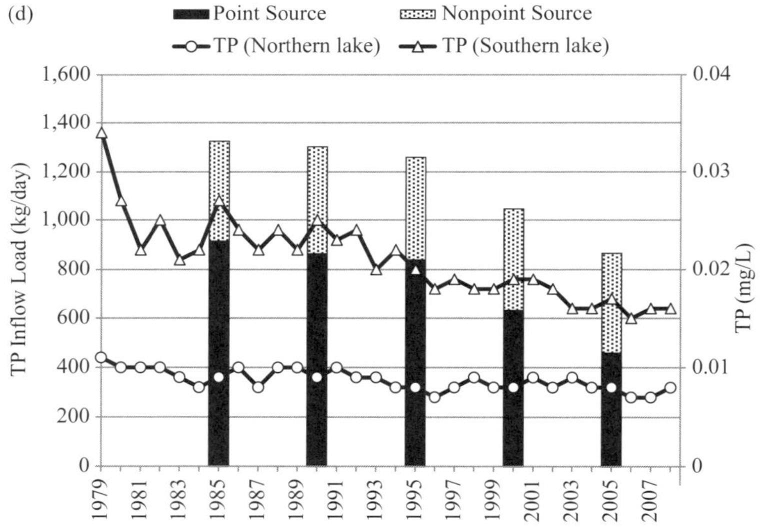

FIGURE 5-2. Trends in the main water quality indices and inflowing load of Lake Biwa.

5.2.2.3 Fish Catches

Fisheries are traditional industries in Shiga prefecture, but fish catches have been rapidly decreasing since the 1950s. In particular, shellfish catches (most of them Seta shellfish, which is an indigenous species) in recent several years have been about 1/100 the amount during peak periods. Fish catches have also been

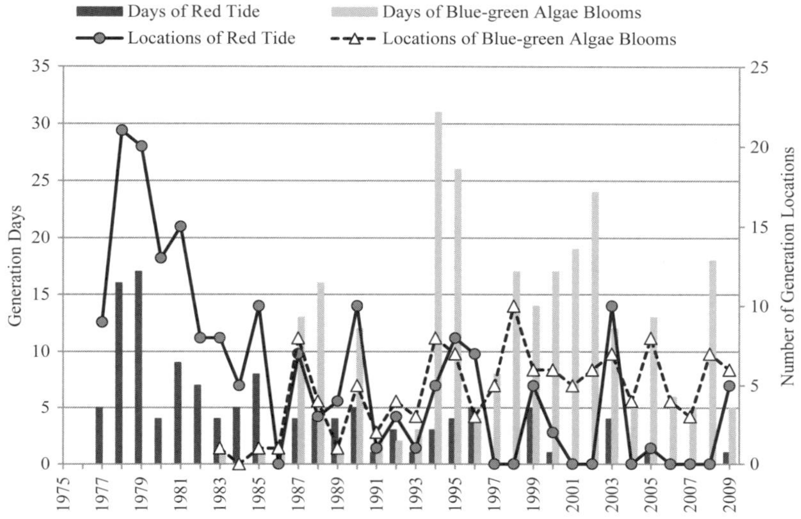

FIGURE 5-3. Occurrence of red tides and blue-green algal blooms.

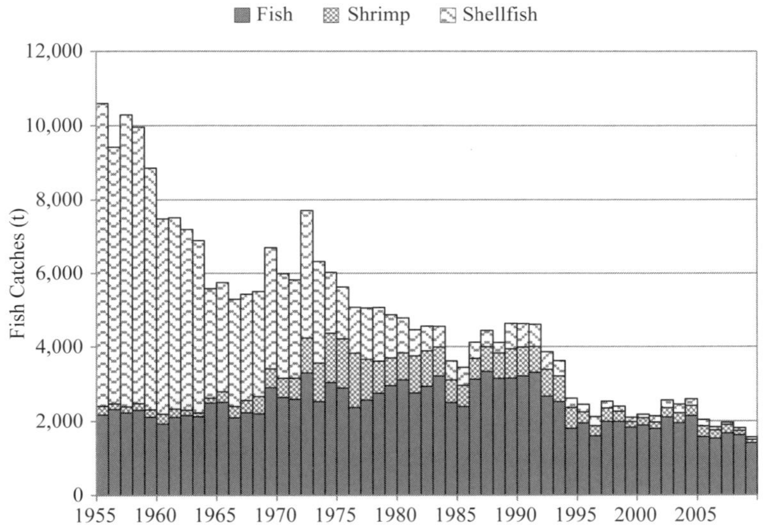

FIGURE 5-4. Fish catches in Lake Biwa.

reduced by about 50% compared to the peak catches (**Figure 5-4**). While many causes of this deterioration have been pointed out, such as the development of the lakeshore, a loss of linkage between the land and lake, and an increase in foreign species, the quantitative relationships have not been made clear.

5.2.2.4 Surrounding social situation

The population in Shiga prefecture has increased from about 800,000 to 1,400,000 since the 1970s (**Figure 5-5**). This has lead to an increase in the load generation from households, but the estimated inflow load into the lake has been decreasing since 1990. One reason for this is the sudden expansion of sewage systems since the 1980s. Moreover, most of these sewage systems have adopted advanced treatment in which nitrogen and phosphorus are highly eliminated. Nowadays, the sewage system adoption rate is up to 85% (**Figure 5-5**).

According to the population increase, the land use in Shiga prefecture has been changing (**Figure 5-6**). In particular, the amount of housing and road areas in recent years has become twice that during the 1960s, and farmland and forest areas have been lost by this same degree.

5.2.3 *History of environmental conservation in Lake Biwa*

In this section, the environment of Lake Biwa over the past fifty years is divided into five phases, and the characteristics of each phase are described. Many of the government efforts (mainly laws and ordinances) are also explained. **Table 5-1** shows the history of Lake Biwa and its environmental conservation measures from the 1960s to the 2010s.

5.2.3.1 First phase: agrichemical pollution (1960s)

The trigger of the pollution of Lake Biwa was agrichemical discharge in the 1960s. The discharge of agrichemicals has severely damaged fish and freshwater clams in rivers and lakeshores. In 1962, in particular, the damage to fisheries from agrichemical discharge was up to 400 million yen. For this reason, the Shiga prefectural government set up the usage standards for agrichemicals and

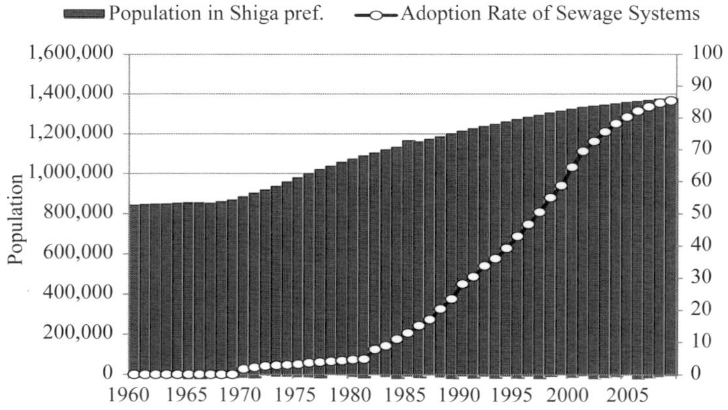

FIGURE 5-5. Changes in the population and rate of sewage installation.

has prohibited the use of agrichemicals with strong toxic or residual characteristics. Moreover, agrichemicals with less effect on the environment have been developed, and this problem is decreasing.

5.2.3.2 Second phase: environmental pollution by industrial effluent (1960s–1970s)

During this period of high economic growth, the amount of industry and production rapidly increased. Antimony, PCB, and chromium hexavalent pollution was detected in Lake Biwa and its watershed. Mold odors and local red tides also started to occur. Therefore, the government of Japan enacted the Basic Law for Environmental Pollution Control for the promotion of planned and comprehensive environmental pollution control in 1967. Under this law, the National Water Pollution Control Act was enacted in 1970, which aims at creating environmental standards for effluent control and the monitoring of public waters. This law was revolutionary because the pollution control was changed from restoration measures to proactive prevention. Responding to this movement, the Shiga prefectural government enacted the Prevention of Public Pollution Ordinance in 1969. This ordinance was drastically revised in 1972, and target facilities and items beyond those of the national law were added (called 'side putting out facility' and 'side putting out item'). Moreover, the Shiga prefectural government enacted Extra Ordinances of the Prefectural Water Pollution Control Act and set the effluent standard to 2- to 10-times stricter than the national law. In this way, Shiga prefectural government has set uniquely strict standards.

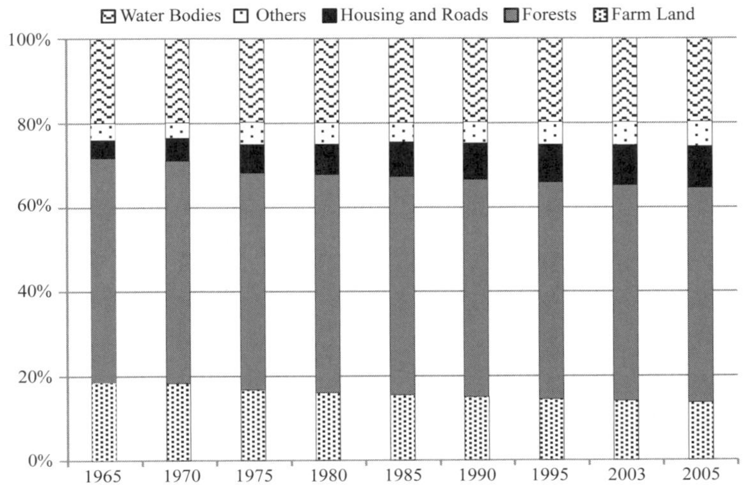

FIGURE 5-6. Land use in Shiga prefecture.

TABLE 5-1. History of Lake Biwa and its environmental conservation measures.

Period	Topics about Lake Biwa		Control measures for water environment: Shiga prefectural government		Control measures for water environment: Government	
1960	• Fish and freshwater clam damage by agrichemical PCP	(1960)	• Enactment of Prevention of Public Pollution Ordinance	(1969)	• Enactment of *Basic Law for Environmental Pollution Control*	(1967)
	• Overgrowth of Elodea Nuttallii in Northern lake	(1963)	• Otsu city Sewage System started in service	(1969)		
	• Antimony pollution in Maibara city	(1968)				
	• Mold odor in Lake Biwa and complaint in Kyoto city	(1969)				
1970	• PCB pollution in Kusatsu city	(1972)	• Environmental standard was introduced to Lake Biwa	(1972)	• Enactment of *National Water Pollution Control Act*	(1970)
	• Local red tides in Shiga town offshore	(1972)	• Formulation of *Environmental Conservation Measures in Lake Biwa*	(1972)	• Establishment of Environment Agency	(1971)
	• Overgrowth of Brazilian Elodea in Lake Biwa	(1973)	• Drastically Revision of *Prevention of Public Pollution Ordinance*	(1972)	• *Start of Lake Biwa Comprehensive Development Project*	(1972)

	• Chromium hexavalent was detected in Kusatsu City's well in high concentration	(1976)	• Enactment of *Prefectural Water Pollution Control Act Extra Ordinances*	(1972)	• Enactment of *Law Concerning the Examination and Regulation of Manufacture, etc of Chemical Substances Chemical Substances Control Law*	(1973)
	• Occurrence of Uroglena red tides in large area of Lake Biwa	(1977)	• Association of *Residential Activities for Promoting Soap Usage for Lake Biwa*	(1978)	• Environmental standard and effluent standard of PCB were introduced	(1975)
			• Regular water quality monitoring was started on 47 points in Lake Biwa	(1979)		
			• Enactment of *Eutrophication Prevention Ordinance*	(1979)		
1980	• Occurrence of blue-green algae blooms in Southern lake at the first time	(1983)	• Formulation of *Lake Biwa Access the Blue and Clean (ABC) Strategy*	(1980)	• Environmental standard of nitrogen and phosphorus was introduced in lakes	(1982)
	• Groundwater pollution caused by organochlorine compound such as trichloroethylene was detected	(1983)	• River basin sewage system in Konan areas started in service	(1982)	• Enactment of *Law Concerning Special Measures for Conservation of Lake Water Quality*	(1984)

(*Continued*)

TABLE 5-1. Continued.

Period	Topics about Lake Biwa		Control measures for water environment: Shiga prefectural government		Control measures for water environment: Government	
			• Enactment of *Conserving and Bringing up the Shiga's Landscape Ordinance*	(1984)	• First World lake Conference was held in Otsu, Shiga	(1984)
			• Enactment of *Prefectural Water Pollution Control Act Extra Ordinances of nitrogen and phosphorus about effluent*	(1985)	• Effluent standard of nitrogen and phosphorus was introduced in lakes	(1985)
			• Lake Biwa was made one of the *Designated Lake and Reservoir*	(1985)		
			• Formulation of *Water Quality Conservation Plan for Lake Biwa*	(1987)		
1990	• Outbreak of picoplankton in Northern lake	(1991)	• Enactment of *Conservation Reed Pipes Ordinance*	(1992)	• *National Water Pollution Control Act* was revised for domestic wastewater measures	(1990)
	• Occurrence of blue-green algae blooms in Northern lake at the first time	(1994)	• Enactment of *Antilittering Ordinance*	(1992)	• Enactment of *Basic Environment Law*	(1993)
	• Water level of Lake Biwa went down to –123 cm under the base level	(1994)	• Enactment of *Shiga prefecture Basic Environmental Ordinance*	(1996)	• Formulation of *Basic Environmental Plan*	(1994)

			• Enactment of *Domestic Wastewater Measures Promotion Ordinance*	(1996)		(1997)
			• Enactment of *Prefectural Water Pollution Control Act Extra Ordinances for small business establishments*	(1996)		(1997)
			• End of *Lake Biwa Comprehensive Development Project*	(1997)	• Enactment of *Environmental Impact Assessment Law*	(1999)
			• Formulation of *Shiga prefecture Comprehensive Environmental Plan*	(1997)	• The 3rd Session of the Conference of the Parties to the United Nations Framework Convention on Climate Change (COP3) was held in Kyoto	
			• Enactment of *Environmental Impact Assessment Ordinance*	(1998)	• Enactment of *Pollutant Release and Transfer Register (PRTR) Law*	
2000-	• Recovery of dissolved oxygen (DO) at the bottom of Lake Biwa was delayed two months on account of warm winter	(2007)	• Formulation of *Mother Lake 21 Plan*	(2000)	• Environment Agency was graded up to Ministry of the Environment	(2001)

(Continued)

TABLE 5-1. Continued.

Period	Topics about Lake Biwa		Control measures for water environment: Shiga prefectural government		Government	
	• Dissolved oxygen (DO) at the bottom of Lake Biwa recorded an all-time low since observations began	(2008)	• Enactment of *Proper Utilization of Lake Biwa by Leisure Activities Ordinance*	(2002)	• *Law Concerning Special Measures for Conservation of Lake Water Quality* was revised for nonpoint source pollution etc.	(2005)
			• Enactment of *Environmental Friendly Farming Promotion Ordinance*	(2003)		
			• Enactment of *Bringing up Lake Biwa Forests Ordinance*	(2004)		
			• Enactment of *Environmental Education Promotion Ordinance*	(2004)		
			• Formulation of *the 5th Water Quality Conservation Plan for Lake Biwa*	(2007)		
			• Formulation of *Sustainable Shiga Social Vision*	(2008)		
			• Enactment of *Shiga prefecture Low Carbon Society Promotion Ordinance*	(2011)		
			• Formulation of *the 2nd Mother Lake 21 Plan*	(2011)		

Sources: Shiga prefecture (2007), Tsuno (2000), and other documents

The Lake Biwa Comprehensive Development Project was started in 1972, and was implemented for twenty-five years. The project plan consisted of three pillars: environmental conservation concentrated on the development of sewage systems, flood control, and water utilization. This project enabled the eutrophication to be curbed, prevented flooding around Lake Biwa and the downstream region, and supplied water for downstream prefectures in later years.

5.2.3.3 Third phase: Eutrophication and nitrogen and phosphorus control (1970s–1980s)

Lake Biwa experienced outbreaks of freshwater red tides by Uroglena Americana in 1977, which came as a shock to many people. As the use of soap was promoted by women groups for health reasons beginning around 1970, this act was strongly advanced for Lake Biwa, as the phosphorus in detergent soap is one of the reasons for the existence of red tides. These women's groups played a central role in the association of Residential Activities for Promoting Soap Usage for Lake Biwa in 1978. Such actions were called a 'soap movement.' Responding to these activities, Shiga prefecture enacted the Eutrophication Prevention Ordinance in 1979, which was the first regulation for nitrogen and phosphorus drainage in Japan. Moreover, Shiga prefecture formulated the Lake Biwa Access the Blue and Clean (ABC) Strategy to promote comprehensive conservation measures that were not implemented by the above ordinance. This movement in Shiga prefecture affected the national government of Japan. Nitrogen and phosphorus were added to the environmental standards for lakes in 1982 and to the effluent standards in 1985. Moreover, to correspond to lake-specific problems, the Law Concerning Special Measures for the Conservation of lake Water Quality was enacted in 1984. Lake Biwa was made one of the designated lakes and reservoirs under this law in 1985, and the Water Quality Conservation Plan for Lake Biwa was formulated in 1987. Water quality conservation programs, such as the development of sewage systems, the implementation of measures to control factory and agricultural effluents, environmental monitoring, and research projects, as well as the water quality conservation target, have been determined. This plan was revised every five years.

5.2.3.4 Fourth phase: Expansion of Lake Biwa environmental conservation measures (1980s–1990s)

In this phase, the adoption rate of sewage systems was rapidly expanded from 4.6% in 1980 to 58.8% in 1999 (**Figure 5-5**). The river basin sewage system in the Konan area started service in 1982. One of the main reasons for a decreasing inflow load was the government promotion for an expansion of the sewage systems. The unit load from the sewage systems (advanced treatment) is smaller than that from other sewage treatments, especially in TP (**Figure 5-7**).

In addition, Shiga prefecture enacted the Domestic Wastewater Measures Promotion Ordinance in 1996, in which the installation of joint wastewater

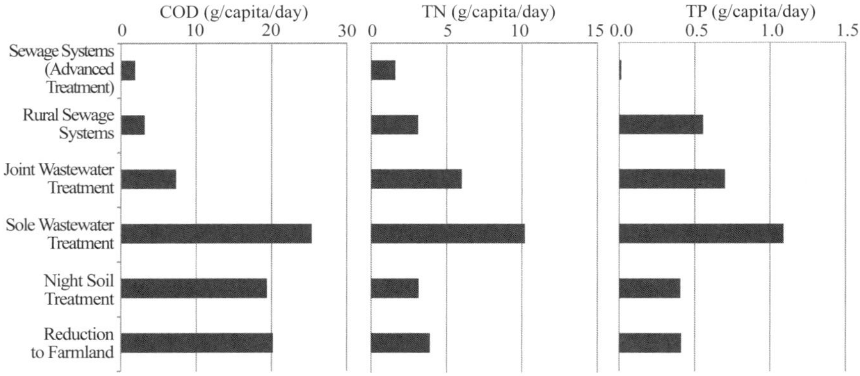

FIGURE 5-7. Unit load from sewage treatment in Shiga prefecture.

treatment tanks became mandatory. Shiga prefecture also enacted the Prefectural Water Pollution Control Act Extra Ordinances for small business establishments (with a daily mean effluent displacement of 10 m^3) in the same year.

These measures and ordinances lead to water quality improvements and a decrease in red tides, but other problems regarding water quality and ecosystems appeared. Blue-green algal blooms first occurred in Southern lake in 1983 and in Northern lake in 1994. An outbreak of picoplankton also occurred in Northern lake in 1991. The COD concentration in Lake Biwa has been increasing since 1985, and fish catches have been gradually decreasing.

A diversity of problems was addressed in the ordinances enacted during this phase. For example, the Conserving and Bringing up the Shiga's Landscape Ordinance was enacted in 1984, and the Conservation Reed Pipes Ordinance and Antilittering Ordinance were enacted in 1992.

5.2.3.5 Fifth phase: Global environmental problems and the comprehensive conservation of Lake Biwa (1990s–2000s)

Japanese environmental government was promoted under the framework of the Basic Law for Environmental Pollution Control, and achieved the recovery from public pollution. While this law has concentrated on pollution regulation, there is a current need to respond to urban and residential problems caused by mass production, mass consumption, and mass disposal, as well as global environmental problems. Therefore, the Basic Environment Law was enacted in 1993, which aims at a comprehensive environmental policy. The Shiga prefecture also enacted the Shiga prefecture Basic Environmental Ordinance in 1996, which aims at securing healthy and high-quality environments. In 1997, the Shiga prefecture Comprehensive Environmental Plan was formulated under this ordinance.

In 1997, the Lake Biwa Comprehensive Development Project ended. This project improved the safety and convenience for residents, but has had a negative influence on ecosystems, such as the loss of habitats for many types of creatures. Therefore, the Shiga prefecture formulated the Mother lake 21 Plan (a comprehensive preservation and restoration plan for Lake Biwa) in 2000 based on reports made by the government of Japan as a guideline to bridging current conservation efforts with next-generation efforts. In this plan, the period between 2000 and 2050 was divided into three stages, and specific objectives were set for each stage. These objectives consist of 'maintaining water quality,' 'improving the recharge capacity of the soil,' and 'preserving the natural environment and scenic landscapes.'

5.2.4 *New challenges in Lake Biwa*

As described above, most of the water quality indices have shown a recovery in spite of the increasing population and other developments. This mainly depends on the ordinances of the Shiga prefecture and the laws of the government of Japan, along with the various measures entailing these institutions. On the other hand, in recent years, many additional problems have been pointed out.

In terms of water quality, for example, the concentration of COD in Lake Biwa is not decreasing as much as before, and the minimum annual amount of dissolved oxygen (DO) in the deep parts of Lake Biwa is decreasing. Recovery of DO at the bottom of Lake Biwa was delayed two months owing to the occurrence of a warm winter in 2007, and global warming is therefore a suspect in the decrease of DO. Other problems regarding ecosystems and the public have appeared, such as forest damage by deer feeding, an overgrowth of waterweeds in Southern lake, and a lack of linkage between our life and Lake Biwa.

For these new challenges, a second Mother lake 21 Plan was scheduled for formulation in 2011, which consists of two pillars, i.e., the 'preservation and restoration of ecosystems in Lake Biwa Basin' and the 'restoration of the relationship between our life and Lake Biwa' (**Figure 5-8**). In addition, the concept of this plan is 'adaptive management.' Ecosystems are complicated, continuously changing, and much more difficult to predict than water quality. Therefore, this plan and its many measures need to be reconsidered from a long-term perspective to respond to these circumstances.

5.3 NEW DEVELOPMENTS REQUIRED

Many lakes have been beset by a myriad of problems affecting their sustainable use. Further, lakes and their drainage basins are fundamentally linked, and interactions between humans and their water and land resources are critical factors influencing lake health and potential long-term uses. Just as the impacts of unsustainable lake uses are felt within the water body, along its

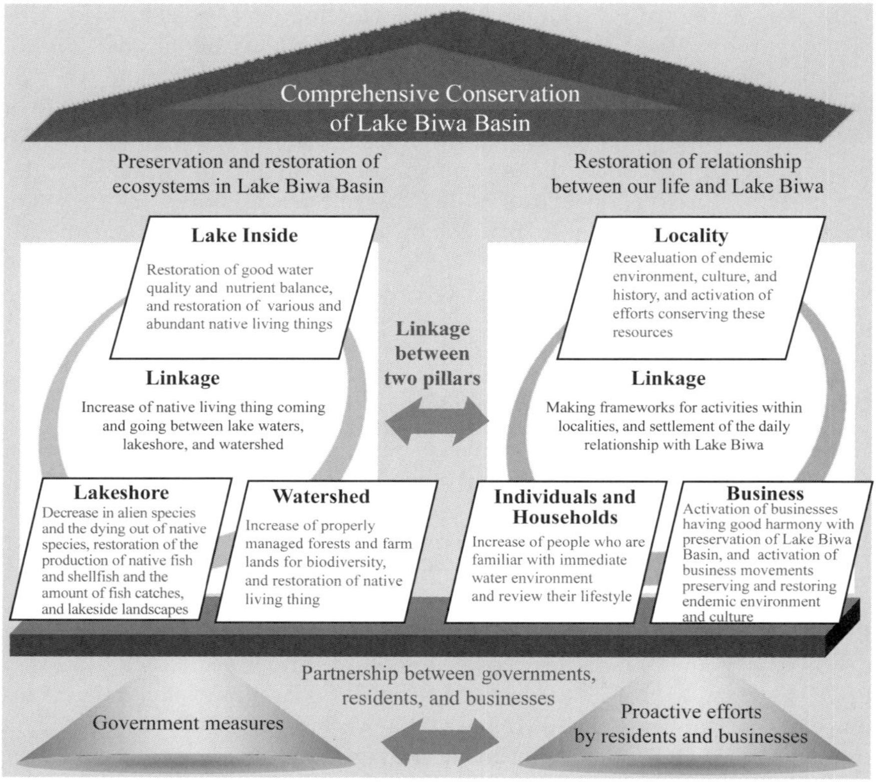

FIGURE 5-8. Pillars and goals of the second Mother lake 21 Plan.

shoreline, or in other parts of its drainage basin, the causes of the problems may lie along their shorelines, and elsewhere within and even outside their drainage basins. The use of water and land resources within a drainage basin, therefore, determines the types and magnitude of its environmental stresses. Further, because many of the world's lakes are being impacted simultaneously by multiple problems, their remediation is often more difficult and costly than addressing a single problem alone. lake problems do not necessarily affect only the people living along its shorelines, but can also have significant economic, health, and/or environmental impacts on inhabitants further afield, both within and outside the drainage basins.

Major environmental or water use problems generally accompany a substantial human settlement of a lake's drainage basin, resulting in the need for large quantities of drinking water and an increase in economic development. In most developing countries, lakeshore communities are heavily or completely dependent upon lakes for their livelihoods through such activities as open-water fishing and intensive aquaculture. Many of the problems facing lakes are deeply rooted in socioeconomic issues, and a major causative factor for the variety of problems facing lakes, in fact, is their multiple roles within human society. Factors contributing to a decrease or degradation of lake uses range from insufficient scientific knowledge and understanding, to technical shortcomings; inadequate intellectual, financial, and/or technological resources; and inappropriate development and governance. There has been no disagreement, however, that the excessive stresses placed on lakes to meet human water needs are a major factor. Such stresses, as well as the issues identified below, have the potential to fundamentally affect the livelihoods of people that depend directly on lakes for their food and basic economic well-being, particularly indigenous peoples and lakeshore communities.

5.3.1 *Sustainable use and preservation of lake watersheds*

The world's population is predicted to grow from six billion today to about nine billion by the year 2050, putting an increasing pressure on local authorities and planners to supply water to satisfy growing agricultural and urban water and sanitation demands. Untreated or inadequately treated sewage is already a major water pollution issue in virtually all developing countries, particularly those undergoing explosive urbanization. Industrial water demands will also continue to grow with increasing economic development. It has been predicted that, if present water use trends continue, approximately two out of every three persons will live under conditions of water stress by the year 2025. Since most of the water taken from lakes and rivers is eventually returned directly or indirectly to these sources, it is necessary that increased lake water withdrawal and usage be accompanied by the development of basic sanitation and wastewater treatment. Excessive water withdrawals from a lake also can cause significant water-level fluctuations directly affecting lake ecosystems, and in some cases, the very existence of a lake itself.

Growing human populations will tend to increase agricultural water demands. It has been predicted that global food production during the next thirty years must double to keep pace with population needs. The need for increased food supplies is causing farmers in many locations to increasingly cultivate marginal lands, resulting in a greater use of fertilizers and other agricultural chemicals, and increasing the potential for lake eutrophication and the accumulation of pesticides, along with the accompanying human and ecological problems.

In certain regions of the world, population-related stresses on the availability of freshwater resources may occur against a background of intensifying adverse impacts of climate change on regional hydrology. This phenomenon could have a profound impact on water inflows and lake levels, especially in inland drainage basins. For lakes that normally receive significant quantities of water from snow melt, reduced snowfield areas associated with climate change can also lead to lower hydrological inputs.

Limited public awareness of the human impact on lakes has contributed to a degradation of their values and uses. Inadequate public awareness may result from insufficient knowledge, data, and/or understanding on the part of citizens, local authorities, decision-makers, the media, industry, and other players of their roles, either individually or collectively, in causing lake problems or in helping solve them. Lake scientists and experts can do more to conduct applied lake research and inform the public and policymakers about their results. Further, in certain cases, governmental agencies and/or decision-makers may believe that the only appropriate role for the public in such matters is to provide the required funds for the programs and activities to address lake problems, in contrast to the proactive approach of working with the public to identify and resolve current problems and/or avoid similar problems in the future. On the other hand, citizens may think that they need to rely exclusively on governmental agencies and/or decision-makers for solutions to such problems. Experience around the world, however, suggests that, where feasible, the involvement of the public can be beneficial in identifying lake problems and in developing sustainable and publicly supportable solutions.

A major contributor to the lack of understanding and awareness by the public and decision-makers regarding lake degradation is the subtle nature of the many types of lake problems. Such problems can manifest themselves very slowly, often across generations. They may only become evident after the degradation has become very severe, and even potentially irreversible. This subtle nature of lake degradation makes it harder to create awareness of lake problems among the public and decision-makers, and to initiate a needed remediation or restoration activities in a timely manner.

5.3.2 *Governance of lake watersheds*

Assuming that an institutional framework for lake management already exists, a lack of proper accountability on the part of citizens and governments is one of the most significant root causes of unsustainable lake use. Poor public consultation, inadequate stakeholder participation, and a lack of appropriate and effective governmental institutions and regulatory mechanisms are the major hindrances to sustainable lake use. The lack of clear policy frameworks that recognize lakes as important aquatic resources, and that specifically address lake management issues, is another threat to sustainable lake use. Further, many countries around the world suffer from inadequate legal expertise in the fields of environmental law and management, resulting in an inconsistent enforcement of environmental regulations. Many also lack sufficient primary and secondary level teachers with necessary education and training regarding important environmental issues and their relevance to human conditions and well-being, and hence fail to convey the urgency of the need for awareness and participation by youth and communities in lake management efforts.

Available avenues for seeking governmental and stakeholder accountability are often limited by fragmented governmental jurisdictions and competing or overlapping responsibilities. As a result, public policies and practices often appear to be insensitive to lake use issues, especially at the local level.

A lack of transparency in the decision-making process is surprisingly common. Further, in emerging democracies, there often is an asynchrony between the process of developing good governance and the urgency of implementing accountable environmental stewardship. This situation can be exacerbated by not recognizing the linkages between the concerns of environmental managers on one hand, and water managers on the other.

Effective training for local and national governmental and non-governmental staff members, particularly for building coalitions, managing projects, and increasing monitoring and evaluation skills, is also lacking in many countries. Further, although many countries may have central environmental authorities, effective institutions for initiating and overseeing the development and implementation of comprehensive, long-term plans for the sustainable use of lakes and their drainage basins are often lacking.

Factors such as these strongly constrain the development and implementation of environmentally compatible and cost-effective management plans for the sustainable use of lakes and their resources.

5.4 SUMMARY

Containing over 90% of the readily available liquid freshwater on the surface of our planet, lakes are the key components of global water-resource systems. Whether natural or artificial (reservoirs), they provide an enormous number and range of resource values that facilitate sustainable human livelihoods and economic development (drinking and industrial water supply, irrigation, fisheries, hydropower, recreation, transportation, aesthetics, and cultural and religious significance, among others). Their development, use, and conservation have therefore been major human elements at both the national and international scales. They also serve as essential habitats for an amazing variety of flora and fauna. In developing countries, they are often the centers of livelihood for small-scale local fishers and lakeshore communities.

Many lakes around the world are shared by two or more countries. Although some riparian countries have discussed the management of international river systems, far fewer people are aware of the implications of lakes as international water systems.

Some involve upstream-downstream relationships and issues between countries sharing a lake, while others involve lakes that serve as international boundaries between countries. This lack of international awareness can be manifested in the unsustainable use of water from shared lakes by one or more of the countries in their drainage basin, as well as in changes in water quantity and quality, wetland ecosystems, aquatic plants, animal communities, and so on. Cooperation and collaboration between countries sharing an international lake will obviously facilitate the identification and implementation of solutions to lake problems.

In regard to lake management issues in general, whether at an international or national scale, it is further noted that we have not yet developed a clear picture on what has worked, what has not, and under what conditions. The lack of such information, data, and experience limits our insight and capabilities about what can be done in the future to improve lake drainage basins from lake management programs. A systematic evaluation of the effectiveness of previous lake management efforts throughout the world, particularly those directed at sustainable lake use, is urgently needed to provide such information and guidance.

REFERENCES

Tsuno, H. (2000), Social Trend and Policy, in Isao Somiya (eds), *Lake Biwa*, Gihodo Press, pp. 7–18 (in Japanese).
Shiga Prefecture (2007), Environment of Shiga.

6

Water Reuse and Risk Management

In this chapter, risk management is discussed from the viewpoint of water reuse. First, reuse methods for water, intended and unintended, are addressed, and risk materials are presented. Then, monitoring methods are described, including a chemical determination method, a bio-assay, and bio-monitoring. Finally, the control and management of risks, including risk evaluation and removal technologies for hazardous chemicals, are discussed.

6.1 INTENDED AND UNINTENDED REUSE OF WATER

With an increasing global population, economic boom, and urbanization, ecological environments, particularly aquatic ecological environments, are under great pressure. The discharge of different kinds of wastewater and the overuse of water resources is causing a water shortage in many lakes and rivers, consequently making the deterioration of aquatic ecological environments the most important environmental issue worldwide.

Water shortage is a worldwide problem, and is particularly severe in China. The amount of water resources per capita is a quarter of the world's average (Ministry of Land, Infrastructure, Transports, and Tourism 2009). At the same time, discharged wastewater causes the pollution of aquatic environments. Water pollution makes the shortage of water resources even more serious. Japan also has a lack of water resources, and the amount of available water is not sufficient in many areas. Annual rainfall per capita in Japan is one-third of the world's average (Ministry of Land, Infrastructure, Transports and Tourism 2009). This is mainly because Japanese territory is small and its population density is large. Changes in weather conditions by global warming are making and will continue to make the condition more severe in both China and Japan.

How to release the pressure caused from societal and economic developments on ecological environments, and how to maintain the environment in its natural state with a sufficient amount of sound quality water are very important issues and great challenges for both China and Japan. Recently, reclaimed water as a new and non-regular water resource has been utilized to resolve water shortages and water pollution. In many cities in Japan, such as Tokyo, Fukuoka, and Osaka, reclaimed water from municipal wastewater treatment plants is reused after advanced treatment as non-portable municipal water, landscape water, and supplied to bodies such as rivers or ponds for restoration of ecological environments. The Tamagawa Jousui and Nobidome Yousui channels are typical examples of recovered landscape through the use of reclaimed water after ozonation in the Tama Joryuu Water Reclamation Plant. In China, more and more attention is being paid to the reuse of reclaimed water, particularly in cities suffering from water shortages. Reclaimed water after advanced treatment is utilized for landscape water, eco-environment water, and so on. For example, in Beijing, treated wastewater from the Gaobeidian Wastewater Treatment Plant is replaced into landscape water bodies including rivers and lakes of Longtanhu Park, Beijing Amusement Park, Temple of Heaven Park, Taoranting Park, and so on.

In many rivers, treated wastewater from wastewater treatment plants is repeatedly used unintentionally. Yodo river is a typical example of a repeated reuse of water. Wastewater discharged from Shiga prefecture flows into water sources of Kyoto city, and furthermore, wastewater discharged from Kyoto city flows into water sources of Osaka and Hyogo prefectures.

6.2 RISK MATERIALS

Along with various pollutants, general organics and nutrients such as nitrogen and phosphorus are included in the effluent standard and are removed well in general advanced wastewater treatment plants. Heavy metals and hazardous chemicals are also included in the effluent standard and have been restricted to discharge. However, other pollutants have been known to exist in natural water bodies and have the possibility to threaten human health and healthy ecosystems. The critical point of reclaimed water for ecological environments, and even for human health for repeated (unintended) reuse, is safety and low environmental risks. Recently, with the advancement of analysis technologies, more and more micro-pollutants have been detected to exist in water environments, such as endocrine disrupting chemicals (EDCs), persistent organic pollutants (POPs), pharmaceuticals and personal care products (PPCPs), and so on. Consequently, the toxicity of these trace pollutants has been getting more and more attention. Disinfection byproducts should also be checked as chlorination is used to secure safety from pathogens for water reuse; however, these issues are discussed in Chapter 3. EDCs, POPs and PPCPs are the focus here.

6.3 MONITORING

6.3.1 *Determination*

Determination methods are fundamental for proper monitoring. Measurement with high level of apparatuses and bio-assays are the major technologies used in the determination of chemicals. These are combined with concentration and extraction techniques as the concentrations of these chemicals, including EDCs, POPs and PPCPs, are as low as μg/L and/or ng/L in natural water. Liquid-liquid extraction and solid-phase extraction methods are broadly used. A liquid-liquid extraction method for determination of POPs in water, and a solid-phase extraction method for EDCs determination, are shown in **Figures 6-1** (Takabe *et al.*, 2011) and **6-2** (Graduate school of Engineering, Kyoto University and Department of Environmental Science and Engineering, Tsinghua University 2010) as respective examples. The solid-phase extraction method is also applied for PPCPs determination.

A concentrated sample of an extracted sample is input into GC/MS for EDCs and POPs determination and into LC-MS/MS for PPCPs determination. The pollution state of these chemicals is explained in Chapter 4. The concentrated sample is used for a bio-assay, which is explained in the next section.

6.3.2 *Bio-assay*

A bio-assay is a determination method using specially designed microbes or cells to detect a target effect of chemicals on a living thing. It has also been

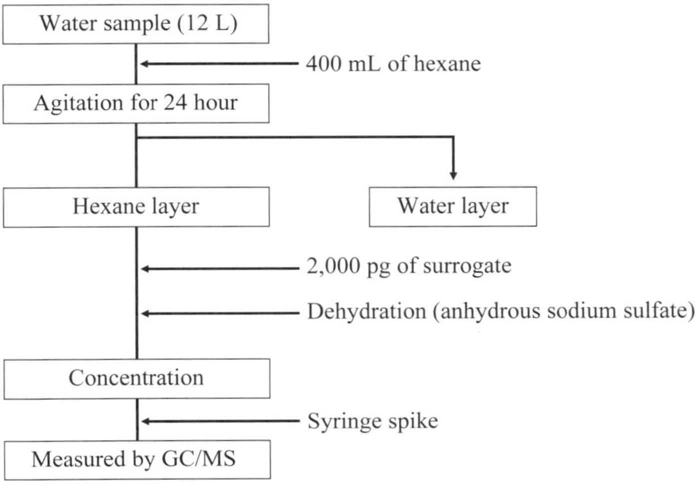

FIGURE 6-1. Pretreatment of POPs in water (Takabe *et al.*, 2011).

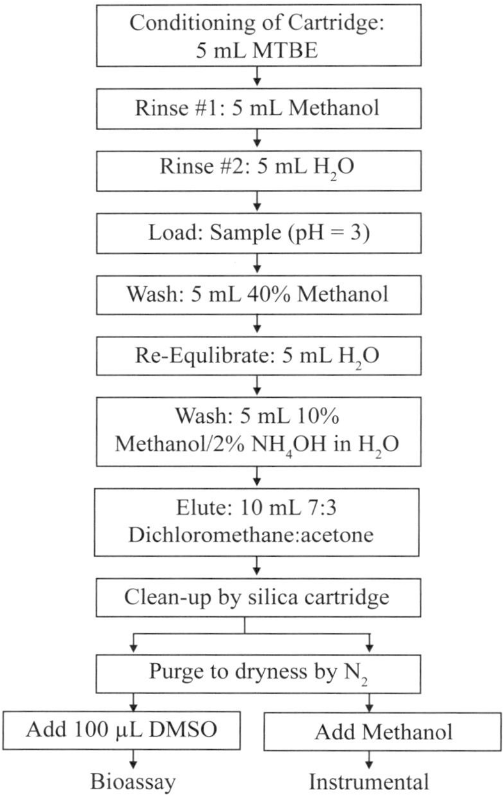

FIGURE 6-2. SPE procedure for dissolved matter (Graduate School of Engineering, Kyoto University and Department of Environmental Science and Engineering, Tsinghua University, 2010).

developed to detect the overall effects of both known and unknown chemicals, and for use as a first screening test with a simple, sensitive, and rapid property. Chemicals that have a target effect bind to a receptor designed in the microbes or cells, and the amount of bound chemicals is measured through absorbance, for example, which is associated with the receptor activity. The mechanism of the NRL/TIF2-BAP assay, which is used for EDCs determination, is schematically shown in **Figure 6-3** (Tsuno *et al.*, 2009) as an example.

The results from the bio-assay are expressed by the strength of the target effect, which is expressed as the concentration of the standard chemical, that is, the effect-equivalent concentration (EEQC). The calculation concept of an EEQC sample is shown in **Figure 6-4** and Equation (6-1).

$$\text{EEQC (g/L)} = \text{RBA} \times \{\text{molecular weight of the standard chemical (g/mol)}\} \quad (6\text{-}1)$$

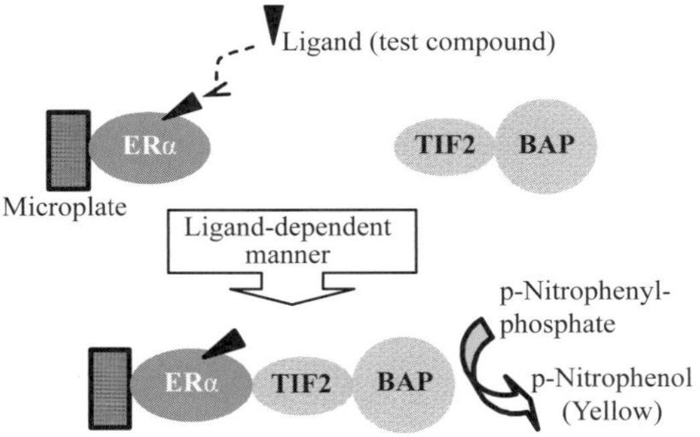

FIGURE 6-3. Schematic of an NRL/TIF2-BAP assay (Tsuno *et al.*, 2009).

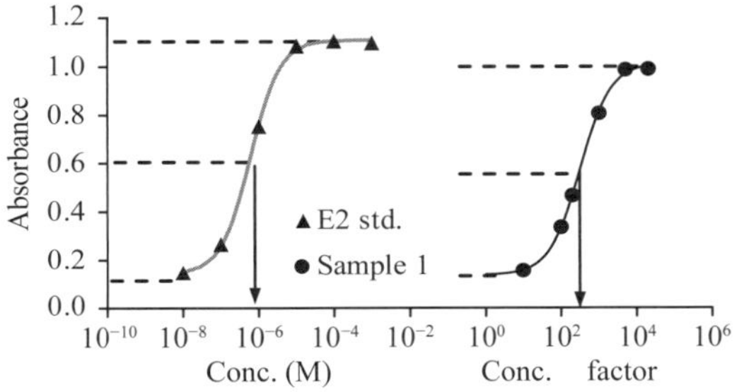

FIGURE 6-4. Dose-response curves of actual samples in an NRL/TIF2-BAP assay.

$$\text{RBA} = \{\text{concentration of the standard chemical at a 50\% effect}\}/\ \text{concentration factor of sample at a 50\% effect}\} \quad (6\text{-}2)$$

RBA values for the endocrine disrupting effect of certain EDCs determined using an NRL/TIF2-BAP assay are shown in **Table 6-1** (Tsuno *et al.*, 2009), and a comparison of the EEQC between the bio-assay results and calculated results with the measured concentration and RBA value of each EDC is shown in **Figure 6-5** (Tsuno *et al.*, 2009). The effects of unknown chemicals are shown at the earlier stage of treatment, and the EEQC values by both methods coincide well.

6.3.3 *Bio-monitoring*

Heavy metals and recalcitrant organic chemicals are concentrated at the biota. When a recalcitrant material is transported from phase A to phase B, the concentration factor (γ) is expressed using the following equation:

$$\gamma = CB^*/CA^* \quad (6\text{-}3)$$

where CA* and CB* are equilibrium concentrations of the target material in phase A and phase B, respectively. The values of γ increase 1,000 to 10,000,000 times depending on the properties of the target material and biota. Therefore, we can determine the extent of pollution through a measurement of the concentration of the target material in the biota using a simple extraction, and without the use of an annoying sampling of larger volumes of water and difficult concentration procedures.

TABLE 6-1. RBA of standard estrogens in different bioassays (Tsuno *et al.*, 2009).

Compound/ Method	RBA (mole/mole)			
	NRL/TIF2-BAP	NRL/FP[1)]	Two-hybrid[2)]	Sumpter[3)]
E2	1	1	1	1
E1	1.6	0.47	5.3×10^{-2}	0.21
E3	2.7	–	1.4×10^{-3}	1.3×10^{-3}
NP	1.5×10^{-3}	5.9×10^{-3}	4.1×10^{-4}	1×10^{-3}
BPA	1.8×10^{-2}	7.0×10^{-4}	1.4×10^{-4}	2.7×10^{-4}

1: Zheng Heging; Behavior of Endocrine Disrupting Chemicals in Wastewater Treatment process and their removal by ozonation, Doctoral Thesis of Kyoto University, 49, 2006.
2: Nishikawa, J. *et al.*, New screening methods for chemicals with hormonal activities using interaction of nuclear hormone receptor with coactivator, Toxicol. Appl. Pharmacol. 157, 76–83, 1993.
3: Routledge, E.J. and Sumpter, J.P., Estrogenic activity of Surfactants and some of their degradation products assessed using a recombinant yeast screen, Environ. Toxi. Chem. 15, 241–248, 1996.

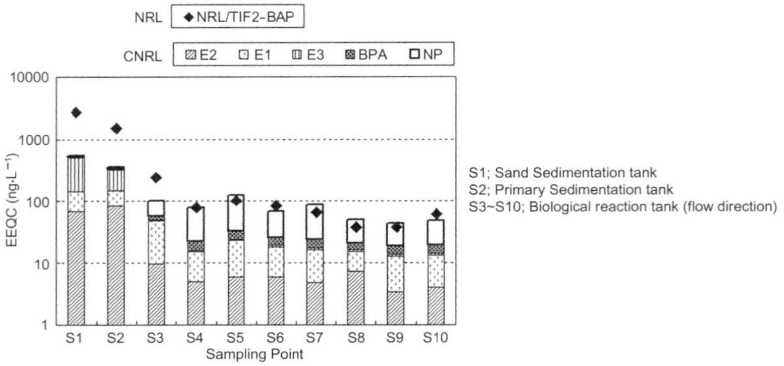

FIGURE 6-5. Comparison between NRL and CNRL of estrogenicity in filtrate during the treatment of Step AO Process (Tsuno *et al.*, 2009).

The biota used for bio-monitoring is called a biological indicator. The essential factors required for a biological indicator are as follows: (1) high γ value; (2) constant γ value independent from size, age, sex, physical state, and so on; (3) worldwide existence; (4) low mortality; (5) easy to collect; (6) easy to culture in a laboratory; and so on. Mussels are widely used as biological indicators and are known to be excellent indicators in seawater. Examples of PCB concentrations in mussels are shown in **Figure 6-6** (Takabe *et al.*, 2009). The concentrations are within the range of 100 to 100,000 pg/g-wet.

To understand local pollution, detect pollution at early stages, and identify pollutant sources, the monitoring of recalcitrant chemicals in fresh and brackish waters is also important, especially for organochlorine pesticides (OCPs). *Corbicula*, which lives worldwide in fresh and brackish waters, has been proposed as a biological indicator (Takabe *et al.*, 2011). Concentrations of DDTs and chlordanes in *Corbicula* as determined in western Japan and the Pearl river delta in China are compared in **Figure 6-7** with those in mussels captured in sea water areas (Takabe, *et al.*, 2011). The results show that these concentrations determined in fresh, brackish, and sea waters are comparable. The compositions of DDTs, chlordanes, and HCHs in *Corbicula* can well reflect those in water and sediment, as shown in **Figure 6-8** (Takabe *et al.*, 2011). The distribution of *cis*-chlordane/*trans*-chlordane ratio in *Corbicula*, water, and sediment as determined in western Japan and the Pearl river delta in China is shown in **Figure 6-9** (Takabe *et al.*, 2011). The solid line in the figure shows the ratios of technical chlordane in Japan, which are similar to those internationally. The ratio

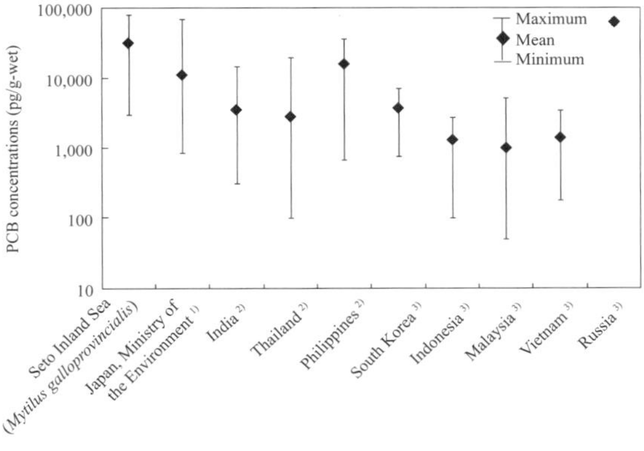

1) Ministry of the Environment: Results of monitoring surveys in 2008.
2) Tanabe, S., Prudente, M.S., Kan-atireklap, S., Subramanian, A. (2000). Mussel watch: marine pollution monitoring of butyltins and organochlorines in coastal waters of Thailand, Philippines and India. Ocean and Coastal Management, 43, 819–839.
3) Monirith, I., Ueno, D., Takahashi, S., Nakata, H., Sudaryanto, A., Subramanian, A., Karuppiah, S., Ismail, A., Muchtar, M., Zheng, J., Richardson, B.J., Prudente, M., Hue, N.D., Tana, T.S., Tkalin, A.V., Tanabe, S. (2003). Asia-Pacific mussel watch: monitoring contamination of persistent organochlorine compounds in coastal waters of Asian countries. Marine Pollution Bulletin, 46, 281–300.

FIGURE 6-6. PCB concentrations in mussels in seawater areas (Takabe *et al.*, 2009).

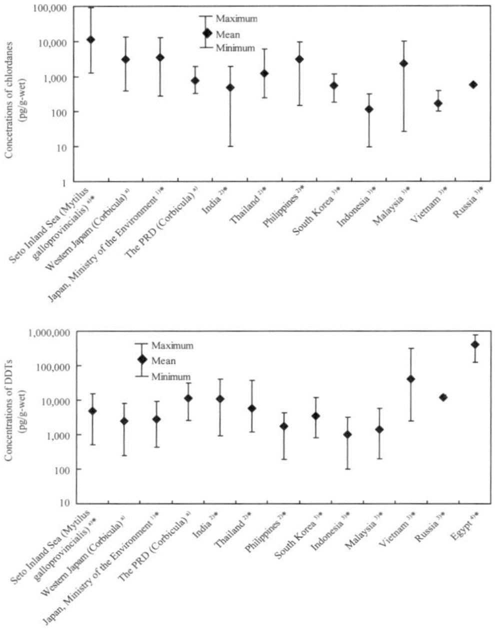

a) Water quality engineering, Kyoto University.
1) Ministry of the Environment: Results of monitoring surveys in 2008.
2) Tanabe, S., Prudente, M.S., Kan-atireklap, S., Subramanian, A. (2000). Mussel watch: marine pollution monitoring of butyltins and organochlorines in coastal waters of Thailand, Philippines and India. Ocean and Coastal Management, 43, 819–839.
3) Monirith, I., Ueno, D., Takahashi, S., Nakata, H., Sudaryanto, A., Subramanian, A., Karuppiah, S., Ismail, A., Muchtar, M., Zheng, J., Richardson, B.J., Prudente, M., Hue, N.D., Tana, T.S., Tkalin, A.V., Tanabe, S. (2003). Asia-Pacific mussel watch: monitoring contamination of persistent organochlorine compounds in coastal waters of Asian countries. Marine Pollution Bulletin, 46, 281–300.
4) Khaled, A., Nemr, A.E., Said, T.O., El-Sikaily, A., Abd-Alla, A.M.A. (2004). Polychlorinated biphenyls and chlorinated pesticides in mussels from the Egyptian Red Sea coast. Chemosphere, 54, 1407–1412.
* Surveys with mussels in seawater areas

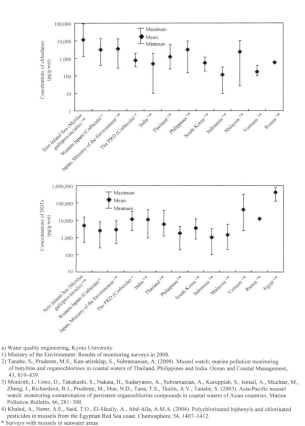

FIGURE 6-7. Concentrations of chlordanes and DDTs in bivalves (Takabe *et al.*, 2011).

in western Japan is higher than for technical chlordane, while the ratio in the Pearl river delta in China is close to the ratio found in technical chlordane. The difference between the two regions reflects the time when chlordane was used, as *cis*-chlordane is known to be more persistent than *trans*-chlordane.

Bioaccumulation is caused by the uptake of chemical compounds by the biota from either water or food. Lipids are known to serve as the primary storage compartment of hydrophobic organic chemicals, and bioaccumulation factors on a lipid basis (BAFLs, pg/g lipid/(pg/mL)) are commonly used to show the bioaccumulation levels of target chemicals from water to biological indicators, and to analyze bioaccumulation characteristics in aquatic organisms. In addition, an octanol-water distribution coefficient (Kow) is used as an indicator of a chemical's hydrophobicity. The relationship between logBAFLs and logKow for OCP and PCB are plotted in **Figure 6-10** (Takabe *et al.*, 2009; Takabe *et al.*, 2011; Tsuno *et al.*, 2007a; Tsuno *et al.*, 2007b), and the

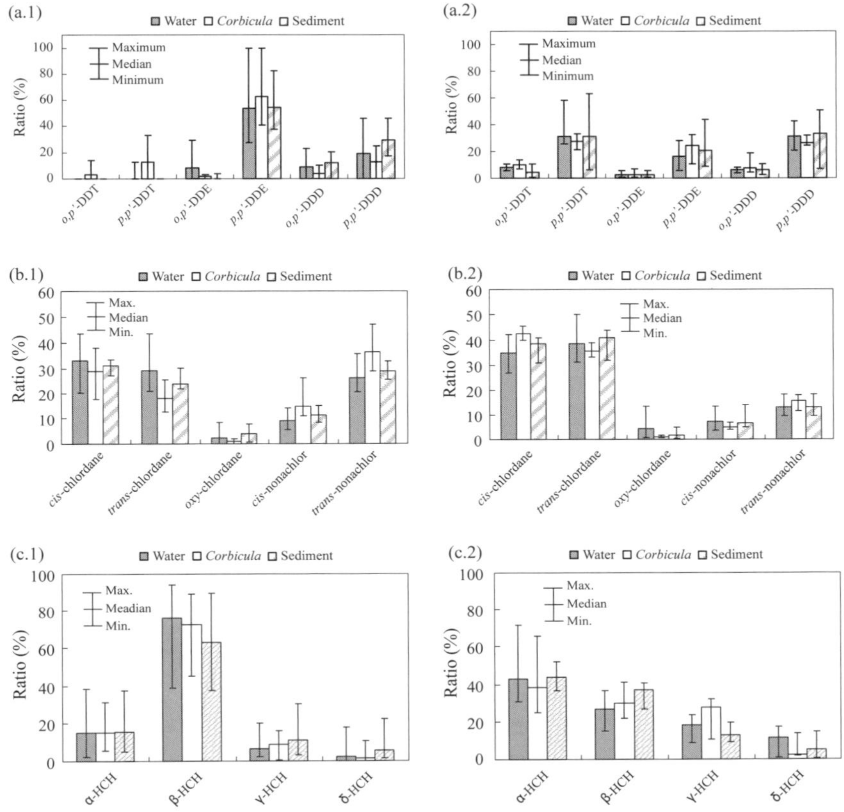

FIGURE 6-8. Compositions of (a) DDTs, (b) chlordanes, and (c) HCHs in water, *Corbicula*, and sediment (Takabe *et al.*, 2011). (1) indicates the compositions in western Japan, and (2) shows the compositions in the Pearl river delta in China.

following relationship equations are obtained when bivalves (*Perna viridis*, *Corbicula* and *Mytilus galloprovincialis*) are used as the indicator biota:

$$\mathrm{logBAFL} = 0.557 \times \mathrm{logKow} + 2.28 \quad (R = 0.569)\ (\mathrm{logKow} \leq 6.8) \tag{6-4}$$

$$\mathrm{logBAFL} = -1.35 \times \mathrm{logKow} + 15.3 \quad (R = 0.404)\ (6.8 \leq \mathrm{logKow}) \tag{6-5}$$

Although a polluted state can be discussed directly using the concentrations of target chemicals in the biological indicator, the concentrations of the target chemicals in water can be calculated using the concentrations in the biological indicator based on these equations.

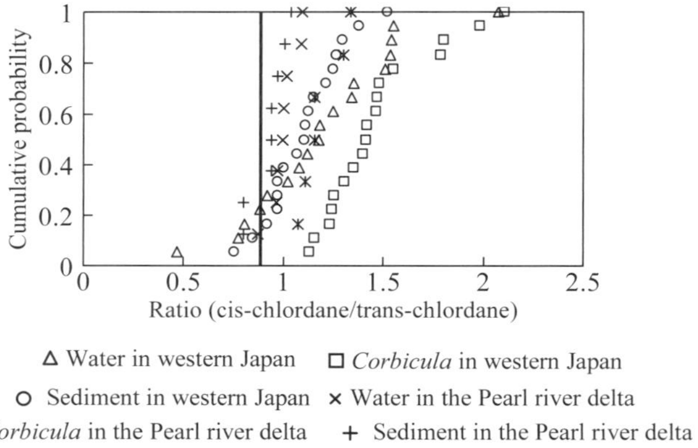

FIGURE 6-9. Ratios of *cis-chlordane/trans*-chlordane (Takabe *et al.*, 2011). The solid line shows the ratios of technical chlordane in Japan found in this study.

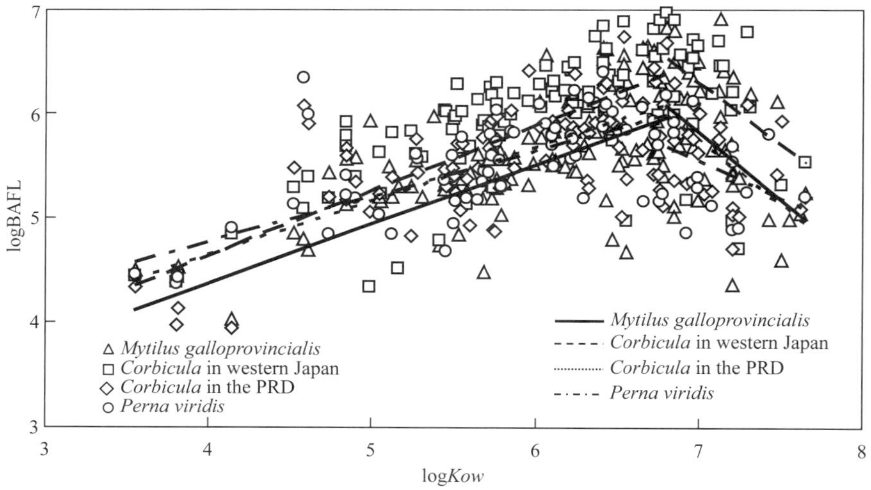

FIGURE 6-10. Relationship between logBAFL and logKow for OCP and PCB (Takabe *et al.*, 2009; Takabe *et al.*, 2011; Tsuno *et al.*, 2007a; Tsuno *et al.*, 2007b). Each plot shows the logarithmic mean value of BAFL of each chemical.

6.4 CONTROL AND MANAGEMENT

6.4.1 *Risk evaluation*

Fundamental risks are evaluated as non-carcinogenic and carcinogenic risks, which are expressed by the following equations:

non-carcinogenic risk

$$HQ = AD/RfD \tag{6-6}$$

carcinogenic risk (excess carcinogenic risk)

$$\Delta R = LAD \times CSF \tag{6-7}$$

where AD indicates the mean daily uptake amount (mg/(kg•day)), RfD is the reference uptake capacity (mg/(kg•day)), LAD is the lifelong mean daily uptake amount (mg/(kg•day)), and CSF is a carcinogenic slope coefficient (1/(mg/(kg•day))).

In addition, HQ is a risk indicator that shows whether AD exceeds RfD, and an HQ value greater than 1.0 indicates a high possibility for the occurrence of a non-carcinogenic hazard effect. A value of 10^{-5} is generally used for ΔR. The data associated with RfD and CSF can be obtained from the USEPA (U.S. Environmental Protection Agency). The HQ and ΔR values, calculated under an assumption of body-weight of 57.6 kg in Japan and 60.0 kg in China, and that *Corbicula* is eaten at a rate of 4.6 g/day in Japan and at 26 g/day in China, which is equal for shellfish such as short-necked clams and regular clams, based on data in the Yodo river in Japan and the Pearl river

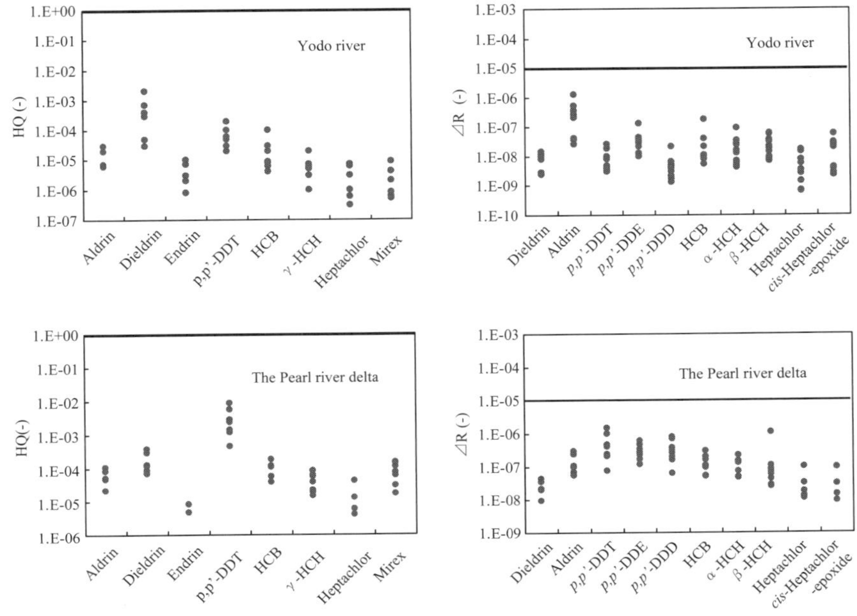

FIGURE 6-11. HQ and ΔR values in Yodo river and the Pearl river delta (Takabe, 2008).

delta in China, are shown in **Figure 6-11** (Takabe, 2008) as examples. All values are under the risk level when evaluated by individual chemical data.

6.4.2 *Removal of hazardous chemicals from wastewater*

Safe and clear effluent from a municipal wastewater treatment plant is required for intended and unintended reuse, and for maintaining safe and healthy water quality for aquatic ecosystems. Important factors for safe and clear effluent are shown in **Figure 6-12**. To accomplish these factors, ozonation is expected to be a promising core technology, as ozonation has strong oxidation ability and is effective in the oxidation of organic and inorganic materials; the removal of odors, color, and micropollutants; the transformation of organics into biodegradable compounds; disinfection; and an increase in transparency. Ozonation has been applied to more than 50 municipal wastewater treatment plants in Japan for effluent reuse (see **Figures 6-13** (Japan Sewage Works Agency, 2009) and **6-14** (Japan Sewage Works Agency, 2009)).

It is very important to remove EDCs and PPCPs in municipal wastewater plants, as they are also discharged from each house as well. An example of a semi-batch experiment is shown in **Figure 6-15**, in which a effluent of a secondary treatment was treated through ozonation under several conditions of ozone concentration in inflowing gas (Kim, 2005). The removal of EDCs and UV254 was accomplished during the ozone consumption per initial TOC of 1.0 mgO_3/mgC, after which the dissolved ozone concentration exceeded 0.5 mg/L, and bromate and organic brominated compounds began being produced with bromide ion included in the secondary treatment effluent. The amount of ozonation that occurs before the dissolved ozone concentration reaches 0.5 mg/L is proposed as the initial ozonation (IO), and the initial ozone consumption per initial TOC (mgO_3/mgC) during IO is used as an operational

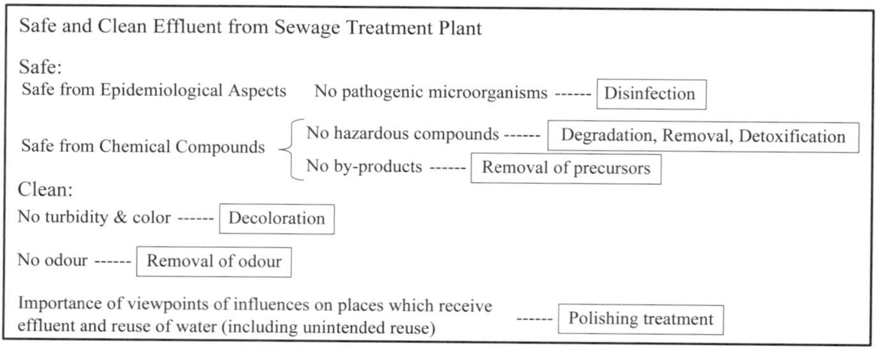

Safe and Clean Effluent from Sewage Treatment Plant
Safe:
Safe from Epidemiological Aspects
No pathogenic microorganisms ------
Disinfection
Safe from Chemical Compounds
No hazardous compounds ------
Degradation, Removal, Detoxification
No by-products ------
Removal of precursors
Clean:
No turbidity & color ------
Decoloration
No odour ------
Removal of odour
Importance of viewpoints of influences on places which receive effluent and reuse of water (including unintended reuse)

Polishing treatment

FIGURE 6-12. Safe and clean effluent from a sewage treatment plant.

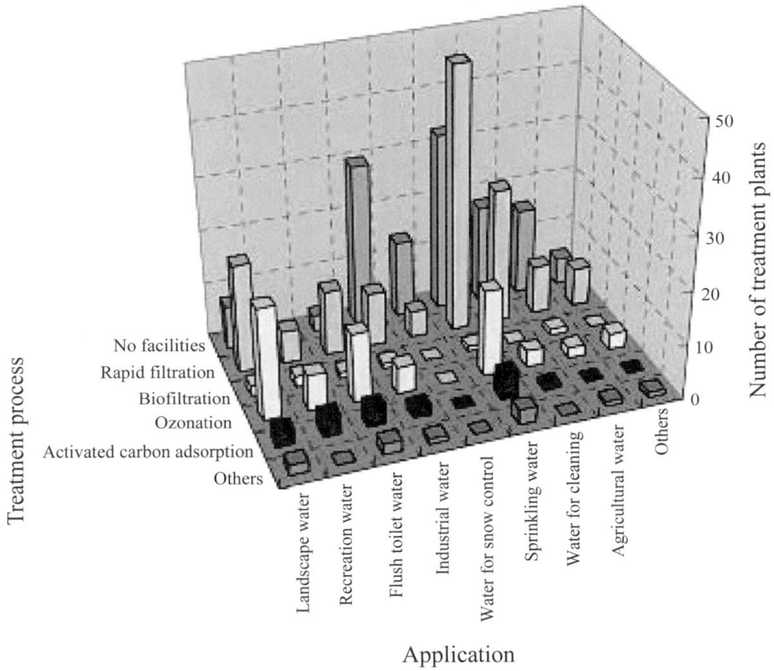

FIGURE 6-13. Relationship between the application of reclaimed water and the treatment process in 2002 (Japan Sewage Works Agency, 2009).

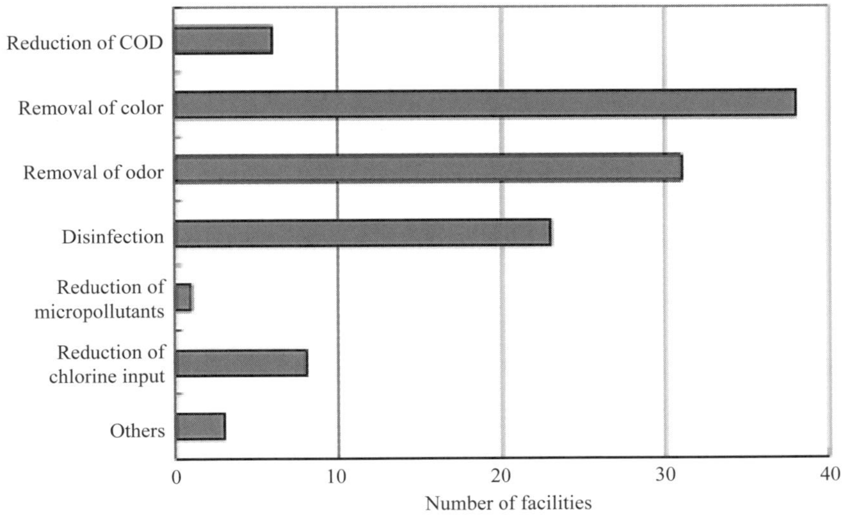

FIGURE 6-14. Objectives of the application of ozonation (Japan Sewage Works Agency, 2009).

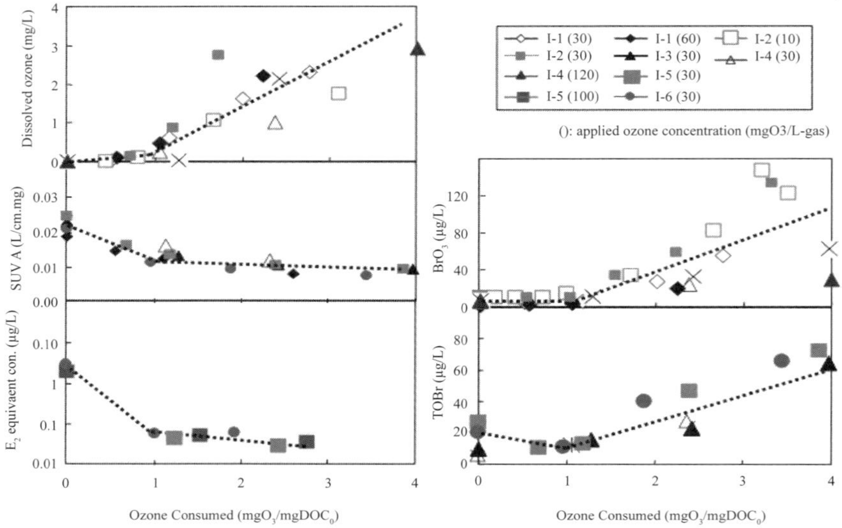

FIGURE 6-15. Characteristics of ozonation in the effluent of a secondary treatment (Kim *et al.*, 2005).

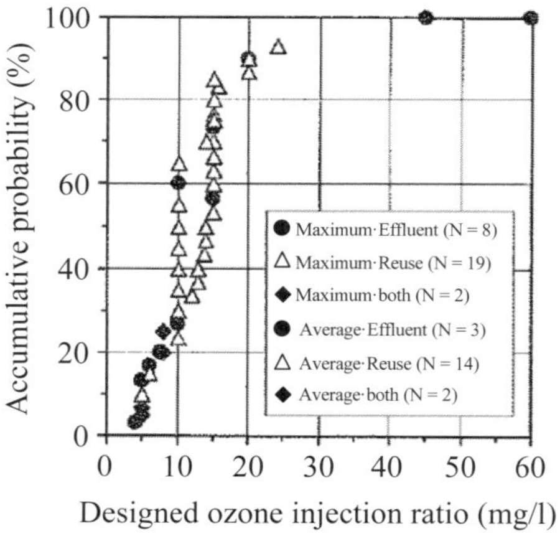

FIGURE 6-16. Accumulative probability distribution of designed ozone injection ratio (results from surveys) (Japan Sewage Works Agency, 2009).

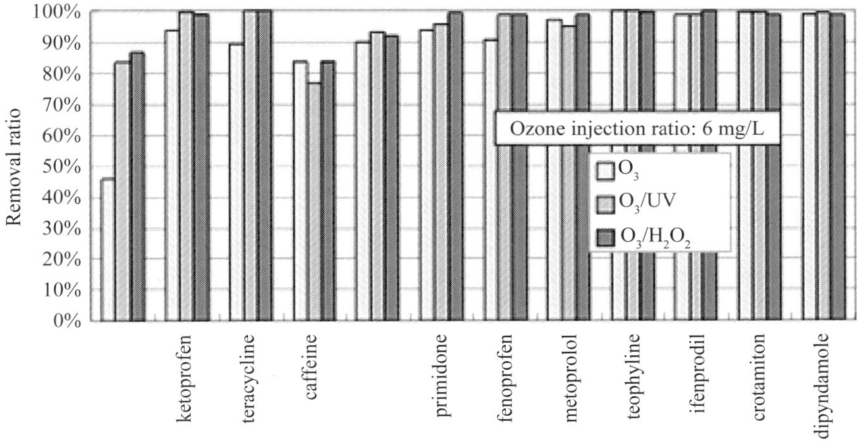

FIGURE 6-17. Effectiveness of ozonation and advanced oxidation processes against PPCPs (Kato *et al.*, 2007).

parameter. The ozone consumption per initial TOC of 1.0 mgO_3/mgC corresponds to an ozone dose of around 10 mgO_3 per L of treated water in a full-scale ozonation reactor. Actually, full-scale ozonation reactors are operated at ozone doses of around 5 mgO_3/L (see **Figure 6-16** (Japan Sewage Works Agency, 2009)). Ozonation and advanced oxidation processes with ozone are also effective in the removal of PPCPs, as shown in **Figure 6-17** (Kato *et al.*, 2007).

REFERENCES

Graduate school of Engineering, Kyoto University and Department of Environmental Science and Engineering, Tsinghua University (2010) *Study on monitoring, assessment and treatment technologies for reduction of risk caused by water resources.*

Japan Sewage Works Agency (2009), *Reports on technical assessment of ozonation.*

Kato, Y., *et al.*, (2007), Decomposition characteristics of Pharmaceuticals by Ozonation and Advanced Oxidation Process, *Proceedings of 17th Annual Conference on Ozone Science and Technology in Japan*, pp. 23–26.

Kim, H. (2005), *Behavior and control of by-products during ozone and ozone/hydrogen peroxide treatments of sewage effluent*, doctoral thesis, Kyoto University.

Ministry of Land, Infrastructure, Transports and Tourism (2009), *Water resources in Japan.*

Takabe, Y. (2008), *Monitoring and primary risk assessment of POPs with bivalves as a bioindicator*, Master Thesis, Kyoto University.

Takabe, Y., *et al.*, (2009), Applicability of POPs monitoring method with bivalves in areas with different environmental conditions, *Environmental Engineering Research*, 46, pp. 553–563.

Takabe, Y., *et al.*, (2011), Applicability of Corbicula as a bioindicator for monitoring organochlorine pesticides in fresh and brackish waters, *Environmental Monitoring and Assessment.*

Tsuno, H., *et al.*, (2007a), PCBs distribution and bioconcentration to mussel in Seto Inland Sea, *Journal of Environmental Systems and Engineering*, 63(2), pp. 149–158.

Tsuno, H., *et al.*, (2007b), Accumulation characteristics of POPs from water to bivalves, *Journal of Environmental Systems and Engineering*, 63(2), pp. 179–185.

Tsuno, H., *et al.*, (2009), Measurement of Estrogenic Activity with NRL/TIF2-BAP Assay, *Environmental conservation engineering*, 38(2), pp. 111–119.

U.S. Environmental Protection Agency, *Integrated Risk Information System (IRIS)*, http://www.epa.gov/IRIS/ (October 29th 2010).

Index